1004721161

EMERGENCE AND CONVERGENCE:
QUALITATIVE NOVELTY AND THE
UNITY OF KNOWLEDGE

MARIO BUNGE

Emergence and Convergence: Qualitative Novelty and the Unity of Knowledge

UNIVERSITY OF TORONTO PRESS
Toronto Buffalo London

© University of Toronto Press Incorporated 2003
Toronto Buffalo London
Printed in Canada

ISBN 0-8020-8860-0

Toronto Studies in Philosophy
Editors: Amy Mullin and Donald Ainslie

Printed on acid-free paper

National Library of Canada Cataloguing in Publication

Bunge, Mario, 1919–
  Emergence and convergence : qualitative novelty and the unity of
knowledge / Mario Bunge.

(Toronto studies in philosophy)
Includes bibliographical references and index.
ISBN 0-8020-8860-0

1. System theory.   2. Evolution.   3. Knowledge, Theory of.
4. Interdisciplinary approach to knowledge.   5. Social epistemology.
I. Title. II. Series.

Q175.B8223 2004   003'.01   C2003-903208-6

University of Toronto Press acknowledges the financial assistance to its publishing program of the Canada Council for the Arts and the Ontario Arts Council.

University of Toronto Press acknowledges the financial support for its publishing activities of the Government of Canada through the Book Publishing Industry Development Program (BPIDP).

Contents

PREFACE xi

Introduction 3

Part I: Emergence

1 Part and Whole, Resultant and Emergent 9
    1 Association and Combination 10
    2 Emergence and Supervenience 12
    3 Levels and Evolution 15
    4 Structure and Mechanism 19
    5 Emergence and Explanation 21
    Concluding Remarks 24

2 System Emergence and Submergence 26
    1 System Emergence 27
    2 Emergence *ex nihilo*? 30
    3 Submergence: System Dismantling 31
    4 System Types 33
    5 The CESM Model 34
    Concluding Remarks 38

3 The Systemic Approach 40
    1 The Systemic Approach 40
    2 Conceptual and Material Systems 42
    3 The Systemic Approach to Physical and Chemical Processes 43

vi   Contents

    4 The Systemic Approach to Life   45
    5 The Systemic Approach to Brain and Mind   49
    Concluding Remarks   52

4 Semiotic and Communication Systems   53
    1 Words, Ideas, and Things   54
    2 Semiotic System   58
    3 Languages as Semiotic Systems   60
    4 Speech and Language   62
    5 Speech Learning and Teaching   65
    6 Communication System   67
    Concluding Remarks   68

5 Society and Artefact   70
    1 The Systemic Approach to Society   71
    2 Microsocial and Macrosocial, Sectoral and Integral   73
    3 Emergence by Design   75
    4 Social Invention   76
    5 Philosophical Dividends of the Systemic Approach   77
    Concluding Remarks   80

6 Individualism and Holism: Theoretical   82
    1 Individual and Individualism, Whole and Holism   84
    2 Ontological   85
    3 Logical   87
    4 Semantic   89
    5 Epistemological   91
    6 Methodological   93
    Concluding Remarks   95

7 Individualism and Holism: Practical   97
    1 Value Theory, Action Theory, and Ethics   97
    2 Historical and Political Individualism   100
    3 First Alternative to Individualism: Holism   101
    4 Hybrids   105
    5 The Systemic Alternative   106
    Concluding Remarks   109

8 Three Views of Society   112
    1 The Two Classical Views of Society   112
    2 The Systemic Approach   114

3 From Statistics to Theoretical Models   115
4 The Science-Technology-Market Supersystem   119
5 Implications for Social-policy Design   121
6 Social Studies Are about Social Systems   122
7 The Competitive Advantage of Systemism   124
Concluding Remarks   126

Part II: Convergence

9 Reduction and Reductionism   129
1 Reduction Operations   130
2 Micro-levels, Macro-levels, and Their Relations   133
3 Same-level and Inter-level Relations   135
4 Inter-level Hypotheses and Explanations   136
5 From Physics to Chemistry   137
6 Biology, Ecology, and Psychology   139
7 From Biology to Social Science: Human Sociobiology and the IQ Debate   141
8 Kinds and Limits of Reduction   144
9 Reduction and Materialism   146
10 Concluding Remarks   147

10 A Pack of Failed Reductionist Projects   149
1 Physicalism   149
2 Computationism   150
3 Linguistic Imperialism   152
4 Biologism I: Sociobiology   154
5 Biologism II: Evolutionary Psychology   156
6 Psychologism   162
7 Sociologism, Economism, Politicism, and Culturalism   164
Concluding Remarks   167

11 Why Integration Succeeds in Social Studies   168
1 The B-E-P-C Square   169
2 Social Multidisciplinarity   171
3 Social Interdisciplinarity   176
Concluding Remarks   177

12 Functional Convergence: The Case of Mental Functions   179
1 Informationist Psychology   181

    2 Mark II Model: Connectionism   183
    3 Localization of Mental Functions   185
    4 Functional Interdependence of Neural Modules   187
    5 Consciousness: From Mystery to Scientific Problem   188
    6 Two Convergence Processes   191
    Concluding Remarks   194

13 Stealthy Convergence: Rational-choice Theory and Hermeneutics   196
    1 Divergence and Convergence   197
    2 Methodological Individualism   199
    3 Subjective Process and Observable Behaviour   201
    4 Inverse Problems   203
    5 Figuring Out Mediating Mechanisms   206
    6 Example: The Delinquency–Unemployment Relation   208
    Concluding Remarks   209

14 Convergence as Confusion: The Case of 'Maybe'   213
    1 Logical Possibility   214
    2 Factual Possibility   217
    3 Likelihood   220
    4 Relation between Frequency and Probability   221
    5 Probability, Chance, and Causation   223
    6 Credibility   226
    7 Probabilistic Epistemology   228
    8 Plausibility or Verisimilitude   232
    9 Towards a Plausibility Calculus   233
    Concluding Remarks   235

15 Emergence of Truth and Convergence to Truth   237
    1 The Nature of Truth   237
    2 Towards an Exact Concept of Correspondence   239
    3 Partial Truth   241
    4 The Emergence of the Knowledge of Truth   245
    5 Truth-centred Ethics and Ideology   247
    Concluding Remarks   249

16 Emergence of Disease and Convergence of the Biomedical Sciences   250
    1 Multilevel Systems and Multidisciplinarity   250

    2 What Kind of Entity Is Disease?   252
    3 Diagnosis as an Inverse Problem   253
    4 Knowledge of Mechanism Strengthens Inference   256
    5 Bayesian Number Juggling   259
    6 Decision-theoretic Management of Therapy   261
    7 Medicine between Basic Science and Technology   263
    Concluding Remarks   265

17 The Emergence of Convergence and Divergence   268
    1 Divergence   268
    2 Convergence   270
    3 Caution against Premature Unification   272
    4 Why Both Processes Are Required   274
    5 Logic and Semantics of Integration   277
    6 Glue   278
    7 Integrated Sciences and Technologies   280
    Concluding Remarks   282

GLOSSARY   285
REFERENCES   293
INDEX OF NAMES   315
INDEX OF SUBJECTS   323

Preface

This book is about parts and wholes, as well as about the old and the new – two perennial problems in science, technology, and the humanities. More precisely, it is about systems and their emergent properties, as exemplified by the synthesis of molecules, the origin of species, the creation of ideas, and social innovations such as the transnational corporation and the welfare state. The present work is also about the merger of initially independent lines of inquiry, such as developmental evolutionary biology, social cognitive neuroscience, socio-economics, and political sociology. This book is, in sum, about new coming from old, both in reality and in its study. Shorter: It deals with newcomers, whether concrete or conceptual. Even shorter: It is about novelty.

However, we shall also examine submergence, or the disappearance of higher-level things and their properties, as in the cases of evaporation, forgetting, and the crumbling of social systems. And we shall not forget that one of the emergence mechanisms is splitting or divergence, as exemplified by nuclear fission, cell division, and the division of a field of study into subdisciplines. Hence a more adequate title for this book would be *Emergence and Submergence, Convergence and Divergence*

The following list of topical and intriguing problems involving both emergence and cross-disciplinarity should help elucidate the nature and importance of these categories:

How did they emerge?	*Why did they converge?*
Molecules	Chemical physics
Life	Biophysics
Mind	Biochemistry
Social norms	Cognitive neuroscience
The state	Socio-economics

Along the way we shall also examine such topical problems as the advantages of looking for mechanisms underneath observable facts, the limitations of both individualism and holism, the reach of reduction, the abuses of Darwinism, the rational choice–hermeneutics feud, the modularity of the brain versus the unity of the mind, the cluster of concepts around 'maybe,' the relevance of truth to all walks of life, the obstacles to correct medical diagnosis, and the formal conditions for the emergence of a cross-discipline.

For example, we shall ask whether individualism can account for the emergence of social norms that curb the liberty to pollute, bear arms, and initiate war. And how is human sex to be understood: as a mechanism for shuffling chromosomes, for effecting reproduction, for giving pleasure, or for strengthening social bonding? Since sex involves all of the preceding, is it not reasonable to promote the convergence of genetics, organismic biology, ethology, psychology, and sociology in studying sex, instead of pushing either a micro-reductionist or a macro-reductionist strategy?

Given the reputation for other-worldliness that philosophers have earned, the following warning may not be amiss. What follows is not an idle fantasy about parallel universes, counterfactuals, immaterial minds, knowledge without inquirers, ingenious but barren puzzles, and the like. On the contrary, this book is about topical problems that occur in all the disciplines that study reality. In particular, the present work is partly a belated spinoff of evolutionary biology. Indeed, the latter has spawned at least three key ontological concepts – those of evolution, emergence, and level of organization. Moreover, evolutionary biology has shown the heuristic value of fusing initially independent lines of inquiry, as exemplified by evolutionary biology itself, which joins molecular biology with whole-organism biology, ecology, and the history of life.

Evolutionary biology has also proposed at least three hypotheses central to our concerns: (1) Living beings emerged from the synthesis of abiotic precursors; (2) new levels emerge, and old properties submerge, in the course of evolution; and (3) understanding organisms and their development and evolution requires the merger of several branches of biology. These hypotheses have spilled over many of the factual sciences and technologies. It is therefore high time that philosophers take them seriously – or rather rethink them, because they were discussed by British and American philosophers between the two world wars.

Although this book tackles philosophical problems, it is addressed to the broad community of people who – regardless of their specialties – are interested in intriguing general problems, rather than to professional philosophers only. One reason is that all the really important philosophical problems overflow philosophy. The Glossary at the end of this book may be of help to the reader who, like the author, has had no formal philosophical training. More detailed elucidations of technical philosophical terms may be found in the author's *Philosophical Dictionary* (2003).

I am indebted to the late David Bohm, with whom I held many stimulating discussions on emergence and levels of organization, causation and chance, as well as on the quantum theory, in 1953, at the Instituto de Física Teórica of the Universidade de São Paulo, Brazil. I thank Joseph Agàssi (Tel Aviv University), Omar Ahmad (Rockefeller University), Silvia A. Bunge (University of California, Davis), Bernard Dubrovsky (McGill University), James Franklin (University of New South Wales), Irving Louis Horowitz (Rutgers University), Michael Kary (Boston University), the late Robert K. Merton (Columbia University), Pierre Moessinger (Université de Genève), Andreas Pickel (Trent University), Miguel Angel Quintanilla (Universidad de Salamanca), Dan A. Seni (UQÀM), and Paul Weingartner (Universität Salzburg), for interesting exchanges about some of the problems tackled in this book. I also thank Virgil Duff, Executive Editor of the University of Toronto Press, for having skilfully shepherded the manuscript of this book, and to John St James for his intelligent copy-editing. Above all, I am indebted to Martin Mahner (Zentrum für Wissenschaft und Kritisches Denken, Rosdorf) for his penetrating questions and criticisms. I also thank Michael Matthews and Johnny Schneider, my hosts in stunning Sydney, Australia, during the winter and spring terms of 2001. This book emerged there and then from the convergence of several threads of thought that had been building up and unravelling in my brain for over half a century.

I dedicate this book to my dear friend, valued co-worker, and uncompromising critic Martin Mahner, biologist and scientific philosopher.

EMERGENCE AND CONVERGENCE:
QUALITATIVE NOVELTY AND THE
UNITY OF KNOWLEDGE

Introduction

The term 'emergence' refers to the origin of novelties, as in the emergence of a plant out of a seed and the emergence of a visual pattern from the juxtaposition of the tiles in a mosaic. And the convergence discussed in this book is that between initially separate approaches and fields, as in the interdisciplinary studies of mental processes and of the creation and distribution of wealth.

At first sight emergence and convergence appear alien to each other, if only because, whereas the former is an ontological category, the latter is an epistemological one. On second thought they are not mutually alien, because the understanding of emergence often requires the convergence of two or more lines of research. Thus, the attempt to explain chemical reactions generated physical chemistry; the wish to understand speciation prompted the union of evolutionary and developmental biology; the urge to understand mental processes led to the merger of psychology with neuroscience and sociology; and the need to understand and control the distribution of wealth gave rise to socio-economics.

The time for a philosophical study of emergence and convergence seems to be ripe. This is because both subjects, earlier neglected or misunderstood, are now moving to the fore. Indeed, the term 'emergence' – in the sense of the coming into being of qualitative novelty – which until recently languished in the jail of forgotten or vilified words, has now joined the ranks of other popular buzzwords, such as 'system,' 'self-assembly,' 'chaos,' 'fractal,' 'complexity,' 'module', and 'consciousness.' As for convergence (or cross-disciplinarity, or unification, or merger, or integration), once the preserve of dilettanti and research-grant officers, it is being practised ever more often in science, technology, and the humanities. Witness the increasingly interdisciplinary approach to such conceptual problems as the emergence of drug-resistant strains of viruses and micro-organisms, the emergence mechanisms of new ideas, the origin of *Homo sapiens*, the diffusion of agriculture, and the emergence of novel organizational forms.

Much the same is happening with practical problems, such as the governance of large corporations and the design of control systems, and such social issues as poverty, illiteracy, crime, overpopulation, desertification, underdevelopment, war-mongering, and the persistence of superstition. Given the systemic and multifaceted nature of such issues, only a cross-disciplinary approach to them may succeed in understanding and managing them.

Though only now fashionable, the concepts of emergence and convergence are far from new. However, the concept of emergence is still often misunderstood or resisted. And the unity of science is usually taken to be either utopian or possible only through reduction – of, say, sociology to biology, and the latter to chemistry.

The two categories in question are intimately related. Indeed, some novelties result from the self-organization of a collection of disparate entities; and every merger of ideas involves the emergence of new ones – those that bridge the initially disconnected items. Thus, when two disciplines converge, a whole new interdiscipline emerges. And when a new broad viewpoint or approach emerges, some previously disconnected fields of inquiry are likely to converge. Consequently, the widespread beliefs that emergence precludes convergence, and that emergence must be rejected because it is an obstacle to the unity of knowledge, are mistaken.

For instance, the scientific study of the origin of life requires a close collaboration of biology with chemistry and geology; the study of the relation between morbidity and mortality, on the one hand, and socio-economic status, on the other, is central to epidemiology and medical sociology; and the investigation of the links between big business and politics is crying for the emergence of econopolitology. In general, emergence calls for convergence because only multidisciplinarity and interdisciplinarity can explain multifaceted and multilevel events. In turn, convergence requires the emergence of new bridge or glue concepts and hypotheses.

Another example: When it was realized that evolutionary novelties emerge in the course of individual development (ontogeny), it became clear that evolutionary biology must be amalgamated with developmental biology. (This movement, often called evo-devo, is currently in full swing.) Again, when it was found that mental processes, such as emotion, vision, speech, reasoning, and decision-making, are brain functions, it became obvious that psychology had to coalesce with neurobiology: this is how cognitive neuroscience, one of the most exciting sciences of the day, was born. When it became clear that neither economics nor sociology by themselves can cope with such cross-disciplinary problems as income distribution and national development, socio-economics emerged. Likewise, understanding the emergence and evolution of the state calls for a synthesis of anthropology, archaeology, sociology, economics, politology, and history (Trigger 2003: 6).

Why devote an entire book to the concepts of emergence and convergence? Because they are often misunderstood or even overlooked altogether. Indeed, many students dislike talk of emergence, suspecting it to be obscure or even obscurantist. (The standard dictionary definition of 'emergence' mistakenly identifies it with the impossibility of understanding a whole through analysis into its components and their interactions.) And just as many students are justifiably suspicious of interdisciplinarity, because it is often preached by bureaucrats fearful of what they see as wasteful research duplication. Others rightly resist the temptation of achieving unification through reduction and concomitant conceptual impoverishment.

Although many hard-nosed scientists mistrust the very idea of emergence, there is nothing strange or murky about it. Indeed, emergence happens every time something qualitatively new arises, as when a molecule, a star, a biospecies, a business, or a science is born. And it consists in a new complex object having properties that none of its constituents or precursors possess. As for cross-disciplinarity, or border trespassing, it has been a common enough research strategy in the sciences and technologies for nearly two centuries: think of physical chemistry, biochemistry, psychophysics, neurolinguistics, or medical sociology. Still, both concepts, the ontological and the epistemological, can do with some additional clarification, if only because emergence is often confused with the ill-defined notion of supervenience, multidisciplinarity is likely to be mistaken for interdisciplinarity, and both seen as dilettantism.

Moreover, it is worthwhile insisting that emergence is important, not just in itself but also because its mere recognition jeopardizes many an attempt at radical micro-reduction – which is far rarer than reductionists realize. Likewise, the merits of cross-disciplinarity are worth emphasizing at a time when increasing specialization narrows down perspectives and hinders the successful tackling of such systemic issues as inequality, ignorance, and violence that defy the one-thing-at-a-time strategy. The only way to prevent the runaway proliferation of subdisciplines is to discover or construct bridges among them.

However, it is not enough to emphasize the importance of emergence and convergence. We should also attempt to explain them. For example, how and where did the first organisms originate from abiotic precursors? How did new health-care organizations emerge? And why did the need for medical sociology arise? Of course, these and similar questions are beyond the ken of philosophers. Still, these scholars can suggest that those are legitimate and important scientific problems that involve some deep philosophical ideas.

Moreover, philosophers may be competent to judge whether certain proposed syntheses are successful or at least promising. For instance, I will argue that the unification of cosmology and quantum mechanics has so far been unsuccessful because it has been effected at the expense of certain fundamental physical laws, such as energy conservation. I will also suggest that current evolutionary psychology has proved to be unsuccessful because it has joined a mistaken conception of evolution with an erroneous psychological theory, and a methodology that does not include the need for empirical tests.

A few other key notions will be elucidated along the way, among them those of system, mechanism, explanation, probability, partial truth, medical diagnosis, and inverse problem. Their importance may be gauged by the following sample of topical questions where those concepts occur: Why is medical diagnosis so often mistaken? Is it true that markets are self-regulating systems in equilibrium because they are ruled by the supply-and-demand mechanism? Which is the mysterious mechanism that explains why globalization is good for some and bad for others? Is it legitimate to assign a probability to anything other than chance events? In particular, is it scientifically and morally legitimate to gamble with truth? Can we dispense with half-truths in the (hopefully convergent) search for total truths? And how are inverse problems – such as those of guessing disease from syndrome and intention from behaviour – best tackled?

PART I

Emergence

1

Part and Whole, Resultant and Emergent

It is a truth of logic that every object is either simple or complex – in some respect or at some level. For example, words are composed of lexical units such as letters or simple ideograms, sentences of words, and texts of sentences; the integers other than 0 and 1 are sums of two or more numbers; polygons are constituted by line segments; theories are composed of propositions, which in turn are fusions of concepts; atomic nuclei, atoms, and molecules are composed of elementary particles such as protons and electrons; droplets, liquids, gels, and solids are composed of atoms or molecules; light beams are composed of photons; cells are composed of molecules and organelles; organs are composed of cells; families, gangs, businesses, and other social systems are composed of persons; machines are composed of mechanical or electrical modules – and so on and so forth.

There are then modules, simple or complex, on every level of organization – or rung in the ladder, hierarchy, or chain of being, as it used to be called (see, e.g., Lovejoy 1953; Whyte, Wilson, and Wilson 1969). In the simplest cases, or in the early phase of a research project, only two such levels of organization, micro and macro, need be distinguished. This distinction is involved in such disparate studies as those concerning signal amplification, the macrophysical effects of tiny crystal impurities, landslides caused by small perturbations, the macroeconomic conditions of microeconomic transactions, and political crises resulting from court intrigues (see, e.g., Lerner 1963). The same distinction is involved in the two strategies for handling micro–macro relations: the bottom-up (or synthetic) and the top-down (or analytic) approaches – of which more later.

However, composition is not everything: the mode of composition, or structure, or organization of a whole, is every bit as important. For example, the words 'god' and 'dog,' and the numerals 13 and 31, are composed by the same elementary symbols, but they are different because the order in which their constituents occur is different. Likewise, when a group of previously unrelated people converge into a classroom, or are drafted into a regiment, the previously shapeless group acquires a structure. (This structure is constituted by the hierarchical relations imposed from above and by the peer relations of friendship and rivalry.) And when atoms combine into molecules, their elementary constituents become drastically rearranged rather than juxtaposed or glued in the manner that the Greek atomists and Dalton had imagined. In these cases one speaks of *integral* as opposed to *modular* structure.

Thus, whereas the architectures of a Lego toy and a computer are modular, those of a cell or a brain are integral. Again, whereas a pyramid can be built stone by stone – though not in an arbitrary order – the assembly of a cell cannot proceed directly from modules to whole. (In the case of the pyramid, the external force – gravity – is overwhelming, whereas in the case of the cell the myriad interactions among the cell components are dominant.) Presumably such assembly must have climbed up, rung by rung, a ladder of levels – atoms, molecules, small molecular assemblies, organelles, membrane, whole cell – every one of which is characterized by simultaneous and interdependent processes, such as the energy fluxes and chemical reactions that go on during the entire assembly process. The upshot is that we must look at two qualitatively different assembly modes: association or mixture and combination or fusion. Let us do so.

1 Association and Combination

Objects can join in various ways. The most common union of items, whether of the same or different kinds, is association, juxtaposition, concatenation, clumping, or accretion, as in the formation of a sand pile or a crowd. Such wholes are characterized by a low degree of cohesion. Consequently, they can alter or be altered rather easily, to the point of dismantling, due to either internal rearrangement or external shocks. Their structure (or organization or architecture) can be said to be *modular*.

The concept of association will allow us to clarify the ever-present if somewhat elusive relation of part to whole. If a and b are two objects, then a can be said to be *a part* of b, or $a < b$ for short, if a adds nothing to b. For example, in ordinary arithmetics 1 is a part of any number b, because $1 \cdot b = b$. Likewise, the cell membrane is part of the cell because 'adding' (juxtaposing) a membrane to its own cell just results in the same cell.

(Both association and the part–whole relation can be formalized with the help of semigroup theory, one of the simplest mathematical theories. An arbitrary set *S*, together with a binary operation ⊕ constitutes a *semigroup* if ⊕ is associative; that is, if, for any elements *a*, *b*, and *c* of *S*, *a* ⊕ (*b* ⊕ *c*) = (*a* ⊕ *b*) ⊕ *c*, where neither *S* nor ⊕ is specified except for the associativity property. According to our verbal definition, *a* is a part of *b* if *a* adds nothing to *b*. In symbols, our definition reads: $a < b =_{df} (a ⊕ b = b)$. In words: *a* is a part of *b* equals, by definition, the concatenation – or juxtaposition or mereological sum – of *a* and *b* is equal to *b*. Thus, thanks to a bit of abstract algebra, the entire rather esoteric discipline of mereology gets compressed into a single paragraph: see Bunge 1977a.)

The above definition of the part–whole relation allows us to define the notion of composition of an object thus:

Definition 1.1 The *composition* of an object is the collection of all its parts. (In formal terms: $C(s) = \{x | x < s\}$.)

Association results in novelty of the combinatory type. This is the most common type of novelty because it is the least demanding, energetically or culturally. In fact, it takes only two solid bodies to constitute a lever; two different liquids to form a mixture; a handful of strangers to get a party going; and two 'atomic' propositions p and q to form a 'molecular' proposition such as p or q. Furthermore, association does not change the nature of the constituents. Still, sometimes it ensues in qualitative novelty of the combinatorial type. For example, the lever may be in equilibrium, and the party may turn into a brawl.

Combinatory novelty is the only kind of novelty that associationist (or 'elementist') psychology admitted. Consequently, it had to postulate that the basic ideas must be either innate (as Socrates, Leibniz, Chomsky, and Fodor have held) or gotten from sense perception (as Aristotle, Locke, Hume, and Mill thought). How radically novel ideas, such as those of vacuum, atom, field, gene, evolution, emergence, logical entailment, zero, or infinity could ever have arisen in either fashion was not explained.

By contrast, the combination of two or more modules of the same or different kinds results in a radically novel thing, that is, one characterized by properties that its constituents lack. For instance, a proton and an electron combine to form a hydrogen atom; two hydrogen atoms combine to form a hydrogen molecule; the collision of an electron with a positron results in a photon; egg and spermatozoid combine into zygote; neurons assemble into neuronal systems or circuits capable of having mental experiences; people organize themselves into families, bands, business firms, or clubs; letters combine to form words, and words phrases; and sets of propositions get systematized into theories (or hypothetico-deductive systems).

Combinations differ from mere aggregates in at least three regards. First, the original items alter in the process, so that they are precursors rather than constituents of the whole. For example, the molecule resulting from the combination of two atoms is not the association of the latter, since their electrons get drastically rearranged; likewise, when two people love and live together they transform each other. Second, combinations, as in the cases of chemical compounds, bodily organs, and social systems, are more stable than mere aggregates, because they are more cohesive. Third and consequently, combination takes more energy, longer time, or rarer circumstances, as the case may be. Think of the emergence of the solar system, the RNA molecule, the cell, the human brain, the school, or the state. These are all cases of emergence. However, this concept deserves a new section.

2 Emergence and Supervenience

Typically, the wholes resulting from combinations of lower-level units have properties that their parts or precursors lack. No mystery here. For example, two switches connected in series mimic the logical connective 'and,' and they mimic 'or' if connected in parallel. Again, a valid reasoning (argument) is a system whose conclusion is not included in its separate premises. Indeed, the conclusion emerges (results) from their combination in a valid manner, as in 'If p, then q, p ⊢ q,' where ⊢ is read 'entail.'

As a rule, then, wholes are not similar to their parts. Only fractals, such as snowflakes and coastlines, are self-similar: that is, their parts are similar in shape to the whole. But shape is neither a universal property nor a source one. Thus electrons and families are shapeless. And organs are not similar to their cells, just as nations are not similar to their citizens. Moral: Let the fractal fashion pass.

Typically, then, wholes possess properties that their parts lack. Such global properties are said to be *emergent*. Every whole has at least one such property. And, so far as we know, there is only one property common to all concrete existents regardless of their complexity, namely energy (Bunge 2000c). Moreover, energy is distributive: the energy of a whole is distributed additively among its parts.

The emergent properties are not distributive but global. Think, for instance, of the validity of an argument, the consistency of a theory, the efficiency of an algorithm, the style of an artwork, the stability (or instability) of an atomic nucleus, the solidity (or fluidity or plasticity) of a body, the flow pattern of a river, the synchrony of a neuron assembly, the body plan of an organism, the self-regulation of an organism or a machine, the cohesion of a family, the organization (or structure) of a business firm, the stability (or instability) of a government, the equilibrium (or disequilibrium) of a market, the division of labour in a factory or in a society, or the level of development attained by a nation. These global (systemic) properties originate in the interrelations among the constituents of the systems concerned.

Other systems are materially indecomposable, though conceptually analysable. For instance, a light wave is constituted by two different intertwining fields, an electric and a magnetic one, but not by two different waves: there are no purely electric or purely magnetic waves. And the intertwined fields are described by electrodynamics, a synthesis whose historical precursors were electrostatics, magnetostatics, optics, and mechanics.

Two types of emergent must be distinguished: absolute and relative. The former are 'firsts': they refer to the earliest occurrence of individuals of a new kind, such as the very first bacterium that emerged on Earth about three billion years ago, the beginning of agriculture, the first car, or the first laboratory in history. This kind of emergence is different from later instances of the same kind, such as newly manufactured cars, which may be called 'relative' emergents. However, save when dealing explicitly with 'firsts,' we shall not draw the absolute/relative distinction. (Incidentally, this terminology is not quite felicitous.)

Another distinction worth drawing is that between natural (spontaneous) and artificial (or made) assembly. The former is also called *self-assembly*. Examples: the solidification of a body of water; the formation of a group of cells that oscillate synchronically; and the coalescence of a street gang or a sports team around a task or a leader. By contrast, car assembly and personnel recruitment are artificial processes. But, of course, natural and artificial emergence can combine, as in the following familiar process:

Seed → Seedling → Sapling → Tree → Log → Pulp → Paper → Book

Some philosophers, such as Ernest Nagel (1961) and Carl G. Hempel (1965), have rightly rejected the holistic construal of 'emergence' as an ontological category, in being imprecise. They have admitted 'emergence' only as an epistemological category equivalent to 'unexplainable (or unpredictable) by means of contemporary theories' – the way Broad (1925) had proposed.

Yet that is not the sense in which 'emergent' is used, say, by biologists when stating that life is an emergent property of cells. Nor is that the concept involved in a social scientist's explanation of stability (or its dual, instability) in some respect as an emergent property of certain social systems. All of which suggests that the concept of emergence is worth rescuing from holistic metaphysics. It is also necessary to clarify the rather popular confusion between the two concepts of emergence:

ontological: emergence = occurrence of qualitative novelty
and
epistemological: emergence = unpredictability from lower levels
– a confusion that even eminent scientists (e.g., Mayr 1982) and philosophers (e.g., Popper 1974) have incurred.

Other philosophers, such as G.E. Moore, Donald Davidson, and Jaegwon Kim, have admitted the occurrence of qualitative novelty but have preferred 'supervenience' to 'emergence.' They have said, for instance, that mental properties 'supervene' upon physical properties, in the sense that the former somehow 'depend' upon the latter – without however being themselves properties of matter. Regrettably, these philosophers have failed to elucidate clearly such dependence.

Jaegwon Kim (1978) tried to define the concept of supervenience in exact terms, but it is generally admitted that he did not succeed. I submit that he failed because he (a) detached properties from the things that possess them; (b) assumed that properties, like predicates, can be not only conjoined but also disjoined and negated; (c) conflated properties with events, thus attributing to the former causal powers; and (d) used the fiction of possible worlds instead of studying cases of emergence in the real world. This amusing fantasy will be examined in chapter 14. The other three ideas are open to the following objections.

First of all, there are no properties in themselves, located in a Platonic realm of ideas: every property is possessed by some individual or n-tuple of individuals. Second, there are neither negative nor disjunctive properties. Of course, we may truly say of a person that she is not a smoker, but she does not possess the property of being a non-smoker, any more than she possesses the property of being a non-whale. Likewise, we can truly say of people that they are people or birds, but such addition confuses without enriching. A true theory of properties will start by distinguishing them from predicates, and thus it will assume that the predicate calculus cannot double as an ontological theory. (More in Bunge 1977a.)

Only changes in concrete things, that is, events, can cause anything. (In other words, the relation of efficient causation holds only among events: see, e.g., Bunge 1959a.) For example, a rise in the temperature of the gas mixture in a car cylinder will cause the gas to expand (or to increase in volume), which will in turn cause the displacement of the piston, which will move the wheels – a familiar enough causal chain.

My own definition of emergence is this (Bunge 1977a). To say that P is an *emergent* property of systems of kind K is short for 'P is a global [or collective

Part and Whole, Resultant and Emergent 15

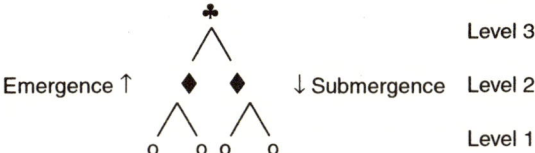

Figure 1.1 Rung-by-rung self-assembly of a complex system from precursors. Every new level is constituted by combinations of lower-level things; every higher-level thing is characterized by emergent properties.

or non-distributive] property of a system of kind K, none of whose components or precursors possesses P.'

No things, no properties thereof. Hence, to ask properly how properties emerge amounts to asking how things with emergent properties arise. In turn, this question boils down to the problem of emergence mechanisms, of which more below. (For the superiority of 'emergence' over 'supervenience,' see Mahner and Bunge 1997.)

3 Levels and Evolution

Whether natural or artificial, assembly may occur step-wise rather than all at once. For example, elementary particles self-assemble into atoms, which combine into monomers, which combine into dimers, which combine into polymers, and so on – the way DNA molecules self-generate from their precursors, not only in the remote past but also in commercially available devices. Some self-assembly processes, such as those that gave rise to stars and organisms, have lasted billions of years. This disposes of the argument from intelligent design, according to which every highly complex system, even if natural, calls for a Designer. Whereas instant self-assembly through chance encounters of zillions of things is indeed extremely unlikely, step-by-step self-assembly through forces of various kinds is nearly unavoidable – if sometimes helped by chance and at other times hindered by it. See figure 1.1.

A *level* is not a thing but a collection of things, namely the collection of all the things that have certain properties in common – such as the collection of all living things or the collection of all social systems. The set L of levels of organization is ordered by the relation < of level precedence, which can be defined as follows (Bunge 1977a). Level L_{n-1} *precedes* level L_n if every element of L_n is composed of entities of level L_{n-1}. This suggests defining the *level hierarchy* as L together with <, or $\mathcal{L} = <L, <>$ for short. (Caution: this modern concept of hierarchy, born in evolutionary biology, must be distinguished from the traditional one, which connotes both sacredness and domination.)

What holds for things also holds, *mutatis mutandis*, for processes and, in particular, for the specific functions of systems, such as the radiation of antennas and manufacture in factories. In some cases we must distinguish several phases or time frames of emergence. For example, the following stages in the emergence of language must be distinguished (MacWhinney 1999: xi): evolutionary, embryological, developmental, online (activities of speakers and listeners), and diachronic (linguistic changes across centuries).

Novelty can be quantitative, as in the cases of lengthening and heating, or qualitative, as in the cases of freezing and cell fertilization. Emergence is the most intriguing and misunderstood kind of novelty. Still, it occurs every time a qualitatively new whole appears – as when new chemicals, organisms of new species, or artefacts of new kinds appear. However, Holland rightly deplores the fact that '[d]espite its ubiquity and importance, emergence is an enigmatic, recondite topic, more wondered at than analyzed' (1998: 3). Only a handful of contemporary philosophers (e.g., Bunge 1959b, 1977a, 1979a; Blitz 1992; Weissman 2000) have paid any attention to emergence. Far more is being written about the plurality of worlds, universal grammar, and the Liar Paradox.

Yet, the fact of emergence and the importance of the corresponding concept have been recognized by a number of thinkers in the course of the last century and a half. Here is a sample. The economist and philosopher John Stuart Mill (1843) pointed out that a chemical compound has properties different from those of its components. The mathematician and philosopher Bernhard Bolzano (1851) remarked that a machine has properties lacking in its separate parts. The polymath George Henry Lewes (1874) introduced the resultant/emergent distinction. The social scientist and activist Friedrich Engels (1878) made much of qualitative leaps and what he unfelicitously called 'the transformation of quantity into quality' (meaning emergence as a result of a quantitative change). In 1912, the psychologist Max Wertheimer introduced the concept of a perceptual Gestalt. In 1923, the psychologist and philosopher Conwy Lloyd Morgan published a book about emergent evolution. The philosopher Roy Wood Sellars (1922) sketched a materialist ontology around the concepts of evolution, emergence, and level. The mathematician and engineer Norbert Wiener (1948) explained self-correction, a whole-system process, in terms of feedback loops. The concepts of emergence and level re-emerged explicitly in a paper by the biologist Alex Novikoff in 1945. The philosopher Nicolai Hartmann (1949) sketched a new ontology based on the concepts of qualitative novelty and layer. The sociologist James S. Coleman (1964) stated that the central problem of sociology is to find out 'emergent group-level regularities' from individual regularities. At about the same time the psychologist and sociologist Jean Piaget (1965) proposed a clear definition of the concept of emergence. (For some of these and other cases see Boring 1950; Sellars et al. 1949; and Blitz 1992.)

The concept of emergence combines two ideas: those of qualitative novelty and of its occurrence in the course of some process, such as freezing or evaporation, ontogeny or phylogeny, technological invention or social innovation. The two concepts in question can be elucidated as follows.

Definition 1.2 A property of a complex object is said to be *emergent* if neither of the constituents or precursors of the object possesses it.

(If preferred, P is an *emergent* property $=_{df} \exists x \forall y (Px \,\&\, y < x \Rightarrow \neg Py)$, where '<' stands for the part–whole relation defined earlier. This definition holds for things of any kind: material, conceptual, or semiotic.)

There is no emergence in itself or separate from emerging things: whatever emerges does so in some (complex) object. (This is not obvious, since in a Platonic ontology the 'forms,' i.e., properties, exist on their own prior to things.) And there is no emergence *ex nihilo*: everything emerges from something, such as interactions among either the constituents of a system or some of them and environmental items. Thus, refraction emerges in a medium from its interaction with light; and language emerges in the heads of toddlers interacting with other humans.

A thing may have an emergent property from the start, or it may acquire such a property by virtue of being incorporated into a system – as in the cases of a firm that hires a worker (who becomes an employee answering to his superiors) or a bride incorporated into her husband's family (who becomes a daughter-in-law under her mother-in-law's rule). The former may be said to be *intrinsic* (or global) and the second *relational* (or contextual) emergence.

A new thing possessing an emergent property is sometimes called an *emergent*. And the process whereby a thing loses one or more properties may be called *submergence*. For example, a newly formed cell is an emergent, whereas cell death instantiates submergence. Other familiar examples are the formation and dismantling of a neuron assembly that occur in learning and forgetting something, as well as the organization and disintegration of a social system such as a firm.

Things are not the only possible emergents: processes too may emerge and submerge. For instance, if several oscillators interact, they may entrain one another, so that the oscillator population acquires a rhythm of its own: the whole population starts behaving as a unit. There are plenty of physical, chemical, biological, and social examples of such emergence of coherence, such as the synchronic 'singing' of cicadas and the rhythmic clapping of a crowd (see

Figure 1.2 Any long-term history of a concrete thing, such as a developing organism or an evolving society, involves both the emergence of some properties and the submergence of others.

Winfree 1980). As with coherence, so with its dual, or decoherence. For example, a quantum-mechanical system, initially in a superposition of states, decoheres quickly when it interacts with its environment.

All developmental and evolutionary processes are continuous in some respects because of the conservation of certain constituents or features. At the same time, those processes are discontinuous in other respects by virtue of the birth and disappearance of qualitative novelties. Hence both gradualism and saltationism, whether with regard to molecular, biological, social, or intellectual development and evolution, are one-sided.

The long-term history of a thing, then, can be characterized by the properties it gains together with those it loses. Let $P(t)$ and $P(t')$ designate the sets of all properties of a given thing at times t and t' respectively, where $t' > t$. Then the gain γ and the loss λ of properties over the $t' - t$ period can be defined as $\gamma(t, t') = P(t') \setminus P(t)$, and $\lambda(t, t') = P(t) \setminus P(t')$ respectively, where '\' stands for set-theoretic difference ($A \setminus B =_{df} A \cap \bar{B}$). Obviously, the emergent properties during that period are those in the gain set, whereas the loss set includes the submerged properties. The intersection of those two sets equals the totality $\kappa(t, t') = P(t) \cap P(t')$ of conserved properties. One of these, in fact the most important of all, is energy, or the ability to change. See figure 1.2.

The previous definitions allow us to state the following general ontological hypotheses.

Postulate 1.1 All processes of development and evolution are accompanied by the emergence of some properties and the submergence of others.

Postulate 1.2 Only one property is common to all concrete things, and it never submerges: this is the ability to change.

The above assumptions entail informally

Theorem 1.1 All long-term histories are gradual in some respects (properties) and discontinuous in others.

An immediate consequence of this theorem is that, contrary to popular belief, something is conserved even in the most drastic of metamorphoses:

Corollary 1.1 There are neither absolute beginnings nor total revolutions.

The preceding postulates and theorems suggest that ontology need not be impregnable to empirical testing. They also suggest that science stands to gain from science-oriented philosophy. For instance, neglect of the level structure of reality is likely to lead to either error or to a dead end.

The history of behaviour genetics is a case in point. The workers in this discipline have unsuccessfully tried to establish direct relationships between individual genes and behaviour features, such as intelligence, alcoholism, schizophrenia, or even religiosity. But no such one gene–one trait relations have been found. Nor are they likely to be found, and this not only because genes come in clusters or networks rather than separately. In addition, and perhaps principally, the brain must be interpolated between molecule and behaviour, since behaviour is an output of the brain. In other words, to find out how genic changes affect behaviour changes one must investigate how the former alter brain processes. The methodological moral of this story is, Don't skip levels.

4 Structure and Mechanism

People can join to form either crowds, such as political constituencies, or organizations (systems), such as political parties. Whereas crowds are structureless, organizations have definite structures, as well as other systemic or emergent properties. Crowds may gather around external events such as large fires and soccer matches: they may not be held together by strong bonds among their components, whence they may disperse as soon as the attracting event ceases. By contrast, the constituents of a system, from molecule to business firm, are held together by bonds. This is why they last, however shortly.

Moreover, one and the same collection of things can be organized in different ways. For example, carbon atoms can join to form either C_{60} molecules, diamond crystals, or graphite fibres. A C_{60} molecule is a hollow mesoscopic object similar in shape to the geodesic dome; diamonds are translucent, hard, and do not burn easily; and graphite is black, soft, and easily combustible. What explains the differences between these three forms of carbon is structure or organization. Likewise, one and the same group of people can be organized into a study group, a sports team, a religious congregation, a political club, a business firm, or what have you. Again, the difference is one of structure.

Since the word 'structure' has been given several meanings in the literature, we should agree on a definition of it. Here is ours:

Definition 1.3 The *structure* (or *organization* or *architecture*) of an object is the collection of relations among its components. Symbol: $S(s)$.

Now, two or more items may be related in either of two ways: bonding or non-bonding. Whereas in the first case the relation makes a difference to the relata, in the second it does not. Spatial and temporal relations, such as 'to the left of' and 'later than,' are non-bonding. (However, such relations, in particular those of spatial and temporal contiguity, may render bonding relations possible.) By contrast, electromagnetic, chemical, biological, ecological, and social relations are bonding: they make a difference to the relata. However, this does not mean that all bonds are causal: in fact, only some are. For example, chemical and loyalty bonds are not causal.

Bonds are the key to self-organization. In particular, some chemical bonds drive the synthesis of molecules and supramolecules. Unsurprisingly, the strongest bonds, such as nuclear forces, give rise to small systems, whereas the weakest bonds hold the constituents of large systems together; hence, paradoxically, the vast reach of weak bonds, from supramolecules to cells to social networks to nations.

The formation and severance of bonding relations in material systems involves some energy exchanges. Obviously, this does not hold for conceptual systems, such as classifications and theories, or in semiotic systems, such as diagrams and texts: Nothing happens in these, so that the concept of binding energy does not apply to them. The bonds that hold such systems together are logical or semantic. In either case, the structure of a complex object equals the set B of its bonds plus the set $\bar{B}$ of nonbonding relations: $S(s) = B \cup \bar{B}$. The first part may be called the *bonding structure*, or $S_B(s)$, and the second the *non-bonding structure*.

These concepts suggest the definition of a system as an object with a non-empty bonding structure:

Definition 1.4 A *system* is an object with a bonding structure. (Formally: s is a system $=_{df} S_B(s) \neq \emptyset$.)

However, this definition will only help one to recognize a system once its structure has been unveiled. Before this task has been performed we can use this criterion: new systems are characterized by new properties. In other words, emergence indicates system, the knowledge of which calls for the unveiling of its structure.

Let us finally introduce the notion of a mechanism or modus operandi – the process or processes that make a system tick. This feature of concrete (material) systems differentiates them not only from simples but also from both conceptual systems, such as theories, and semiotic systems, such as diagrams, where nothing ever happens. We propose the following:

Definition 1.5 A *mechanism* is a set of processes in a system, such that they bring about or prevent some change – either the emergence of a property or another process – in the system as a whole.

Examples: 1/ The mechanism that characterizes a self-winding watch is this causal chain: Wrist motion → spring compression and the accompanying storing of elastic energy → motion of the watch's hands (transformation of elastic into kinetic energy). 2/ The mechanism of light emission is the random decay of atoms or molecules from higher energy levels. 3/ Catalyzers act by forming short-lived intermediate compounds with one or more of the reagents in a chemical reaction. 4/ There are two biological growth mechanisms: cell growth and cell division. 5/ The main mechanisms of biological evolution are mutation, recombination, and natural selection. 6/ Social systems work through mechanisms of several kinds, among them work, trade, cooperation, competition, and one-sided dependence (in particular parasitism). 7/ Technological innovation is pushed by research and pulled by the market.

Clearly, simple things, such as quarks and electrons, have no mechanisms. Nor do conceptual systems, such as classifications, or semiotic systems, such as languages considered in themselves, have mechanisms. Only concrete systems, such as atomic nuclei, cells, and governments, have mechanisms.

5 Emergence and Explanation

The concept of emergence introduced above is ontological, not epistemological. Therefore, contrary to a widespread opinion, it has nothing to do with the possibility or impossibility of explaining qualitative novelty. Hence, it is mistaken to define an emergent property as a feature of a whole that cannot be explained in terms of the properties of its parts. Emergence is often intriguing but not mysterious: explained emergence is still emergence (Bunge 1959b: 110). For example, quantum chemistry explains the synthesis of a hydrogen molecule, or H_2, out of two hydrogen atoms. Yet H_2 is obviously an emergent relative to H_1: for example, H_2 has a dissociation energy and a band (rather than a line) spectrum.

Emergence processes are far more difficult to explain than are aggregation and dispersion processes. For example, there is no accepted theory of the way(s) organisms emerged about three billion years ago from abiotic materials – surely one of the most sensational of all emergence processes. Much the same holds for the emergence of mind, both in evolution and during development. By contrast, something is known about the emergence, about ten millennia ago, of the domestication of plants and animals: it was necessitated by population growth, which in turn it favoured.

Likewise, we know something about another revolution that occurred five millennia later, namely the emergence of civilization and the concomitant institutions, such as public services, legal systems, taxation, and armies. It seems that all of these systems were set up to manage irrigation, agriculture, urban settlement, trade, defence, and education. In other words, every one of the subsystems of a civilized society is characterized by a specific function or mechanism.

Some of the most interesting and toughest problems, in any science, are to discover mechanisms of emergence and submergence. This is the task of figuring out and, if possible, effectively finding the processes that end up in the assembly (or the dismantling) of a system characterized by one or more emergent properties. Think of the following open problems. What mechanisms are 'responsible' for the self-assembly of nucleotides into genes? What is the mechanism of protein synthesis? (Don't answer that it is a matter of templates and information transfer, for these are just didactic props. A precise explanation calls for a precise, i.e., mathematical, theory.) What are the mechanisms causing the assembling of neurons into systems capable of perceiving a figure or uttering a word? How and why do people get together to push for or prevent a social reform? What has caused the recent decline of the family, both nuclear and extended, in so many advanced countries?

Hegel and his followers claimed that there is a universal emergence mechanism, namely the thesis-antithesis-synthesis sequence. However, the dialectical metaphysicians did not disclose the secret of this alchemy: they only offered a couple of alleged examples, such as nothingness-being-becoming, and acorn–oak tree–new acorn. Even assuming that these are genuine instances of the alleged law, the following counter-examples destroy the generality claim: sixty carbon atoms combine to form one fullerene molecule; a zillion water molecules condense to form a water droplet; a hundred like-minded citizens join to form a political party. Where is the 'anti' in these fusion processes?

An adequate and general definition of the conditions for emergence is elusive, if not impossible, given the large variety of emergence mechanisms. What mechanism might be common to freezing and magnetization, nuclear fusion and cell clumping, chemical combination and political alliance, herd formation and the merger of corporations? Water droplets emerge from water molecules as a result of hydrogen bonds; magnets emerge from the alignment of atoms whose spins were initially distributed at random; social groups are constituted by people with similar interests or under external pressure; business mergers arise from the wish to quash competition – and so on and so forth.

Thus, we need different theories to account for widely different emergence mechanisms. This is why scientific explanations are specific: because mechanisms are specific. In other words, there are no all-encompassing explanations because there are no one-size-fits-all mechanisms. This alone should render the universality claims of dialectical, psychoanalytic, natural-selection, and rational-choice explanations suspect.

When a mechanism in a system is hypothesized and found, it can be claimed that the behaviour of the system in question has been explained. Otherwise one only has either a description or a subsumption under a generalization. For example, to say that the vending machine dispensed a candy bar because a coin was inserted only describes superficially (functionally) the way the machine works. In general, input-output (or black-box) models and functional accounts are purely descriptive, and therefore shallow – descriptive rather than explanatory. Likewise, to say that so and so died of old age does not explain why he did not die a year earlier or a year later. A genuine explanation of life and its end, just like an explanation of the coin–candy bar correlation, requires hypothesizing mechanisms, that is, building translucent boxes. (For black and translucent boxes, see Bunge 1964, 1967a, 1999; for the limitations of functional accounts, see Mahner and Bunge 2001.)

The preceding motivates proposing the following convention:

Definition 1.6 To *explain* X is to propose the mechanism(s) that give(s) rise to (or else maintains or destroys) X.

Caution: Mechanisms can be modelled by translucent boxes but are not to be confused with them, as they sometimes are (as in Hedstrom and Swedberg 1998). There are two main reasons for avoiding such confusion. One is that there were mechanisms long before any of them were modelled. The second reason is that one and the same real mechanism may be modelled in different ways, not all of them equally true or deep. For example, economics students are taught that all markets are always in or near equilibrium – a global property. The underlying clearing mechanism would be negative feedback. That is, a market is modelled as a black box with supply as input and demand as output. Any change in demand is fed back into supply: increases (decreases) in demand would cause increases (decreases) in supply, so that the end result would be the ideal condition where supply equals demand. This is the standard mechanismic explanation of market equilibrium. True, it fails to explain the all-too-frequent market disequilibria. But this only shows that real mechanisms must be distinguished from their models.

Another example is this. Our economic and political leaders have assured us that globalization, or worldwide trade liberalization, is the automatic answer to the poverty of individuals and the backwardness of nations. However, they have not deigned to explain why this should be so. That is, they have not described the mechanism that would translate free trade into universal prosperity and equality. Worse, socio-economic statistics falsify that claim. Indeed, under globalization, inequality rose significantly nearly everywhere over the last two decades of the twentieth century (Galbraith and Berner 2001; Streeten 2001). One cause of this increase is that high-skilled jobs are vanishing in high-income countries while they multiply in the others; another is that social services have been reduced everywhere; a third is that the more powerful nations subsidize some industries while exporting others. In general, there can be no real freedom of any kind among unequals for, if some agents are more powerful than others, they will tend to use their power to their own benefit. The only way free trade can work to the benefit of all the partners concerned is when an impartial umpire sees to it that the wealthier help the poorer attain a comparable level, the way the European Union has been doing successfully for decades.

Contemporary biologists tend to try and explain everything in terms of genes and natural selection. But they seldom succeed in this ambitious endeavour, because a large number of mechanisms are neither genetic nor selective. For example, the growth of bones, such as teeth, is regulated by certain genes, and natural selection results in most of us having teeth that are neither cat-like nor sabre-tiger-type teeth. The explanation of teeth growth must be sought elsewhere, with the gradual emergence of layers of enamel, dentine, nerve, and other components. Genes and the proteins they help synthesize make some things possible, and natural selection makes others impossible, but neither of them creates anything. The detailed emergence mechanisms, whether physical, chemical, or biological, are specific. And as long as we do not understand or at least guess at such mechanisms, we cannot claim to understand anything about the corresponding processes.

Although our elucidation of explanation is likely to sound familiar to any scientist, it is at variance with the standard philosophical account of explanation as subsumption under a generalization (Mill 1952 [1843]; Braithwaite 1953; Popper 1959 [1935]; Hempel 1965). According to this, the so-called covering-law model of explanation, a fact is explained if its description is deducible from a law statement together with the pertinent circumstances, such as the initial or the boundary conditions. (In short: Law & Circumstance $\Rightarrow$ Explanandum.) Clearly, this account covers the logical aspect of explanation, but misses its ontological kernel – mechanism. However, some philosophers have since realized that in the sciences only the description of a mechanism counts as an explanation (Bunge 1959a, 1964, 1967a [1998], 1996, 1997c, 1999; Athearn 1994; Machamer, Darden, and Craver 2000).

Concluding Remarks

The first methodological maxim we learn is, Analyse! The second is, Synthesize! This is because, to understand how a complex item works, we must first decompose it, and then connect its parts and place the whole in a wider context. In addition, the world is made up of interconnected systems. If the world were just a pile of items, analysis would suffice; and if it were a solid block, only the pre-analytic intuition of the whole could help. Fruitful methodology follows, inspires, and checks ontology. All of this points to systems and their emergence and submergence – subjects of the next chapter.

2

System Emergence and Submergence

How do new systems emerge and dismantle? Can there be emergence out of nothing? How many basic types of system are there? And how can systems be modelled? These are some of the problems to be tackled in the present chapter. Regrettably, very few contemporary philosophers are interested in these problems. Worse yet, the very concept of a system is absent from 'official' ontology or metaphysics (see, e.g., Lowe 2002). Yet, all scientists and technologists deal with systems and face those questions once in a while, though seldom in general terms. For instance, astronomers keep looking for extrasolar planetary systems; biologists are intrigued by the problem of the origin of living systems; physiologists investigate the interactions among the nervous, endocrine, immune, and muscular systems; historians investigate the emergence and decline of social systems such as welfare capitalism and statist socialism; and engineers design new artificial systems such as nanomotors and electroneural prostheses.

All of these specialists might presumably benefit from some philosophical ideas about the emergence and submergence of systems of any kind, since a general idea is expected to strike at the nucleus of the matter, and thus to suggest a potent strategy for its study or control. Imagine where science would be today if Newton, instead of crafting the earliest general scientific theory in history, had applied himself to the industrious but mindless observation of the fall, rotation, oscillation, and collision of bodies of various kinds – bone, steel, wood, rubber, and so on. Full-blown science only emerges when natural history becomes spent or gets bogged down in uncounted unassimilated details. And good philosophy helps ask deep questions and build deep scientific theories, that is, theories capable of guiding the search for patterns underneath apparently stray particulars, and of explaining facts in terms of mechanisms.

1 System Emergence

As we saw in chapter 1, there are two ways a whole may come into being: by association or by combination. The accretion of dust particles and the coalescence of droplets exemplify association; so do the formation of garbage dumps, water pools, sand dunes, clouds, crowds, and columns of refugees fleeing from a disaster. What characterizes all of these wholes is the lack of a specific structure constituted by strong bonds: such wholes are neither cohesive nor, consequently, lasting.

However, when an accretion process keeps going beyond a certain threshold, it can give rise to qualitatively new things, as in the sequence: Dust → Pebbles → Boulders → Planetesimals → Planets. A more familiar example is: Cotton wool → Thread → Fabric → Dress. These are illustrations of what dialectical philosophers call 'the law of the transformation of quantity into quality' – which, taken literally, is an oxymoron.

Unsurprisingly, any account of the emergence of qualitative nova calls for new ideas. Consider, for instance, the emergence of laser beams from photons, of crystal from atoms, or, in general, of what may be called classons from quantons. Such emergence, still only partially understood, necessitated the replacement of classical physics by quantum physics, which is characterized by the concepts of quantization, superposition of eigenstates, spin, anticommutation, and vacuum polarization. Likewise, the account of speciation and species extinction required the emergence of evolutionary biology, with its peculiar concepts, such as those of natural selection and exaptation.

However, let us go back to the coalescence processes. When two or more things get together by interacting strongly in a specific way, they constitute a system. This is a complex thing possessing a definite structure. Atomic nuclei, atoms, molecules, crystals, organelles, cells, organs, multicellular organisms, biopopulations, ecosystems, human families, business enterprises, and other organizations are systems. They may all be said to emerge through combination or self-organization rather than aggregation – even though, once generated, some of them may grow by accretion or decline by attrition.

What holds for things also holds, *mutatis mutandis*, for events (changes of state) and processes (sequences of states). For example, random molecular movements aggregate into macrophysical regularities; likewise, some of the actions of mutually independent persons give rise to social statistical regularities – for instance, average numbers of marriages, accidents, and suicides. In particular, accident on one level may translate into pattern on a higher level:

Macro-level   Global regularities (e.g., constant averages and variances)
↑
Micro-level   Individual irregularities (e.g., chance encounters and errors)

Self-organization, in particular biological morphogenesis, is a wondrous yet ubiquitous and still poorly understood process. No wonder that it has been the object of much pseudo-scientific speculation peppered with high-sounding but empty expressions, such as 'building force,' 'entelechy,' ' vital impulse,' 'morphogenetic field,' and the like. Such factors have been invoked by vitalists, who have regarded them as immaterial, hence beyond the reach of physics and chemistry. They are neither described in any detail nor manipulated in the laboratory. Hence, talk of such factors is just hand-waving, when not magic-wand-waving

By contrast, the scientific approach to self-organization is down-to-earth yet imaginative. Let us peek at a recent instance of this approach: the work of Adams et al., (1998). Colloids consisting of tiny rods and spheres were randomly suspended in a buffer sealed in glass capillaries, then left to their own devices, and observed under a microscope. The rods were viri, and the spheres plastic balls; the former were negatively charged, and the latter positively charged. After some time, the mixture separated spontaneously into two or more homogeneous phases. Depending on the experimental conditions, a phase may consist of layers of rods alternating with layers of spheres; or the spheres may assemble into columns.

Paradoxically, these demixing processes of various types are explained in terms of repulsions between the charged particles – which, intuitively, should preclude the crowding of particles with the same charge. And the equally paradoxical decrease in entropy (increase of order) is explained by noting that the clumping of some of the colloids is accompanied by a raise in the translational entropy of the medium. In any event, the entire process is accounted for in strictly physical terms. At the same time, the authors warn that their results are at variance with the pertinent theory – though of course not with any general physical theories. Such incompleteness is typical of factual science in the making, by contrast with pseudo-science, where everything is cut and dried from the start. (However, see Ball 2001 for a large catalogue of stunning self-organization processes, biological as well as physical, that are rather well understood in terms of standard physical and chemical laws.)

Unlike mere aggregates, systems are more or less cohesive. However, they may break down either as a result of conflicting relations among their parts or in response to external forces. That is, a system may end up as an aggregate. And conversely. For example, a star starts as a cloud of dust and gas; it becomes a star when this aggregate condenses; and when it does, its density and temperature increase to the point when thermonuclear fusion starts.

The concept of a system is bound to occur in the very statement of any scientific problem dealing with wholes of some kind. Think of the problem of solving a system of equations of some kind. Such solution requires tackling the system as a whole, not equation by equation, because every variable in one equation is related to variables in other equations of the same system. Or think of solving any non-trivial problem in mechanics, such as a many-body problem. One does not investigate the motion of each interacting body and then somehow joins the individual solutions. Rather, one attempts to solve the system of $3n$ equations of motion that describe the motion of each of the n bodies in relation to the others. Something similar happens with the equations describing the fields generated by two or more electric charges or currents. Contrary to the methodological-individualist prescription, one starts with the system, though not as a sealed unit but as a complex thing made up of distinct interacting constituents.

In accounting for the emergence and dismantling of aggregates, we focus on their composition and environment, in particular the external stimuli that favour the aggregation (or the dispersion) process. In this case structure matters little: a heap does not cease to be a heap if its constituents exchange places. Therefore, basically we explain aggregates (and their dispersion) in terms of their composition and environment. By contrast, structure, in particular internal structure, is essential to systems. Indeed, to account for the emergence of a system we must uncover the corresponding combination or assembly process and, in particular, the bonds or links resulting in the formation of the whole. The same holds, *mutatis mutandis*, for any account of a system's breakdown.

In other words, we explain the emergence, behaviour, and dismantling of systems in terms not only of their composition and environment, but also of their total (internal and external) structure. Nor is this enough: we should also know something about the system's mechanism or modus operandi: that is, what process makes it behave – or cease to behave – the way it does.

A way to find out the mechanism that makes a system tick is to look for the specific functions of the system – that is, which processes are peculiar to it (Bunge 2003b). Indeed, we define a mechanism as a process necessary for the emergence of either a property or another process – the specific function. See table 2.1.

In some cases, a given specific function can be accomplished by systems with different mechanisms. In these cases, the systems in question can be said to be *functionally equivalent*. For example, transportation can be effected by car, ship, or plane; some computations can be carried out by either brains or

30 Emergence

Table 2.1 Specific functions and associated mechanisms of familiar systems

System	Specific function	Mechanism(s)
River	Drainage	Water flow
Chemical reactor	Emergence of new molecules	Chemical reactions
Organism	Maintenance	Metabolism
Heart	Blood pumping	Contraction-relaxation
Brain	Behaviour and mentation	Interneuronal bonds
Time piece	Time-keeping	Several
School	Learning	Teaching, study, discussion
Factory	Production of merchandise	Work, management
Shop	Distribution of merchandise	Trade
Scientific laboratory	Growth of knowledge	Research
Scholarly community	Quality control	Peer review
Courthouse	Seeking justice	Litigation
NGO	Public service	Voluntary work

computers; and the redress of grievances can be sought by collective bargaining, litigation, violence, or bribing. (To find the function given the mechanism is a direct problem. By contrast, going from function to mechanism is to work on an inverse problem – one that, if soluble at all, has more than one solution given that the functions-mechanisms map is one-to-many.) A common fallacy is to infer system identity from functional equivalence. This fallacy, called *functionalism*, is the core of the computationalist view of the mind, on which more in chapter 9, section 3.

2 Emergence *ex nihilo*?

Whatever emerges does so from some pre-existing thing: this is one of the ontological presuppositions of all science and all technology. For example, the earliest organisms are assumed to have been the final products of a process of stepwise self-assembly starting with prebiotic materials. (Yes, there must have been spontaneous generation, but it is likely to have taken a billion years or so.)

And yet there is one rather fashionable theory, namely quantum cosmology, which postulates that the universe originated *ex nihilo* through what quantum physicists call a tunnelling process (see, e.g., Atkatz 1994). Clearly, this hypothesis contradicts Lucretius's famous principle *Ex nihilo nihil fit*, exemplified by the energy-conservation principle, and that has always been regarded as a cornerstone of every naturalist cosmology, whether philosophical or scientific. Let us face this challenge.

This is not the first time cosmologists have questioned Lucretius's principle. Half a century ago, in a heroic attempt to save their now defunct 'steady-state' theory of the universe – an alternative to the Big Bang conjecture – Hermann Bondi and Fred Hoyle postulated a 'law' of continuous creation of matter. I dubbed it 'magic' because it contradicts all of the conservation theorems of physics (Bunge 1962b). Popper rebuked me for this criticism: he regarded the theory as scientific because it is refutable.

In my view, refutability, though highly desirable, is neither necessary nor sufficient for scientificity. First, existential hypotheses, such as those of the existence of gravitational waves, extrasolar planets, and the neural system that 'binds' the various features of language comprehension (syntactical, semantical, and phonological), are only confirmable. Second, hypergeneral theories – such as classical mechanics, the synthetic theory of evolution, and information theory – are untestable without further ado: only specific theories (theoretical models) are empirically testable (Bunge 1973b). (In other words, any basic general equations must be enriched with subsidiary assumptions and data to yield precise predictions and explanations.) Third, compatibility with the bulk of science (external consistency) is a much weightier criterion of scientificity than refutability (Bunge 1967a).

Quantum cosmology is guilty of a similar mistake as the steady-state theory. Indeed, although it is based on two solid theories – quantum mechanics and general relativity – it contradicts both of them, since it violates all of the conservation principles of the founding theories. Still, contrary to what some of its followers claim, at least the theory does not assume the sudden emergence of the universe out of nothing. Indeed, it postulates that before the Big Bang there was the so-called vacuum field. This fluctuating field has neither mass nor electric charge, and its average intensity is zero. But, since it has a positive energy density, that field is material according to the definition of 'material' as whatever possesses energy, or the ability to change (see Bunge 2000c). However, quantum cosmology is still far too speculative to either undermine ontology or support theology (see Stenger 1995). In any event, it is not a paragon of successful theory merger. Besides, the most recent astronomical observations seem to support the hypothesis that the universe is infinite, eternal, and flat rather than curved (Tegmark 2002).

3 Submergence: System Dismantling

The loss of higher-level properties may be called *submergence*. Since properties have no independent existence, but are possessed by things, property submergence is just a feature of the (partial or total) dismantling of systems of some kind. For example, it occurs when a molecule dissociates into its atomic precursors, and when the members of a family or a political party disband.

Only physicists, chemists, and engineers have studied intensively submergence processes, such as ionization, nuclear fission, chemical dissociation, and the breakdown of solids. Biologists have started recently to deepen their understanding of the aging and death mechanisms, such as oxidation, telomere shortening, unrepaired damage, and programmed cell death. So far, social scientists have only been fascinated by a few very large-scale spectacular dismantling processes, notably the fall of the Roman Empire and the French Revolution. The crumbling of the Soviet Empire took them all by surprise, and has not yet been satisfactorily explained. I attribute this failure to the adoption of sectoral approaches (purely economic, political, or cultural) to what was actually a systemic crisis that had been brewing over several decades (Bunge 1998).

One of the features of the collapse of the so-called Socialist camp is the submergence of the legal and moral order. All of a sudden millions of people accustomed to being told what to do were left to fend for themselves, and in particular to invent and try out new social and moral norms in a normative vacuum. The persistent social disarray of the former Soviet societies suggests that this submergence process is far from over. Yet it does not seem to have attracted, as it should have, the attention of an entire army of social scientists. Thus, the latest work on the emergence of social norms (Hechter and Opp, eds 2001), authored by fourteen scholars, several of them renowned, omits the subject altogether.

I submit that a major reason for this lack of interest in a massive process of normative crisis that has been unfolding in front of us since 1989 is the following. The system in question was more than macrosocial: it was megasocial, and its dissolution affected all the aspects of life, from day-to-day survival to work places to modes of thinking, in particular ideological allegiances. Any serious study of such a process calls for plenty of data that are hard to procure, as well as the adoption of a systemic and multilevel approach instead of the individualist one that prevails nowadays among the more lucid students of society. Indeed, a system of social and legal norms is a code for an entire society, and it covers all manner of individual and concerted conduct. Its scientific study requires far more than anecdotes about a handful of leaders and ingenious applications of the Prisoner's Dilemma game.

Philosophers won't remain satisfied with examples of system dismantling: they will seek general patterns. However, there is but one general dismantling mechanism, namely the weakening of the internal bonds that hold the system together. Such weakening can happen in various ways. The most common of them is the intrusion of an external agent, as in the case of the lover who disrupts a married couple, and of the soap we use to wash our hands. The case of soap merits mention because of its familiarity, simplicity, and generality.

The surface of pure water is hard to penetrate, because of the strength of the hydrogen bonds that hold the water molecules together: this is the source of the surface tension that allows certain bugs to skate on water surfaces. The effect of soap is to weaken the hydrogen bonds and thus to make the contact of body parts with water more intimate. The mechanism is this. Soap contains stearic acid molecules; these are, schematically, rods with two tips: one hydrophilic, or water-friendly, and the other hydrophobic, or water-repellent. When in water, the hydrophilic tip of the molecule dips into the liquid between water molecules, thus weakening or even breaking the hydrogen bonds.

In sum, ironically, to understand the dismantling of a system we must understand the bonds that gave rise to it and have held it together. Shorter: Emergence explains submergence.

4 System Types

There are systems of several different kinds. A first classing of them is the ideal/material dichotomy: whatever is ideal is not material, and conversely. Idealists as well as materialists uphold this dichotomy. However, whereas idealists attribute independent existence to ideal objects, materialists contend that they exist to the extent that they are thinkable by someone.

However, the ideal/material dichotomy is insufficient, since some material systems, such as the social, technological, and semiotic ones, incorporate or express ideas. A somewhat finer distinction of systems is as follows:

1 *Natural*, such as a molecule, a river network, or a nervous system
2 *Social*, such as a family, a school, or a linguistic community
3 *Technical*, such as a machine, a TV network, or a high-tech hospital
4 *Conceptual*, such as a classification, a hypothetico-deductive system (theory), or a legal code
5 *Semiotic*, such as a language, a musical score, or a blueprint for a building

Note the following points. First, this typology belongs in an emergentist (or non-reductionist) materialist ontology. It makes no sense in alternative ontologies. In particular, it is as unacceptable to idealism (in particular Platonism and phenomenalism) as it is to vulgar materialism (in particular physicalism).

Second, our typology is not a partition, let alone a classification, because (a) most social systems are artificial as well as social: think of schools, businesses, or armies; (b) some social systems, such as farms and factories, contain not only people but also animals, plants, or machines; (c) all semiotic systems, even the natural languages, are artefacts, some of which – such as scientific formulas and diagrams – designate conceptual systems; and (d) activities in all social systems involve the use of semiotic systems. Still, the above typology does represent in a rough manner some salient objective features of the systems that compose the world.

Quick (and therefore vulnerable) definitions of the five above concepts follow.

Definition 2.1 A *natural* system is one all of whose components, and the bonds among them, belong in nature (i.e., are not man-made).

Definition 2.2 A *social* system is one some of whose components are conspecific animals, and others are artefacts (inanimate like tools or living like domestic animals).

Definition 2.3 A *technical* system is one constructed by people with technical knowledge.

Definition 2.4 A *conceptual* system is one composed of concepts.

Definition 2.5 A *semiotic* system is one composed of artificial signs (such as words, musical notes, and figures).

Definition 2.6 An *artificial* system is one some of whose components are made.

Obviously, the class of artificial systems equals the union of technical, conceptual, and semiotic systems, as well as the formal social organizations, such as schools, business firms, and governments. All languages are artificial in being made. The difference between 'natural' languages, such as English, and 'artificial languages,' such as predicate logic (when used as a language, not as a calculus), is that the latter are designed instead of evolving more or less spontaneously.

5 The CESM Model

Three of the more common definitions of a system found in the systems-theory literature are the following:

- D1 A system is a set, or collection of items, that behaves as a whole.
- D2 A system is a structured set or collection.
- D3 A system is a binary relation on a set of items of some kind, such as the input-output pairs in a black box.

None of these definitions is suitable for scientific purposes. D1 is defective because (a) it does not point to the features that makes a collection behave as a unit – namely the emergent properties; and (b) it identifies 'set' with 'collection' – a mistake because, whereas sets are concepts, and their composition is fixed once and for all, that of a concrete collection or aggregate, such as a biospecies, may change over time. D2, though not wrong, is incomplete, in failing to specify the structure of a system, that is, the collection of relations that hold the constituents together. And D3 is flawed too because it only holds for a black box, which is the coarsest representation of a complex material thing, and moreover one assuming that the system only changes in response to external stimuli, while actually the internal forces are at least equally important.

Because of these objections we proposed earlier our own definition of a system as a structured object. Still, this alternative definition, though correct, is far too coarse, because it fails to include the enviroment and the mechanism of a system. The following characterization, to be called the *CESM model*, is more comprehensive. It states that any system *s* may be modeled, at any given instant, as the quadruple

$$\mu(s) = \langle C(s), \mathcal{E}(s), S(s), \mathcal{M}(s) \rangle,$$

where
  $C(s) = $ *Composition*: Collection of all the parts of *s*;
  $\mathcal{E}(s) = $ *Environment*: Collection of items, other than those in *s*, that act on or are acted upon by some or all components of *s*;
  $S(s) = $ *Structure*: Collection of relations, in particular bonds, among components of *s* or among these and items in its environment $\mathcal{E}(s)$.
  $\mathcal{M}(s) = $ *Mechanism:* Collection of processes in *s* that make it behave the way it does.
  Examples: 1/ A two-member semigroup, $C(s) = $ set of nondescript elements a and b; $S(s) = $ concatenation (as in $a \oplus b, b \oplus a, a \oplus a, b \oplus b, a \oplus b \oplus a$, and $b \oplus a \oplus b$); $\mathcal{E}(s) = $ predicate logic; $\mathcal{M}(s) = \emptyset$. 2/ A sentence is a (semiotic) system, for it results from concatenating some words. 3/ A text may or may not be a system, depending on whether its component expressions 'hang together' in some way – either for referring to the same subject or for being connected by the relation of implication. 4/ An atom, where $C(s) = $ the constituent particles and associated fields; $\mathcal{E}(s) = $ the things (particles and fields) with which the atom interacts; $S(s) = $ the fields that hold the atom together, plus its interaction with items in its environment; $\mathcal{M}(s) = $ the processes of light emission and absorption, combination, etc. 5/ A linguistic community, where $C(s) = $ collection of people who speak the same language; $\mathcal{E}(s) = $ the culture(s) where the language is used; $C(s) = $ the collection of linguistic communication rela-

tions; $\mathcal{M}(s)$ = the production, transmission, and reception of symbols. 6/ A firm, where $C(s)$ = personnel and management; $\mathcal{E}(s)$ = market and government; $C(s)$ = the work relations among the firm's members, and among these and the environment; $\mathcal{M}(s)$ = the activities that end up in the firm's products. 7/ Finally, a grab bag of non-systems: an arbitrary set of unspecified elements devoid of structure; an arbitrary collection of symbols picked at random from one or more languages; the pile of parts of a dismantled machine; a clan or a village whose members have migrated to the four corners of the world.

Note the following points. First, a collection may or may not have a constant membership; only if it does may it be called a set. Because concrete systems are always in flux, their composition may change over time: think of a natural language or of a linguistic community. Second, everything except for the universe as a whole has an environment with which it interacts. Third, the word 'bond' (or its synonym 'tie') stands for a relation that makes some difference to the relata. For example, an interaction is a bond, whereas the relations of being larger than and to the left of are not. Fourth, the structure of a system may be divided into two: (a) the *endostructure*, or collection of bonds among members of the system; and (b) the *exostructure*, or collection of bonds among the system components and environmental items. The exostructure of a system includes two particularly important items: the *input* and the *output*. Whereas the former is the collection of actions of environmental items upon the system, the output is the action of the system upon its environment. Any system model that includes only the input and the output is called a *black box*, whereas a model that also represents the endostructure and the mechanism may be called a *translucent box*. Fifth, the subset of the exostructure that contains only the system members that hold direct relations with environmental items may be called the system *boundary*. Note that (a) this concept is wider than that of shape or geometric form; (b) the explicit mention of the boundary or edge is required whenever the system mechanism depends on it, as in the cases of quantum-mechanical systems and continuous media confined in finite regions; and (c) the universe has no boundary.

Note also that the input-output or black-box model is a special case of the CESM model. Indeed, the box with input and output terminals is a CESM model in which the composition is a singleton, the environment is only sketched, the structure is the set of inputs and outputs, and the internal mechanism is specified in purely functional (behavioural) terms. This is why behaviourism is sometimes called 'the empty organism model.' Cybernetics is another example of emphasis on structure at the expense of composition, since it focuses on control systems regardless of the 'stuff' they are made of (see, e.g., Wiener 1948, Ashby 1963).

As simple as it looks, a CESM model is unwieldy in practice, for it requires the knowledge of all the parts of the system and of all their interactions, as well as their links with the rest of the world. In practice we use the notions of composition, environment, structure, and mechanism *at a given level*. For example, we speak of the atomic composition of a molecule, the cellular composition of an organ, or the individual composition of a society. Except in particle physics, we never handle the ultimate components of anything. And even in particle physics one usually neglects a number of interactions, particularly with environmental items.

More precisely, instead of taking the set $C(s)$ of all the parts of s, in practice we only take the set $C_a(s)$ of parts of kind a; that is, we form the intersection or logical product $C(s) \cap a = C_a(s)$. We proceed similarly with the other three coordinates of the quadruple $\mu(s)$. That is, we take $E_b(s)$, or the environment of s at level b, $S_c(s)$ or the structure of s at level c, and $M_d(s)$, or the mechanism of s at level d. In short, we form what may be called a *reduced CESM model*:

$$\mu_{abcd}(s) = \langle C_a(s), E_b(s), S_c(s), M_d(s) \rangle.$$

For example, when forming a model of a social system (or group) we usually take it to be composed of whole persons; consequently we limit the internal structure of the system to interpersonal relations. However, nothing prevents us from constructing a whole sheaf of models of the same society, by changing the meanings of 'a,' 'b,' 'c,' and 'd.' We do this when we take certain subsystems of the given social system – for instance, families or formal organizations – to be our units of analysis. Similar sheaves of models can be constructed, of course, in all fields of knowledge.

The above model of a system should be supplemented with a model of emergence and submergence, that is, of system generation and disintegration. The most general approach to modelling quantitative and qualitative changes of systems of any kind is the state-space approach. This is used or utilizable in any discipline, from quantum physics to genetics to demography. We proceed to sketching it (for details see, e.g., Bunge 1977a).

Consider a process involving only three quantitative properties, called X, Y, and Z, such as concentrations of chemicals in a chemical reactor, vital signs of an organism, densities of populations in an ecosystem, or what have you. Each of the three properties is a function of time, and all three can be combined into a single function $F = \langle X, Y, Z \rangle$. This is called the *state function* of the system, because the value $F(t) = \langle X(t), Y(t), Z(t) \rangle$ at time t represents the *state* of the system at t. $F(t)$ is a snapshot of the processes that occur in the system. $F(t)$

can also be imagined as the tip of a vector that describes a trajectory in the state (or phase) space. This trajectory, the ordered sequence of states $H = \langle F(t) | t \in T \rangle$, represents the *history* of the system over the period T in question. This history is confined within a box representing all of the really possible (or lawful) states of the system. This is a finite subset of the entire state space, because no real properties of a finite system can attain infinite values. When these singularities are taken seriously, as they often are in cosmology, the model in question ceases to be scientific.

Suppose now that, up to a certain time t_e, the vector $F(t)$ lies on the $X - Y$ plane, and that its Z-component starts to grow at that time. In other words, the system undergoes changes that lead at t_e to the emergence of property Z, which up until then was merely possible. (Think, e.g., of a chemical reaction of the form $X + Y \rightarrow Z$ that only starts when the environmental temperature attains a certain value.) From then on, and as long as all three properties last, the tip of the state vector will move in the three-dimensional state space. Just as emergence is representable as the sprouting of axes in a state space, so submergence can be represented as their pruning. And the entire history of the system within a given time interval, complete with its quantitative and qualitative changes, is representable by its trajectory in the state space characteristic of its kind. These state spaces must not be confused with physical space, if only because in general their dimensionality is larger than three. (The state spaces in quantum mechanics are infinitely dimensional Hilbert spaces, and in some cases their axes constitute a continuum.)

Concluding Remarks

In this chapter and the preceding we have sketched a world view and an approach that we have called sometimes *systemism*, and at other times *emergentism*, for its foci are the concepts of system and emergence. Systemism, or emergentism, is seen to subsume four general but one-sided approaches:

1. *Holism*, which tackles systems as wholes and refuses both to analyse them and to explain the emergence and breakdown of totalities in terms of their components and the interactions among them; this approach is characteristic of the layperson and of philosophical intuitionism and irrationalism, as well as of Gestalt psychology and of much of that passes for 'systems philosophy.'

2. *Individualism*, which focuses on the composition of systems and refuses to admit any supra individual entities or properties thereof; this approach is often proposed as a reaction against the excesses of holism, particularly in the social studies and in moral philosophy.

3. *Environmentalism*, which emphasizes external factors to the point that it overlooks the composition, internal structure, and mechanism of the system – the behaviourist view.

4. *Structuralism*, which treats structures as if they preexisted things or even as if things were structures – a characteristically idealist view.

Each of these four views holds a grain of truth. In putting them together, systemism (or emergentism) helps avoid four common fallacies.

3

The Systemic Approach

As we saw in the previous chapter, systemism is the view that every thing is a system or a component of one. In this chapter and the next I will argue that systemism holds for atoms, ecosystems, persons, societies, and their components, as well as for the things they compose. It holds for ideas and symbols too: there are no stray ideas or isolated meaningful symbols, whether in ordinary knowledge, science, technology, mathematics, or the humanities. Indeed, it is hard to understand how an idea or a symbol could be grasped, worked out, or applied except in relation to other ideas or symbols. Only the universe is not connected to anything else – but it is a system rather than a mere aggregate. In fact, any component of the universe interacts with at least one other component, whether directly (as in face-to-face social interactions) or indirectly (e.g., via physical fields).

Systemism is the alternative to both individualism (or atomism) and collectivism (or holism). Consequently, it is also an alternative to both micro-reductionism ('Everything comes from the bottom') and macro-reductionism ('Everything comes from the top'). Individualism sees the trees but misses the forest, whereas holism sees the forest but overlooks the trees. Only the systemic approach facilitates our noticing both the trees (and their components) and the forest (and its larger environment). What holds for trees and forests applies, *mutatis mutandis*, to everything else as well, as will be seen below and in subsequent chapters.

1 The Systemic Approach

The systemic ontology suggests a systemic approach to all problems, whether epistemological or practical. Let us see how it works in a simple case: how does one choose a car make? Ordinarily one looks for the best car compatible with one's budget, regardless of the service facilities. But this approach courts disaster in the case of imported cars, because the parts and the expertise are expensive and harder to get. A systemic approach to the problem will examine the four possible solutions of the car problem, collected in the following manner:

$$\left\| \begin{array}{ll} \langle \text{Good car, good service} \rangle & \langle \text{Good car, bad service} \rangle \\ \langle \text{Bad car, good service} \rangle & \langle \text{Bad car, bad service} \rangle \end{array} \right\| = \left\| \begin{array}{ll} V_{11} & V_{12} \\ V_{21} & V_{22} \end{array} \right\|$$

The values of the four entries of this matrix may be ranked as follows:

$$V_{11} > V_{12} \geq V_{21} > V_{22}.$$

This way of proceeding will be called the *systemic approach*, and its opposite the *sectoral approach*. I submit that the former is more efficient than the latter, because reality itself happens to be systemic rather than either an undifferentiated blob or a loose assemblage of separate items. Neither cars nor people, nor even atoms or photons, exist in a vacuum. (Moreover, there is no such thing as a total vacuum: every place is the seat of physical fields.)

Everything, save the universe, is connected to something else and embedded in something else. However, not everything is tied to everything else, and not all bonds are equally strong: this renders partial isolation possible, and enables us to study some individual things without taking into consideration the rest of the universe. This qualification distinguishes systemism from holism, or the block-universe doctrine.

The systemic approach emerged together with modernity. For example, astronomers did not talk about the solar *system* before the seventeenth century; the cardiovascular *system* was not recognized as such before William Harvey, at about the same time; and talk about the digestive, nervous, endocrine, immune, and other *systems* is even more recent; so is the treatment of machines in relation to their users and their social environment.

As usual, philosophers took their time to note these scientific changes. In fact, the earliest modern systemic philosophy was crafted by the famous Baron d'Holbach. At the very beginning of his *Système social* (1773) he wrote: 'Tout est lié dans le monde moral [social] comme dans le monde physique.' Three years earlier, in his *Système de la nature*, he had made his case for the systemicity (and materiality) of nature. The powers that be were not amused: those influential works were banished in France, d'Holbach's adoptive country. Even today, the entire French Enlightenment is all but ignored in most universities, where systemism, often confused with holism, is every bit as unpopular as materialism, the bugbear of the fainthearted.

However, if a potent philosophical idea is ignored by the philosophical community, it is likely to flourish elsewhere. This is what happened to systemism, which was explicitly advocated by the biologist Ludwig von Bertalanffy (1950), who inspired the general systems theory movement. Like all movements, this one is heterogeneous: it contains tough-minded scientists and engineers (e.g., Ashby 1963; Milsum 1968; Whyte, Wilson, and Wilson 1969; Weiss 1971; Klir 1972) alongside with popular writers who confuse systemism with holism (e.g., Bertalanffy 1968; Laszlo 1972). The latter believe that systems theory is a recipe for tackling problems without engaging in empirical research. Their purely formal analogies and wild claims have earned them the acerbic criticisms of Buck (1956) and Berlinski (1976).

General systems theorists, like philosophers, are then split into the tough-minded and the tender-minded. My own book *A World of Systems* (1979) adopts a sober systemic approach, uses formal tools, and applies it to chemistry, biology, psychology, and social science. I claim that systemism is part of the ontology inherent in the modern scientific world view, and thus a guide to theorizing rather than a ready-made substitute for it.

2 Conceptual and Material Systems

Modern mathematics is the systemic science par excellence. Indeed, the modern mathematician does not handle stray items, but systems or system components. For example, one speaks of the real numbers, manifolds, Boolean algebras, and Hilbert spaces, as *systems*. In all of these cases what turns an aggregate or set into a system is a structure, that is, a set of relations among, or operations on, the system components. (Mathematical systems are sometimes called 'structures.' This is a misnomer, because structures are properties, and every property is the property of something. For example, a set has the group property if its elements are organized by the concatenation and inversion operations.)

Actually contemporary mathematicians tackle systems of two kinds: mathematical objects proper, such as rings, topological spaces, systems of equations, and manifolds; and theories about such objects. A theory is of course a hypothetico-deductive system, that is, a system composed by formulas linked by the implication relation. But a theory may also be viewed as a mathematical object – the object of a metatheory, such as the logic of theories and the algebra of logic. Finally, the whole of contemporary mathematics may be regarded as a system composed of interrelated theories, every one of which refers to mathematical systems of some kind.

All of this is of course well known to mathematicians. As Hardy (1967) stated, the importance of a mathematical idea is somehow proportional to its relatedness to other mathematical ideas. One might even say that, in mathematics, to be is to be a component of at least one mathematical system. Strays do not qualify.

In what follows we shall consider only concrete systems, that is, complex changeable things: we shall not deal in any depth with conceptual systems such as theories. We define a *concrete* (or *material*) system as a composite thing such that every one of its components is changeable and acts on, or is acted upon by, other components of itself. Equivalently: A material system is a complex thing all of whose constituents possess energy. By virtue of these broad notions of materiality and energy, not only persons but also social systems are material – though, of course, social matter has supraphysical emergent properties.

In addition to being changeable, every concrete system – except for the universe – interacts with its environment. However, such system–environment interactions are weaker than the inter-component interactions. If this condition were not met, there would be no system other than the universe, which would be a single bloc.

3 The Systemic Approach to Physical and Chemical Processes

Historically, the first known example of a physical system was the solar system. It took no less than Newton (1667) to recognize it as such rather than as a mere assemblage of bodies, such as a constellation. He hypothesized that the solar system is held together by gravitational attraction, and that the latter does not cause its collapse into a single body, because every component of the system has inertia (mass). If any of them ever stopped, it would fall onto the sun, and the orbits of all the other planets would alter. The removal of any planet would have a similar effect. Thus, the solar system exhibits the property of wholeness. However, this whole can be analysed: the state of the whole is determined by the state of every one of its components. The main task of the planetary astronomers is, precisely, to measure or calculate the variables that characterize the state of the planets and their moons. But they are also interested in finding out how the system as a whole moves relative to other celestial bodies.

However, Newton's original formulation of particle mechanics is not well suited to the study of mechanical systems, because its referent is the single particle subjected to the forces exerted by the remaining particles in the system. Euler, Lagrange, and Hamilton generalized Newton's method (vectorial mechanics), introducing functions that describe the global properties of a mechanical system. One such function, the action, satisfies Hamilton's varia-

Figure 3.1 (a) Vectorial mechanics: the movement of the particle *p* of interest is influenced by the other particles in the system. (b) Analytical mechanics: the referent is the mechanical system as a whole, where the individual particle is just one among several interacting components.

tional (or extremal) principle: the action of a system is either a maximum or a minimum. (Equivalently: Of all the conceivable histories of a system, the actual one corresponds to an extremal – minimal or maximal – value of its action.) In turn, this principle implies the differential equations of motion. See figure 3.1. This is the approach adopted in all the branches of theoretical physics: it combines analysis (differential equations) with synthesis (variational principles).

Solids, liquids, and physical fields, whether gravitational, electromagnetic, or other, provide further examples of wholes. A perturbation in a region of any such continuous medium propagates throughout the whole. Think of a stone dropping on a pond, or of an electron moving through an electric field. It is no coincidence that solids and liquids, unlike gases, are system of atoms or molecules held together by fields.

Another example of wholeness and emergence is the so-called entanglement, intertwining, or non-separability typical of quantum physics. This means that the state of a multi-component microphysical system cannot be decomposed (factored) into the states of its constituents. In other words, when two or more quantons get together to form a system, their individuality gets lost even if the components become spatially separated (see, e.g., Kronz and Tiehen 2002).

Let us finally examine the concept of a chemical system. A system of this kind may be characterized as one whose components are chemicals (atoms or molecules) varying in numbers (or in concentrations) because they are engaged in reacting with one another (see, e.g., Bunge 1979a). Hence, before the reactions set in, and after they are completed, the system is physical, not chemical. For example, a battery is a chemical system only while it is operating.

Chemical reactions have long constituted a prime example of emergence or qualitative novelty. However, the concept of a chemical system became familiar only with the emergence of chemical reactors in the chemical and pharmaceutical industries during the nineteenth century. Since every change in

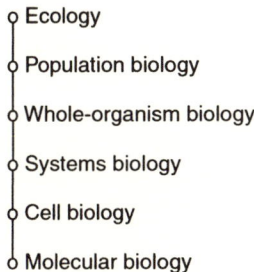

Figure 3.2 The organization of biology reflects the part–whole relation found in nature. The edges should be read as double arrows, which in turn symbolize the flux of problems, concepts, hypotheses, methods, and findings.

chemical composition is caused by interactions among the components, or among the latter and the environment, a true model of the system will incorporate not only its composition, but also its environment, structure, and mechanism: it will then be a specification of the elementary CESM model introduced in chapter 2, section 5.

Let us move on to biosystems, which are supersystems of chemical systems endowed with suprachemical properties.

4 The Systemic Approach to Life

Modern biologists have always studied systems, from cells to organs to large systems to whole multicellular organisms to populations to ecosystems. The very organization of biology reflects this 'chain of being': see figure 3.2.

However, most specialists have isolated their fields from the rest. The generalized recognition of the need for a systemic approach came only recently, when some molecular biologists pointed out the limitation of the traditional endeavour to determine the function of individual genes in the course of development – the popular one-gene-one-trait hypothesis. Indeed, this program fails to analyse the large regulatory systems organized as genic networks.

Even this gene-network approach is insufficient, for genes not only interact with one another, but are expressed (activated) or repressed (inactivated) by enzymes. Hence, gene networks must be combined with protein networks. Presumably, the correct approach to development will focus on the genome-proteome supersystem embedded in its immediate environment. But, of course, such synthesis would be unattainable without the gigantic prior analytic work.

Here, as elsewhere, the right strategy is a combination of micro-reduction (whole → parts) and macro-reduction (parts → whole). However, let us turn to the ontology of biology (on which more in Mahner and Bunge 1997).

A recurrent problem in biology and its philosophy has been the general characterization of its referents, that is, the definition of the concept of life. The systems approach should help here if only because it avoids the extremes of mechanism (and the attendant radical reductionism) and vitalism (a variety of holism often coupled with spiritualism), as well as machinism, or the Artificial Life project.

The mechanist mistakes the living cell for its composition; the vitalist overlooks the latter as well as the structure and the environment, and focuses instead on the emergent (supraphysical) properties of organisms; and the machinist, such as the Artificial Life fan, ignores composition or stuff and remains content with computer mock-ups of morphological features. All three mistakes are avoided by granting that the components of a cell are not alive, and assuming that they self-organize in peculiar ways unknown to physics, chemistry, and computer science.

There are two traditional approaches to the definition of the concept of life. One is to postulate a single peculiar entity contained in every organism and functioning as its first motor, such as the ancient immaterial entelechy or the ultramodern genome. The second is to postulate a single peculiar property, such as the equally ancient purposefulness or teleology, nowadays rechristened 'teleonomy.' Neither of these strategies has worked. Entelechies are imaginary and inscrutable, hence beyond the ken of science. DNA, though necessary, is impotent without enzymes, and neither does much outside the living cell. As for teleology, only highly evolved organisms behave (sometimes) in a purposeful way. We must look for an alternative.

The systemic approach suggests the following characterization of life, couched in terms of the CESM model proposed in chapter 2, section 5. A living system or organism is a semi-open material system s, far from thermodynamic equilibrium with its environment, whose boundary is a semi-permeable lipid membrane, and such that

Composition of s = physical and chemical microsystems and mesosystems, in particular water, carbohydrates, lipids, proteins, and nucleic acids; all of these components are sufficiently contiguous so that they can enter into chemical reactions; and some components are control systems that maintain a fairly constant *milieu intérieur* despite environmental changes within bounds;

Environment of s = a medium rich in nutrients and energy fluxes, but whose variables (such as pressure, temperature, radiation intensity, and acidity) are confined within rather narrow intervals;

Structure of s = all the bonds, direct and indirect, physical or chemical, covalent and non-covalent, that interconnect the components of *s* and thus keep it together; plus all the ties – physical, chemical, and biological – with items in the environment of *s*;

Mechanism of s = all the processes that keep *s* alive, among them the synthesis of some molecules – a process regulated jointly by nucleic acids and enzymes; the transport, rearrangement, assembly, and dismantling of components that accompany metabolism – which allows for the system's maintenance and self-repair; the capture and storage of free energy (*e.g.*, in ATP molecules); the signals of various kinds, electric or chemical, through which components near and far intercommunicate – such as those carried by hormones and neurotransmitters; the expression and repression of genes; the detection of environmental stimuli; and the repair or even regeneration of some parts.

Note the simultaneous occurrence of concepts denoting entities and processes on several levels: atom, molecule, organelle, cell, whole organism, and environment. Hence, the above characterization is systemic rather than either micro-reductionist or macro-reductionist. Note also the tacit occurrence of Lawrence J. Henderson's notion of fitness of the environment – absent from both the mechanist and the vitalist accounts. The mechanism had to be mentioned because two cells originating from a single parent cell may have roughly the same composition, environment, and structure, except that one of them is alive and the other dead. Analogue: A light bulb before and after being switched off. Reproducibility has not been included either, because not all organisms reproduce.

(Nor have we included the popular concepts of genetic information and its kin – genetic program, code, blueprint, transcription, translation, and error correction – because they are imprecise and metaphorical, and therefore sometimes suggestive but at other times misleading [see Mahner and Bunge 1997]. The uncritical use of those terms encourages the illusion that the processes they name are understood, which is false, because nothing is understood until the underlying mechanisms have been unveiled. In any event, we inherit genetic material, that is, DNA, and the laws inherent in it, not immaterial genetic programs.)

According to this elucidation of the concept of a biosystem, chromosomes are not alive, because they do not metabolize. Likewise, viruses are not alive because they do not function at all outside some host cell. (Only the host-virus system is alive, though, alas, often sick.) Nor do robots, however sophisticated, qualify as biosystems, if only because they are made of mechanical and electrical components instead of biochemical ones; and because, far from having evolved spontaneously, they have been designed and assembled by people for the use of people.

Our characterization of the concept of a biosystem suggests a precise definition of another much-discussed concept: that of a biospecies. We stipulate that a species of concrete things is a *biospecies* if (a) all of its members are organisms (present, past, or future); (b) it is a natural kind (rather than either an arbitrary collection or a mathematical set); and (c) it is a member of an evolutionary lineage.

According to this convention, a biospecies is not a thing or concrete individual that behaves as a whole in some respects, but a collection of things. Hence it is a concept, though of course not an idle one but, like the concepts of number and volume, a key component in any conceptual construction purporting to represent life. Moreover, species and all the other taxonomic taxa are more or less hypothetical, whereas living organisms are real; so much so, that some classes, such as 'reptile' and 'rodent,' are no longer accepted even though no one doubts the reality of individual alligators and rats.

However, nowadays several biologists and philosophers claim that a species is an individual, or concrete, system extended over space and time – a description that applies correctly to populations. That view is mistaken for a number of reasons. First, the populations of many species are geographically dispersed, so that, although every one of them is a system, their totality is not. (Think of sparrows or sea lions, alfalfa or wheat.) Second, the concept of a biospecies is necessary to build the concepts of (monospecific) population and (polyspecific) community and ecosystem.

Third, whoever refuses to employ the concept of a class at the species level is forced to introduce it at the next higher taxonomic level, namely genus – otherwise it will be impossible to do systematics. Presumably, in this case one would define a genus as a set of biospecies construed as individuals. But then no organism would belong to any genus, for the members of a genus would be species, not individuals. In particular, no person would be a member of the genus *Homo* and, a fortiori, no person would be a primate, a mammal, a vertebrate, or even an animal. To realize this one needs only to understand that three utterly different relations are involved in these considerations: those of the part-whole, set membership, and set inclusion. For instance, a heart is a part ($<$) of an organism that belongs ($\in$) to a species included ($\subseteq$) in a genus: $h < o \in S \subseteq G$. Regrettably, these elementary distinctions are not usually taught to biology students. Exactness is not just fastidiousness but a component of clear thinking, a condition of theorizing, and a deterrent to barren controversy. An early realization of the fallacious character of the 'species as individual' thesis would have saved entire forests from the logger.

A consequence of the species-as-individual view is that not all biologists are sure which are the so-called units of evolution. That is, they do not know exactly what evolves. Some of them have stated that species evolve. But, although species are natural groupings, they cannot change by themselves because they are concepts, not things. Nor can populations be said to evolve except vicariously, because they do not suffer genic changes. Developmental biology (or embryology) suggests that evolution starts at the level of the individual organism. This is where evolutionary novelties emerge – whence the need for the union of evolutionary biology with developmental biology. (See, e.g., Maynard Smith et al. 1985; Gould 1992; Mahner and Bunge 1997; Wilkins 2002.) However, if we remember that every organism interacts with organisms belonging to several species, we realize that higher-level systems, such as populations, ecosystems, and even the entire biosphere, evolve as well. That is, evolution is a multi-level process. This claim will become clearer upon elucidating the concepts involved in it.

We stipulate that a system is (a) a *biopopulation* if it is composed of individuals of the same biospecies; (b) an *ecosystem* if it is composed of several interacting populations of organisms belonging to different species; (c) a *biosphere* if it contains all the biosystems on a given planet.

Another issue that can be clarified in the light of the systemic approach is that of biofunction, often mistaken for purpose or goal – as, for instance, when it is said that the hand was 'made for' grasping. This is of course the core of teleology, or teleonomy as it is prudishly called nowadays. Thus, instead of saying that organ X *does* Y or that X discharges the *function(s)* Y, many people, even some eminent evolutionary biologists, say that Y is the *purpose* or *goal* of X. I submit that, although the notions of purpose and goal are indispensable in psychology and social science, biology must get rid of them, for they are relics of anthropomorphism and vitalism.

I suggest, moreover, that the notions of purpose and goal must be replaced, in biology and other disciplines, by the concept of specific function or role, which can be given the following definition: The specific biological function (or role) of a subsystem of an organism is the function (or process) that only that subsystem can carry out. For example, the specific function of the human brain cortex is to have cognitive experiences. However, the matter of mind deserves a separate section.

5 The Systemic Approach to Brain and Mind

Psychology is of course the science of behaviour and subjective experience. In psychology, as in every other scientific field, one should begin by identifying the object(s) of study or referents. If one were to believe mainstream philosophy of mind, the object of study of psychology is the immaterial soul, spirit, or mind. In its most recent version, spiritualism holds the mind to be a collection of computer-like programs that can be 'embodied' (or 'instantiated') either in the flesh or in the machine.

Yet if the mind were immaterial, it could not be studied the way concrete things, such as brains, are investigated both experimentally and theoretically. Hence, the science and philosophy of mind would have nothing to learn from brain research or from physiological and social psychology. By the same token, it would be impossible to build and test mathematical models of the kind that have been so successful in science: those involving the state-spaces constituted by the salient properties of the systems in question. Consequently, there would be a chasm between the study of mind and science.

Whoever takes seriously the recent findings of neuroscience, as well as the systemic approach, will reject the foregoing view, without necessarily rejecting the hypothesis that the mental processes have peculiar features that set them apart from all other bodily functions. Although laughing and walking are different, both are bodily functions. If we choose the scientific approach to mind, we shall single out the central nervous system – the organ of mind – as our main (though not sole) system, and shall endeavour to understand its specific properties and functions.

We already know some of these peculiar properties and functions. One is the unusual level of spontaneous (stimulus-independent) activity of neurons discovered (but forgotten) in 1914: there can be output without an input – contrary to what happens in a computer. Another peculiarity of nervous tissue is lateral inhibition, which accompanies every excitation. That is, contrary to what happens in an elastic body or even in a vacuum, excitations do not propagate, but remain confined.

A third specific property of nervous tissue, in particular the mammalian cerebral cortex, is the grouping of neurons into systems, such as the minicolumns, columns, and larger systems discovered by Vernon Mountcastle (1998). These system have (emergent) properties of their own; they act as wholes in certain respects, and often self-assemble in the course of individual development.

A fourth property is the functional differentiation and relative independence of some subsystems of the brain; this accounts for the parallel 'processing' of, say, the shape, colour, texture, and motion of a visual stimulus. This relative autonomy of certain subsystems of the brain would also explain why the loss of one subsystem impairs or erases some mental abilities but not others.

A fifth specific property of the cerebral cortex is the plasticity (as opposed to elasticity) of certain interneuronal junctions. Neural plasticity is the key to behavioural and social plasticity.

Any of the first four properties suffices to relegate stimulus-response psychology to the dustbin of history. And the fifth, the ability to form new systems of neurons whose mutual (synaptic) connections can change in a lasting way, is even more interesting. These neurons are found mainly in the phylogenetically more recent parts of the brain. Tanzi, Lugaro, and Cajal more than a century ago, and half a century later Donald Hebb and Dalbir Bindra, hypothesized that learning an item is identical with the emergence of a specialized system of neurons held together by excitatory plastic synaptic junctions (see Hebb 1980).

We generalize and assume that such plastic multi-neuronal systems are in charge of mental functions (Bunge 1980). In other words, the mind would be the collection of specific functions of the plastic regions of the brain. The animals lacking such plastic neural systems, that is, those having rigid (or 'wired-in') neural circuits, or no neuronal systems at all, would have no mental life. If, on the other hand, the connectivity (or structure) of a neuronal system can change with the strengthening or weakening of synaptic connections, then the system can acquire and lose certain functions in the course of its life. These changes on the cellular and neural system levels are rooted in such subcellular processes as the sprouting and pruning of dendrites, and such molecular changes as altered gene expression and protein synthesis.

In sum, the human brain is plastic rather than either rigid or elastic: it can learn and unlearn, perceive or conceive new items, and sometimes create wholly new ideas. People reconstruct ('rewire') their own brains as they learn and unlearn; and individuals with different experiences and different professions develop correspondingly different brains. By contrast, nearly all the other organs, such as the lungs and the kidneys, have fixed specific functions.

We hypothesize then that an animal endowed with a nervous system undergoes a *mental process* (or performs a mental function) only and while that system has a plastic subsystem engaged in a specific process. We call a *psychon* of kind K the smallest plastic neuronal system capable of discharging a mental function of kind K. Every state or stage in a mental process – or, equivalently, every state of a psychon or of a system of psychons – is called a *mental state* of the animal. For example, the formation of purposes and plans appears to be a specific activity of psychons in the prefrontal cortex.

By contrast, pain, hunger, thirst, fear, anxiety, rage, sexual urge, and the control of circadian rhythms seem to be processes in subcortical neural systems, such as the hypothalamus, that have little or no plasticity. Yet, because these phylogenetically older systems are 'wired' to the cognitive 'areas,' they can be influenced by them: there are neither purely cognitive nor purely affective processes. This is one of the reasons that it is wrong to detach cognitive psychology from the rest of the science of mind and behaviour.

In sum, we can now define the tricky concept of mind. We stipulate that the *mind* of an animal during a given period is the union of all the specific functions (processes) occurring in the plastic part of its nervous system during that period. We shall return to the mind-body problem in chapter 11.

So far we have focused on the brain. However, far from being fully autonomous, the brain is intimately connected with the endocrine and the immune systems, as well as with such support systems as the cardiovascular, digestive, and musculo-skeletal systems. Therefore there are hardly any purely nervous processes. Hence, an adequate understanding of mental functions and treatment of mental dysfunctions can only be attained by studying the neuro-endocrino-immune supersystem. In particular, the so-called psychosomatic phenomena, like blushing, premenstrual stress, and psychosomatic eczemas, only began to be understood when psycho-neuro-endocrino-immunology was born in the early 1980s. Likewise, the social emotions – such as empathy, shame, pride, jealousy, and compassion – are being explained by social cognitive-affective psychology in terms of brain–society interactions – one more vindication of the systemic approach.

Concluding Remarks

The concept of a system is central to mathematics and natural science. There are two reasons for this. One is that the real world is the system of all systems. The other reason is that all ideas, whether or not they refer to real things, come in bundles. An isolated idea would be unintelligible, hence no idea at all. It is therefore amazing that the very concept of a system is absent from most contemporary philosophies – one of the indicators of the crisis in the discipline (Bunge 2001a). The more is the pity, because the explicit recognition of the ubiquity of systems helps to identify and analyse them. It also avoids mistaking systemism for the fuzzy and irrationalist holism characteristic of Romantic philosophy, the New Age cult, postmodernism, and feminist philosophy.

4

Semiotic and Communication Systems

Ordinary language offers one of the simplest and yet most sophisticated illustrations of the concepts of system, module, concatenation, emergence, and level: see table 4.1 (following page).

Only a finite though open subset of the infinitely many concatenates of 0-level modules, such as letters, constitute a vocabulary. Likewise, not all word combinations are phrases, and not all combinations thereof result in sentences, questions, or commands. Signification (the linguistic counterpart of meaning) emerges on level 1, and carries through to all higher levels. However, only a subset of all possible sentences signify: the relation between signs and meanings is many-to-one. And truth (or rather its linguistic surrogate) emerges on level 3. Strictly speaking, truth is a property of propositions, not of their linguistic representatives (utterances and sentences), because one and the same proposition may be designated by different sentences.

However, spoken languages constitute only one species of semiotic system. Other semiotic systems include musical scores, blueprints, graphs, and technological diagrams. A semiotic (or symbolic) system is one composed of signs, such as utterances, gestures, words, and pictures, that signify something to someone by virtue of certain conventions. Sentences, paragraphs, texts, languages (both spoken and signed), diagrams, blueprints, maps, and time-tables are semiotic systems. By contrast, stray gestures, individual words, lines, ideograms, or numerals can only be constituents of semiotic systems. Strictly speaking, taken in isolation they are syncategorematic signs, that is, lacking in signification. They acquire signification only in context or in combination with other signs – that is, by incorporation into a system. For example, 'Go!' can only mean something in the context of a discussion about some action to be taken. It means nothing at all in relation to an algebraic problem, let alone in itself.

Table 4.1 Five lexical levels

Level	Lexical category
4	Text
3	Sentence, question, command
2	Phrase
1	Word
0	Letter, numeral, auxiliary symbol, e.g., ')', '?', and the blank

Although our definition of a semiotic system looks obvious, it becomes problematic as soon as we are asked to elucidate its key components, namely the concepts of system, sign, and signification. The first of them was defined in chapter 1. In this chapter the other two will be characterized. I will also attempt to show the advantages deriving from regarding languages as semiotic systems on a par with, though of course quite different from, natural systems such as bodily organs, and social systems such as linguistic communities.

Regrettably, the very idea of a system is absent from the standard philosophy of language (see, e.g., Martinich 1996). And yet a systemic approach to language has several advantages. One of them is that, when exhibiting a language as a system rather than a mere aggregate, one can account for contextuality, since the signification of a sign depends partly on its context. Another advantage of systemism is that it emphasizes the relations between linguistic items and extralinguistic ones, both cognitive and non-cognitive. A third advantage is that it encourages stressing rather than cutting the links between form, content (meaning), and use. A fourth is that it exhibits linguistics as a multidiscipline straddling the natural science / social science divide erected by idealism (Bunge 1986).

Finally, we will also characterize the concept of a communication system. This, unlike a semiotic system, is constituted by living animals that exchange signals belonging to some semiotic system. Hence, unlike semiotic systems, communication systems must be studied by ethologists, biosociologists, anthropologists, sociolinguists, and sociologists.

1 Words, Ideas, and Things

Grammarians tend to view a language as just a set of words that can be combined in accordance with certain rules. That is, they focus on the composition (vocabulary) and structure (grammar) of a linguistic system, with neglect of the environment in which it is embedded. And yet this environment, in particular culture, is what supplies linguistic expressions with signification, because (a) language is eminently instrumental; and (b) the sign-thing, sign-idea, and sign-sound links are conventional. (See Donald 1991 for a criticism of the view of language in a social void.)

For the preceding reason, different languages are parts of different cultures. This is also why there can be an international ordinary language, such as Latin or English, across cultures with a sizeable overlap. And convention, rather than either genome or reason, is why in English one says *round table* instead of *table round*, which fits better the corresponding thought, as the French *table ronde* does, since adjectives are subsidiary to nouns. Convention is also the reason for saying *trai'bju:nl* instead of *tribju:nl*, which is more reasonable because it matches the root *tribju:n*.

Imagine the following situation. You are addressed by a stranger who is visibly distraught, and who expresses herself in a language that is utterly foreign to you. You wish to help her, so you make an effort to understand what she has said to you. Being neither a formal logician nor a transformational grammarian, you do not attempt to figure out (reverse-engineer) the rules of formation and transformation inherent in the stranger's speech. All you want to find out is the meaning of her utterances. Accordingly, you start by trying to find out what is the referent of the stranger's speech: What was she talking about? If you succeed in guessing the referent, perhaps by making and inviting gestures, you are ready to tackle the second problem: What did your interlocutor say, that is, what is the sense or content of what she spoke about? In short, you try to ferret out the signification or meaning of the speech in question, while paying little if any attention to syntactic niceties. You know that, Chomsky notwithstanding, content drives form, not the other way round. And you tacitly define meaning as reference cum sense – or denotation together with connotation.

Despite the centrality of meaning in speech, linguists have yet to come up with an acceptable theory of meaning, that is, a semantics. It is as if their discipline were a victim to Wernicke's aphasia, which is characterized by a marked impairment of language comprehension. A patient with this condition may be able to speak distinctly with perfect grammar, but without making any sense: he retains his phonology and his syntax, while his semantics is gone. All the other aphasia syndromes are mild by comparison. For instance, a patient with syntactic aphasia may say, 'Water drink want I' instead of 'I want a drink of water,' but he is likely to be understood. By contrast, he will not be understood if he utters syntactically well-formed gibberish such as 'He parsed himself to death' or '[T]he immediate "I," already enduring in the enduring primordial sphere, constitutes in itself another as other' (Husserl 1970: 185).

In short, although the grammatical analysis of utterances starts with unveiling their form or syntax, the formation of utterances starts with their content or meaning. Thus, Chomsky's opinion notwithstanding, syntax is ancillary to semantics. Linguistic theory without semantics is like a planetary astronomy that ignores the sun. Consequently semantics, rather than syntax or phonology, should be at the centre of every linguistic theory. And yet standard linguistic theory does not even contain an exact and generally accepted theory of meaning, or even reference (denotation), that is, a theory elucidating the relation between signs, on the one hand, and the ideas or things these stand for, on the other.

Let us peek at this sign–denotatum relation. We assume that, because they have to be understood to be used by humans – though not by computers – signs denote only through ideas. For example, a Chinese ideogram for a house denotes a house because it evokes the idea of a house. In other words, the sign–thing relation is actually the composition of the sign–concept and concept–thing relations. Again, signs are such to the extent that they convey meanings, and these are properties of constructs – concepts and their combinations. This is why psychology and sociology are relevant to linguistics but not to computer science.

Concepts – such as the predicates 'is clear' and 'is clearer than' – are the units of meaning. They combine into conceptual systems of many degrees of complexity. The simplest of all conceptual systems are propositions, such as 'Humans are sociable.' For purposes of analysis it is convenient to formalize this particular proposition to read: 'For all x, if x is human, then x is sociable,' or '$(\forall x)(Hx \Rightarrow Sx)$' for short. The components of this system are the logical concepts $\forall$ ('for all') and $\Rightarrow$ ('if-then'), the neutral blank x, and the extralogical predicates 'is human' and 'is sociable.' (I assume that predicates are concepts representing properties, whether real or imaginary. Whereas properties are possessed, predicates are attributed, truly or falsely. This distinction holds only for extralinguistic items. See Bunge 1977a.)

The *composition* of a proposition is the set of concepts occurring in it. The *structure* of a proposition is its logical form, best displayed with the help of predicate logic, as in the previous example. And the *environment* of a proposition is the more or less heterogeneous and untidy set of propositions that are or may become logically related to the given proposition. Two of the members of the environment of the above example are 'All humans are animals' and 'Sociability is necessary for welfare.' Without such an environment the given proposition would not be such, for it would be devoid of sense. Indeed, the sentence designating (or expressing) a proposition would be incomprehensible in isolation.

Propositions may have or acquire an emergent property that their constituents lack, namely a truth value. Thus, whereas '.' is neither true nor false, 'The ampersand designates conjunction' is a true proposition. However, propositions cannot be defined as the objects that are true or false, because there are untestable (or undecidable) propositions, not only in mathematics but also in theology and philosophy, such as 'There are worlds other than ours.' A proposition can then be meaningful in some context, yet it may lack a truth value, either pro tem or forever. This suggests the following methodological ranking:

Meaningfulness < Testability < Actual test < Truth-value assignment,

where '<' stands for the precedence relation.

Since propositions are systems, any systems of propositions, such as systems of equations and theories (unlike single hypotheses) are supersystems, that is, systems composed of subsystems. Nor are these the only conceptual systems. We may distinguish at least two further types of conceptual system: context and classification.

A *context* is a set P of (at least two) propositions together with their domain D (or universe of discourse). That is, $C = \langle P, D \rangle$, where D = Set of referents of the P's. Example: Any set of propositions about verb-noun-subject orders. P and D must be kept separate (though related) because one and the same set P of (formal) propositions may be assigned now one reference class, now another. A context is a system proper only if its component propositions have at least one common referent, for in this case they are related by the equivalence relation of having a non-empty reference class in common. This, then, is the structure of a context. (More precisely, if $P = \{P_1, P_2, ..., P_i, ...\}$ and $\mathcal{R}(P_i) = D_i$, then D equals the union of the D_i, and the common referents of the p_i are in the intersection of the D_i. $\mathcal{R}$ is of course the reference function, which maps propositions into their reference classes.) As for the environment of a context, it may be taken as the union of the environments of all the propositions in the context.

Finally, a *classification* of a given domain D of items may be characterized as a partition of D together with the set of relations among the subclasses of D resulting from the partition. For example, the natural languages can be split, with respect to the verb-noun-subject order, into six classes: VOS, VSO, OSV, OVS, SVO, and SOV. (Empirical investigation may show that one of these classes is empty. This poses an intriguing problem to historical linguists, sociolinguists, and psycholinguists, but cannot be counted against the partition itself.)

So much for conceptual systems. Let us now take a look at their linguistic representations or perceptible 'embodiments' (pardon the Platonic idiom), namely semiotic systems.

58 Emergence

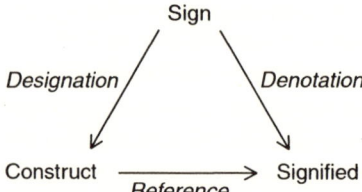

Figure 4.1 A sign either designates a construct or denotes a concrete item. Only in the first case is the sign meaningful, i.e., it possesses both a sense and a reference.

2 Semiotic System

The basic unit of a semiotic system is of course the artificial sign. 'Natural signs,' such as dark clouds, are such only by way of hypothesis. And 'social signs,' such as winks, are such only by virtue of social convention. That is, natural and social signs are not signs proper but rather perceptible indicators of imperceptible things, properties, or events. Hence they do not signify, and therefore talk of their 'meaning' is at best metaphorical, at worst plain wrong. In particular, it is mistaken to think of social life as a text or 'like a text' just because people 'interpret' social behaviour, that is, frame hypotheses concerning its intention or goal. Shorter: Social hermeneutics is basically flawed, in confusing hypothesis with interpretation. (More on this in chapter 13.)

An *artificial sign*, or symbol, may be characterized as a sign produced or used to either designate a concept, such as that of language, or denote an extraconceptual item, such as an individual material thing or another sign. We may call them *designating* and *denoting* signs respectively. Example of a designating sign: a numeral (which designates or names a number). Example of a denoting sign: a proper name. (For details see Bunge 1974a.)

The relations of designation (sign-concept) and denotation (sign-thing) can combine with the relation of reference (or aboutness), as in figure 4.1.

A sign that designates a well-defined construct can be said to be *signifying*: it conveys the meaning (sense cum reference) of its designatum. Thus, signification is linguistic meaning, or meaning by proxy. And synonymy, or meaning identity, translates into equisignificance. That is, two signs are equisignifying if they proxy for the same construct. This is, of course, the concept behind translation: a text is a faithful translation of an original text if every sentence in the former signifies the same as its original homologue. Note that this definition contains the indefinite article 'a' rather than the definite one 'the.' The reason is, of course, that there may be more than one faithful translation of a sentence. What matters in translation is meaning invariance, and meanings lurk behind the signs that stand for well-defined constructs.

Signs are perceptible (visible, audible, or touchable) entities, not abstract ones like concepts and propositions: think of readable sentences, visible drawings, Braille dots, and audible words. However, only iconic (or representational) signs, such as most road signs, are directly interpretable. Non-iconic signs, that is, symbols, cannot be read without an accompanying, albeit often tacit, code. Think of the letters of the alphabet and the words they compose, in contrast to hieroglyphs. Or think of maps, musical scores, graphs, circuit diagrams, organization and flow charts, or even architectural blueprints.

Artificial signs (symbols) can be read only with the help of (explicit or tacit) semiotic conventions, such as 'Letter S → Sibilant sound,' 'Blue patch on a map → Water body,' 'Serrated line on a circuit diagram → Ohmic resistance,' '\$ → dollars,' and 'Money → commodities or work.' In other words, whereas non-symbolic signs are purely material artefacts, symbols are material artefacts together with (explicit or tacit) designation rules. This shows that semiotic systems, far from being self-existing entities, stand or fall along with the people who use them.

Note that the designating-denoting and the symbol–non symbol splitting do not coincide because, whereas some symbols stand for concepts, others don't. Thus, while the numerals '5,' 'V,' and '||||| ' designate the number five, which is a pure concept, proper names and place names denote concrete things. Likewise, the sign '\$5' denotes a five dollar bank note (or checque or money order) or its equivalent in commodities.

(I have just suggested a controversial definition of the concept of money, which so far seems to have eluded economists, namely, 'Paper money is a symbol for commodities or work.' In other words, paper money is a semiotic item, since it is only worth what it stands for. Hence monetary systems, such as the European Monetary System, are social systems with a semiotic feature. However, they are not languages: all languages are semiotic systems, but the converse is false. This construal has the advantage that it avoids the reification inherent in stating that money, or any other symbol, is powerful: only people using certain symbols can wield power, and this by virtue of the things or processes that those symbols stand for.)

Figure 4.1 symbolizes two key semiotic relations, according as the object signified is material or conceptual. If the sign denotes an actual or conceivable concrete thing, such as a table or a sign, then it signifies the item it points to, but lacks a meaning. If, by contrast, a sign designates a construct, as in 'numeral → number,' then it has a meaning proper. We stipulate that in the latter case the (vicarious) *meaning* of a sign is the ordered pair ⟨*sense, reference*⟩, where both members of this couple are defined elsewhere (Bunge 1974b). If the construct in question is defined, then its sense is the set of its definers. If it pertains to a theory, its sense is the set of implying and implied constructs. (That is, $S(c) = \{x \in C \mid x \Rightarrow c \vee c \Rightarrow x\}$.) As for the referents of a construct, they are the object(s) it is about. Thus, the referents of 'Aphasiology studies linguistic disorders' are aphasiology and the set of linguistic disorders.

The objects a construct refers to may be concrete or abstract, actual or potential, possible or impossible. Consequently, to say that a construct (or the sign that designates it) is non-referring, just because its referent is non-existent, betrays ignorance of the role of hypothesis in scientific research. Sometimes explorers of the real world start their journey by hypothesizing things, properties, or events for which they have no evidence: the aim of the research may be to produce evidence for or against the actual existence of such items. In sum, although all signs are real, not all of them are realistic: some of them stand for (either conceptually or materially) impossible items. Thus, the predicate 'is Allah' is perfectly meaningful for any infidel conversant with Islam.

As Stuart Mill noted, since proper names have referents but no senses, they are meaningless. They are only conventional tags, like the slips we get when checking our overcoats in a cloak room. (True, a few proper names, such as Benedict and Argentina, originated as descriptors, but they are nowadays being assigned by stipulation.) Hence, the philosophical essays on the meanings of names are not just boring but also mistaken. Shakespeare (or was it Marlowe?) would have winced at them, for he knew that the shape, colour, and scent of a rose are naming-invariant. Likewise, actions can have no meanings – *pace* Weber and the hermeneutic school in social studies. What actions do have, when deliberate, is intentions or goals. (More on this in chapter 13.) Only signs can signify, and they do so only if they convey meanings of constructs.

The realization of the conventional nature of proper names allows one to avoid Byzantine puzzles, such as whether Shakespeare and The Bard would still be the same person in a different world. The answer is that the very question is wrong, for it has been stipulated that 'Shakespeare' and 'The Bard' denote (refer to) the same person, and stipulations are independent of matters of fact. We are dealing here with words, not worlds.

3 Languages as Semiotic Systems

A language, whether natural (historical), artificial (designed), or mixed, is made up of artificial (or conventional) signs together with rules for combining them and assigning them meanings. A language which, like French, is partly regulated by an academy and a literary elite can be said to be half-natural and half-artificial. Hence, the natural/artificial distinction does not amount to a dichotomy. By contrast, the mathematical/non-mathematical distinction is a dichotomy. A non-mathematical language, such as English, is made up of words, whereas a mathematical language, such as that of the predicate calculus, is constituted by symbols, like $\exists$, x, P, $\Rightarrow$, and $\partial/\partial x$ that are not words, even though they combine to form sentences. Example: '$Pb \Rightarrow \exists x Px$,' reads 'If individual b has the property P, then some individuals have the property P'. Consequently, the syntax of mathematical theories is fundamentally different from that of a natural language. For example, the rule that the expression 'the derivative of a function' is a well-formed formula, whereas 'the function of a derivative' is not, lacks an ordinary-language counterpart. In turn, the root of this difference is the difference between ordinary knowledge (or knowledge of everyday matters) and scientific knowledge. Because standard linguistic philosophy overlooks this difference, it can make no contributions to the philosophy of science, in particular the philosophies of linguistics and mathematics.

Now, we all know that stray signs, such as the ones that one may invent on the spur of the moment to make a linguistic point, or just for fun, are non-signifying. Therefore, when in doubt about the signification of a sign, we place it in some context. That is, we attempt to discover or conjecture the system of signs from which it may have been drawn. We always do so when trying to disambiguate an ambiguous expression, such as 'That was a [wedding, criminal, telephone, algebraic?] ring.' The moral is obvious: A language is a system, so that no sign signifies (that is, is a sign proper) unless it is embedded in a semiotic system. (Incidentally, this is not to be mistaken for Jacques Derrida's infamous dogma that 'there is nothing outside texts.')

Because meaning is contextual, no purely combinatorial or computational linguistic theory can be adequate. In other words, it is not true that, to understand a linguistic expression, all we need to know is the constituents and the rule(s) for computing the meaning of the whole in terms of the meanings of its constituents. We also need to know the context: we must be able to place the linguistic item in an epistemic system. For example, the expression 'The schoolmen were interested in that problem' is a piece of historical information in one context, and an accusation of either obsolescence or futility in another.

The thesis that a language is a system was proposed by Franz Bopp as far back as 1816, and it had been 'in the air' when Saussure adopted it and worked it out in his famous *Cours* exactly one century later (Koerner 1973: 2.2.4). However, at that time it was not yet clear whether a language had to be conceived of as a system of signs or as a system of relations. In any case, the very idea of a system was so fuzzy that it was often equated with that of an organism – as befits the holistic (or organicist) world view dominant in the place and time of birth of linguistics as a separate discipline.

62   Emergence

Recalling the CESM model of a system (chapter 2, section 5), I submit that every language L, whether natural or designed, is a semiotic system with

composition of L = a collection of artificial signs (symbols);
environment of L = the collection of natural and social (in particular cultural) items referred to by expressions in L;
structure of L = the syntactic, semantic, phonological, and pragmatic relations in L.

The syntax of L plus the logical relations among the concepts designated by the signs of L is the internal structure (or *endostructure*) of L. (Syntax is a linguistic category, while the logical structure is independent of the particular linguistic wrapping.) And the *exostructure* of L is the collection of relations that bind the signs of L with the (natural, social, and cultural) world, in particular the speaker and his interlocutor. The relations of designation, denotation (or reference), speaking, and hearing belong in the exostructure of a language: they relate signs to concepts and concrete things. In other words, the exostructure of a language is its boundary, that is, the bridge between the language and the world. It is what makes language a means of communication – on which more anon. (See Bunge 1974a and Dillinger 1990 for formal definitions of a language.)

As defined above, languages – unlike individual signs – are not real, concrete, or material systems. What are real are language users, their speech and writing activities, and the social systems (linguistic communities) they constitute. (The same holds for all semiotic systems.) Consequently, languages do not develop or evolve by themselves, and there are no linguistic mechanism changes, in particular evolutionary forces. Only concrete things, such as speakers and linguistic communities, can develop and evolve. And, of course, as they develop or evolve, they modify, introduce, or jettison linguistic expressions. The history of mathematics is parallel: research mathematicians do come up with new mathematical ideas, which are adopted or rejected by the mathematical community, but mathematics does not evolve by itself.

If we are interested in the dynamics of semiotic change, we must look at communication systems, in particular linguistic communities: that is where languages and other semiotic systems originate, change, and become extinct.

4 Speech and Language

Saussure drew the important distinction, now standard, between *langue* (language) and *parole* (speech). Whereas speech is a process in a material thing –

a head or a social group – a language is a system of linguistic expressions taken in themselves, and therefore a conceptual object. A similar distinction applies to sign languages.

This difference entails that, whereas speech is studied by anthropologists, psycholinguists, sociolinguists, field linguists, experts in speech disorders, and language engineers, languages are studied by general linguists, such as grammarians and philologists. The same speech-language distinction also entails that one should speak of the emergence of speech rather than of language – just as one should refer to the origin of organisms and biopopulations rather than of biospecies.

How does speech emerge in the course of individual development, and how did it emerge in the course of evolution? These problems, banned in the mid-nineteenth century because they had been handled in a wildly speculative way, have returned to the fore (see, e.g., Corballis 2002). Opinion is divided into the innatists (or nativists) and those who believe that all speech is learned. Innatism is untenable for several reasons. One is that all linguistic items are conventional, that is, made and local rather than found and universal. Another is that genes are not complex enough to include phonemes, let alone grammatical rules – on top of which they can do nothing by themselves. A third reason is that, although right-handed people are normally born with intact Broca and Wernicke 'areas,' a young child can learn to speak with his right hemisphere if the former has been damaged. That is, language is not etched in nervous tissue before birth.

Still, it is true that, as Chomsky (1984) and Pinker (1994) have asserted, language is biological. But not more so than other cultural artefacts, such as art, technology, science, or religion. What makes language unique and interesting is that, by contrast to other biological items, such as mastication, it is a tool for both cognition and social interaction. This is why linguistics is neither a natural nor a social science but, as we stated above, an interscience that straddles biology, psychology, and social science. Even the Chomsky school has finally admitted the mongrel nature of linguistics (Hauser, Chomsky, and Fitch 2002).

Furthermore, who invents or learns linguistic signs, and under what circumstances? This problem requires not only smart guessing but also empirical investigation. One of the more recent and interesting findings is that babies and toddlers do not wait to master the rules of the mother tongue: they construct their own primitive semiotic systems (protolanguages) before starting on the mother tongue. Another important finding is that some deaf schoolchildren in Nicaragua created their own language in the mid-1980s: the Nicaraguan Sign Language (Helmuth 2001). In particular, they made up their own grammatical rules.

Since the neuromuscular system activated during signing is quite different from the one that delivers ordinary speech, it is doubtful that we are born with a language instinct. What we are normally born with, on top of the ability to learn, is sociality, which is realized in multiple ways as we grow up, from communicating to playing to mock-fighting to cooperating. The modality we choose to activate depends largely on ability and circumstance.

The above suggests that there are three ways of studying a language:

1 as 'the window on the mind,' an object in itself disconnected from social circumstances and linguistic communities (*internalism*);
2 as a means of communication, and therefore one of the social relations (*externalism*); or
3 as both a mental and a social tool (*interno-externalism*, or *systemism*) that must be learned.

The first approach is that of pure grammarians such as Noam Chomsky, the father of transformational-generative grammar. It comes together with apriorism – which is why it attracts many more philosophers than laboratory or field linguists. The hallmark of this school is the following set of theses: (a) speech is rule-directed rather than spontaneous; (b) the grammatical rules are purely syntactic, that is, they concern the formation and transformation of symbols regardless of their meaning, sound, and use; (c) such rules hover above society rather than being social conventions; (d) although they differ in details from one language to the next, the grammatical rules are basically manifestations of a single universal grammar; and (e) the rules of this grammar are innate rather than made, remade, and learned.

This sophisticated school has made notable contributions to syntax. But it has produced neither a theory of meaning (semantics) nor a theory of use (pragmatics), let alone true theories of language acquisition, emergence, or change. Studies on babies, conducted since the mid-1970s, have falsified the twin hypotheses of innateness and universal grammar. They have shown instead that language, far from 'growing in the mind,' develops in socially embedded brains, whose architecture changes with experience (e.g., Kuhl 2000). And of course geneticists have failed to find the 'grammar gene' that would support the innateness thesis. (What have been found are genes associated with certain specific linguistic disorders, which may turn out to be features of more general cognitive deficits.) In sum, we are born with an ignorant brain – but of course one capable of learning to eat, speak, socialize, hammer, drive, heal, organize, calculate, use credit cards, philosophize, steal, murder, and so on. (See further criticisms in Hebb et al. 1971; MacWhinney 1999.)

Moreover, no such learning is a mere automatic response to environmental stimuli: as Jean Piaget surmised, it is a constructive process – on which more in the next section.

The second approach is that of anthropology, sociolinguistics, and geographical and historical linguistics (e.g., Labov 1972; Newmeyer 1988; Cavalli-Sforza et al. 1994). These interdisciplines, utterly disregarded by the internalists, have been in existence for over a century. They have exhibited the sensitivity of language to such social features as occupation, social class, education, migration, foreign invasion, colonization, and mere fashion. In particular, they have shown how pidgins can evolve into créoles in the course of a single generation; how migratory waves can spread a language across geographical and political barriers; and how invasions (such as that of England by the Normans) may alter the speech of both invader and invaded.

Finally, the third approach is that proposed by Michael Halliday (1985), who calls it *systemic* (or *functional*). This school investigates the function of a linguistic expression in a particular discourse; and it places languages in their social context. Thus, because it studies both the ideational and the interpersonal uses of lexical expressions, it takes meaning as basic and form (syntax) as derivative. In other words, systemic linguistics starts from the realization that one and the same 'meaning' (conceptual item) can be designated by different linguistic expressions: as Halliday puts it, 'the wording "realizes," or encodes, the meaning.' This is why systemic grammar is also called 'functional.'

Like Bopp and Saussure before them, contemporary systemic grammarians regard a language as a system, every symbol of which acquires a meaning by virtue of its relation to other symbols. For instance, the first, second, and third person belong to the 'system of person'; by contrast, 'singular' and 'past' do not belong to the same system because they do not share any meanings (Berry 1975). Last, but not least, systemic (or functionalist) grammarians seek to check their hypotheses. That is, they adopt the scientific method.

However, adopting a scientific and systemic approach is not enough: emergentism too is necessary for stating and investigating the problems of the emergence of language both in the child and in the course of evolution.

5 Speech Learning and Teaching

Language learning and teaching are studied by psycholinguistics and education science. Both fields are arenas of spirited philosophical controversies. Let us glimpse at them. The most notorious of them is a replay of the old nature/nurture debate. The apriorists, like Chomsky and Pinker, hold that language is either innate or 'grows in the mind,' and that the environment can only fine-tune linguistic competence. Moreover, every native speaker would distinguish intuitively, without instruction, the grammatical from the ungrammatical expressions.

Most experimental developmental psychologists, beginning with Jean Piaget in the 1930s, have disputed the innateness hypothesis. Instead, they have adopted the constructivist view, that all learning emerges from interactions between sensorimotor activity and the environment. And nobody has produced evidence that speech is spontaneous, detached from other functions, and that it emerges in a social vacuum. We learn to speak as we learn to walk, eat, play, study, or form moral judgment, namely interacting with our surroundings, in particular our caregivers and peers. Children reared in orphanages, boat houses, or isolated farms, are known to be linguistically (and socially) handicapped.

Empirical research on the precise mechanisms of language emergence in child development is in full swing. One finding of this research is that, contrary to what the Chomsky school contends, there is no autonomous language or specific grammar organ: 'the neural mechanisms that "do" language also do a lot of other things' (Bates and Goodman 1999: 36) – which is not surprising given that meanings emerge in the processes of discovering the world and acting upon it. This is why the blind from birth cannot understand sentences involving colour names.

As for the thesis of the alleged grammaticality intuition, it presupposes that there is such a thing as a single grammaticality standard. But there is no such thing, as revealed by even a superficial examination of the dialects spoken in any given region. For instance, the expressions 'I don't know nothin'' and 'It ain't so,' though unacceptable in academia, are common in American streets. They have survived despite their ungrammaticality because they are intelligible. Meaning is far more important than grammar. So much so that most of us, the present writer included, get by without an explicit knowledge of any grammatical rules.

In any event, there are no objective standards of intuitiveness, linguistic or otherwise: your intuitions may be my counter-intuitions. Hence, appeals to intuition are at best heuristic and at worst misleading (see Sutherland 1995; Hintikka 1999). And in general intuitionist philosophies, such as Bergson's and Husserl's, are subjectivist and irrationalist, hence unscientific (see Kraft 1957; Bunge 1962a, 1984).

The main philosophical controversy raised by language teaching is that between the partisans of two methods: phonics and whole-word, or even whole-text. The former hold that language students, in particular schoolchildren, ought to be taught the correspondences between spoken sounds and the signs that symbolize them. The aim is to learn to combine letters into words, and words into sentences. This is clearly a systemic strategy, since it boils down to integrating a small number of modules (in this case phonemes and elementary symbols) to produce an unlimited number of linguistic expressions. Moreover, the modules are presented in context, not in isolation. For example, the letter *t* can be taught by highlighting it in such words as 'tap' and 'pet.' This is the traditional method.

The opponents of phonics claim that students learn to read and write spontaneously upon being presented with whole words or even whole stories. Like the Gestalt psychologist one century ago, they are holists: they claim that we perceive wholes, and sometimes proceed to analyse them. Neuroscientists and experimental psychologists have falsified this contention, showing that, in perception, analysis precedes synthesis (see, e.g., Treisman and Gelade 1980).

Not surprisingly, phonics works better than whole-language in the classroom too (e.g., Rayner et al. 2002). The fact that nowadays the whole-text method is favoured by political progressives and denounced by conservatives is only an indicator of the low intellectual level of contemporary ideological debates, where the left often adopts philosophies that used to be conservative or even reactionary (see Gross and Levitt 1994).

6 Communication System

A communication system may be characterized as a concrete (material) system composed of animals of the same or different species, as well as non-living things, in some (natural or social) environment, and whose structure includes signals of one or more kinds – visual, acoustic, electromagnetic, chemical, and so on. Communication engineers, ethologists, sociolinguists, linguists, and others study, design, maintain, or repair communication systems, such as TV networks, the Internet, and linguistic communities.

Linguistic communities are of course the units of study of the sociolinguist. The latter, unlike the general grammarian, is not interested in language as an abstract object but as a means of communication among people. Moreover he may not confine his interest to language proper, but may also be interested in other means of communication, such as body language. In other words, sociolinguists deal with semiotic systems embedded in social systems. That is, they study the uses that flesh-and-bone people make of the semiotic systems they use and occasionally alter.

The study of a linguistic community over time is expected to yield not only descriptions and rules, but also laws and explanations of linguistic changes. Now, an explanation proper (unlike a mere subsumption under a generalization) invokes a mechanism, as we saw in chapter 1, section 5. In the case of linguistic changes, the change mechanism is psychosocial: Even if started by an individual, it becomes a linguistic change only if it is adopted by an entire culture or at least subculture. And presumably it is socially condoned in being generally regarded (rightly or wrongly) as convenient, fashionable, or 'cool.' For example, the sound /t/ is slowly being replaced by /d/ in American English, modern Greek, and other languages, presumably because of ease of delivery and thus communication. By contrast, subjunctives are disappearing from English for a different reason – perhaps because of the increase in permissiveness and the vulgarization of culture, both of which impoverish language in cogency while enriching its vocabulary.

In addition to such spontaneous changes there are, from time to time, linguistic decrees such as those issued by governments, political groups, or academies. A good example is the linguistic 'cleansing' operations of the German language conducted first by the Nazi regime – which, among other things, Germanized all the non-Germanic words, such as 'Adresse' and 'Telephon' – and after the war by the authorities in the two German nations – with the result that the West and the East German vocabularies diverged (Drosdowski 1990). A similar, more recent and more familiar case, is the linguistic 'cleansing' demanded by 'political correctness.' A consequence of it is that some illogical and ungrammatical expressions, such as 'When a person is forced to shut up, they feel humiliated,' have been consecrated.

The only point in recalling these cases is to emphasize the idea that languages evolve within linguistic communities rather than by themselves. Hence, the patterns of language evolution must be sought in speakers and their societies. Shorter: Only psycholinguists and sociolinguists can explain linguistic changes. Still, this does not imply that we must admit linguistic innovations just because they are popular or 'politically correct.' LC (logical correctness) is superior to PC if only because rationality is universal.

Concluding Remarks

The upshot of this chapter is the plain thesis that languages are semiotic systems composed of conventional signs (symbols) employed primarily for purposes of communication, and secondarily for thinking. The fruitfulness of the systemic approach is apparent in linguistics, where the relation between syntax and semantics is still unclear, partly because of the individualist dogma that the meaning of a whole is a well-defined function of the meanings of its parts. This dogma is false because, in focusing on composition, it overlooks structure. For instance, 'Gods make brains' is not the same as 'Brains make gods' for, although both sentences have the same constituents, these are ordered differently.

Another widespread dogma that evaporates in a systemic perspective is the belief that the key to understanding language is the study of the syntax of artificial context-free languages – that is, the analysis of linguistic expressions with neglect of their natural and social environment. This dogma is false because in natural languages meaning shifts with context. Thus, the injunction 'Wait !' means one thing if the recipient is to await the occurrence of an external event, and another if the tacit referent is an action that the speaker is expected to perform. Both dogmas are avoided if a composition-environment-structure model of language is adopted, and if linguistics is regarded as a bio-sociological discipline rather than an autonomous one.

However, the systemic approach is insufficient for understanding language: we also need to adopt emergentism if we wish to avoid getting entangled in the nature/nurture debate and want to learn how protolanguage may have emerged from prelinguistic grunts and calls, how proto-Indo-European spread from Asia to Europe and branched out into the contemporary European languages, or how we learn to speak. The resulting evolutionary and developmental views of language will correct the impression of timelessness given by standard liguistic theory: It will show that *parole* or speech is just as changeable an item as commerce, law, or any other institution; and this to the point that languages emerge and submerge along with the people and the communities that use them.

Further mistakes are avoided upon remembering that, in turn, semiotic systems, far from being self-existing, are components of communication systems constituted by live speakers (and e-mailers) who know something about the world and are embedded in changing societies, where they are exposed to such social mechanisms as formal learning, imitation, and the invention or adoption of neologisms necessitated by technological and social innovations. Which brings us to our next topic.

5
Society and Artefact

The concept of a system is unavoidable in the natural sciences, because all natural things are systems or are about to be absorbed or emitted by systems, from atoms to crystals, and from cells to ecosystems. Something similar happens in society, the supersystem of all social minisystems and mesosystems. The orphans Romulus and Remus are only the stuff of legend, and the few feral children that have been found had to be trained to become fully human. Robinson Crusoe was alone only for a while, and he survived largely thanks to the supplies he found in the shipwreck, an unwitting gift of the society he had left.

However, most students of society are wary of adopting a systemic viewpoint. This reluctance seems to have two major sources. One is that much of what passes for systemism is actually holism, long on high-sounding words but short on specifics. The second reason is that most students of society have not yet cut their umbilical cord to traditional philosophy, which is either individualist or holist, and either denies emergence or regards it as mysterious and therefore intractable.

True, a number of social scientists, from Parsons and Merton to Coleman and Boudon, have written about 'systems of actions.' But the point of systemics is that social actions come in bundles because they are performed, either individually or collectively, inside social systems. And these happen to be made up of active people rather than disembodied ideas, intentions, and actions. We shall therefore insist on the centrality of the twin concepts of a system of live people and the emergence of social features.

We shall also preach the systemist gospel with regard to technology. Of course, all artificial things, whether machines or factories, are systems. Yet the systemic viewpoint was adopted explicitly in technology only in the mid-twentieth century, thanks to the emergence of cybernetics and operations research – both of which are cross-disciplinary. In particular, it is now generally recognized that whoever designs a machine or a social system should place it in its natural and social environment. That is, the technologist should adopt a systemic rather than sectoral approach.

1 The Systemic Approach to Society

Every structured human group, from the married couple to the global market, can advantageously be conceived of as a system composed of human beings and their artefacts, embedded in an environment that is partly natural and partly artificial. Such a system is held together by bonds of various kinds: biological (in particular psychological), economic, political, and cultural; and it is the seat of a number of mechanisms, such as those of production, exchange, competition, cooperation, communication, and division of labour.

In addition, every human society, no matter how primitive, can be analysed into four tightly interrelated subsystems: the biological system, characterized by kinship and friendship relations; the economic system, centred on productive work and trade; the political system, characterized by management and the struggle for power; and the cultural system, revolving around cultural activities and relations, such as designing tools and exchanging information.

This systemic view of human society has a number of interesting features and advantages. First, the variety of types of interpersonal bonds, and of the resulting subsystems, suffices to refute all the one-sided views of human society, such as biologism (e.g., sociobiology), psychologism (e.g., symbolic interactionism), culturalism (or cultural idealism), economic determinism (in particular Marxism and the currently fashionable 'economic imperialism'), and political determinism (such as Gramsci's).

To be sure, the ultimate components of human society are organisms who have mental functions, such as feelings, emotions, perceptions, thoughts, and intentions. But society itself is a supra-organic and non-psychological entity: it is not alive and it does not feel, perceive, think, or plan. Moreover, society has emergent (global or non-distributive) properties irreducible to biology and psychology, such as division of labour, distribution of wealth, type of political regime, and level of cultural development. And, as Stinchcombe noted (1968: 67), social emergents are more stable than individual features.

This does not entail that societies are spiritual entities and therefore objects of study of the so-called cultural sciences (*Geisteswissenschaften* or *sciences morales*). Societies and their subsystems are concrete or material systems, because they are composed of material entities; however, they are characterized by global (supra-individual) properties, regularities, and conventions (or norms). Of course, these emergent properties and patterns are rooted in the properties of individuals. For example, human social plasticity may be traced back to behavioural plasticity, which in turn derives from neural plasticity (the ability of neurons to form new neural connections). Still, although individual actions have internal sources, they are conditioned by their natural and social environments. (Remember Marx's famous statement: 'Men make their own history but they do not make it just as they please.') Such micro-macro links are beyond the reach of both individualism and holism.

Another feature of the systemic perspective is that, since the biological, economic, political, and cultural subsystems of a society are components of one and the same system, they are tightly interconnected. An important methodological consequence of this platitude is that no subsystem can be adequately modelled unless the model contains some of the 'exogenous' variables characterizing the remaining subsystems. For example, a realistic model of the political system will have to contain more than politological variables such as the intensities of popular political participation and repression. It will also have to contain biosocial variables such as total population, age distribution ('structure'), and birth and death rates; cultural variables, such as literacy and access to mass media; economic variables, such as GDP; and socio-economic ones, such as the latter's distribution among the various social groups.

Third, because every society has biological, economic, political, and cultural subsystems, it cannot possibly develop unless all four subsystems develop hand in hand. In particular, the purely biological (or ecological or cultural or economic or political) models of development are bound to fail. By the same token, purely economic (or ecological or biological or cultural or political) development or aid programs are bound to miscarry. For example, a high-tech industry cannot be started without a skilled workforce whose members are in reasonably good health, nor in a society lacking the requisite stable institutions. Likewise, the sound development of a society, like that of an organism, is integral, not partial (Bunge 1997a; Galbraith and Berner 2001.) Regrettably, most development experts and agencies still adopt a sectoral approach. They have yet to learn that systemic (or structural) problems must be tackled in a systemic way.

The interlocking of the various subsystems of society suggests the need for a cross-disciplinary (multidisciplinary or interdisciplinary) approach to the study of social facts. Much the same holds for the social technologies, which design management strategies, macroeconomic and social policies, and the like. For example, realistic urban planners are expected to start by learning about the socio-economic and cultural status of the prospective inhabitants of their projects, and the social networks they belong to, as well as about their needs, habits, and aspirations. Contrary to what the architects of the Modernist movement thought, housing is a component of social change rather than its main engine. Any action based on sectoral planning is bound to have perverse consequences, such as the rapid decay of a housing project into a ghetto (see Portes 2000 and Vanderburg 2000).

The systemic approach is gaining adepts in social studies, partly under the increasing pressure of large-scale social issues on policy-makers, and partly in view of the utter inability of all the existing schools to predict or even explain the dismantling of the Soviet empire, the resurgence of nationalism, and the failure of neo-liberal economic policies in the Third World. Witness the proliferation of interdisciplines over the past few decades: economic sociology (and even socio-economics), political sociology, social history, economic history, and legal and medical sociology among others.

Still, obstacles to further integration remain. Some of them are philosophical. An example is the Kantian and hermeneutic claim that the social studies are totally disjunct from natural science. If true, this would condemn all of the biosocial sciences, such as geography, anthropology, demography, epidemiology, and social cognitive neuroscience. Another obstacle to integration is the simplistic thesis that all humans are natural capitalists, so that all social facts can be modelled along the lines of rational-choice theory, which overrates economic 'rationality' while it underrates passions and disinterested motives, as well as tradition and social interaction. Let us peek at the latter.

One can certainly hope that the emergence, maintenance, repair, or dismantling of any social system, and the rules and conventions inherent in it, can 'ultimately ' be explained in terms of individual interests, beliefs, preferences, decisions, and actions. But individual events, in turn, are largely shaped by social interaction, social context, and the social conventions entrenched in tradition. People cultivate the connections and support the systems that they perceive (rightly or wrongly) as beneficial; and they neglect, sabotage, or fight whatever they believe hurts them. In sum, agency and structure are only two sides of the same coin. Consequently, overlooking either of the sides leads to blindness to the whole coin.

2 Microsocial and Macrosocial, Sectoral and Integral

Persons are studied by natural science and psychology. And the latter, along with anthropology, linguistics, demography, epidemiology, and other disciplines, is one of the biosocial sciences. The social sciences proper, such as sociology, politology, and economics, do not study individuals, except as components of social systems. Thus, anthropology studies entire communities, such as villages and tribes. Sociology studies social systems, all the way from the childless couple to the world system. Economics specializes in the study of the social systems engaged in production, services, trade, or finance. Politology studies power and management relations in all systems, particularly the political ones. And history studies social (structural) changes on all scales.

The real problem is not to try and reduce the social to the individual, but to relate them and, in particular, to explain how the former emerges from the latter and in turn shapes it (Coleman 1964). Now, to relate the macrosocial and microsocial levels is not just to point out the social context or circumstance of a fact. Social scientists are expected to study *social* facts, a social fact being one that occurs in a social system – such as work or a strike in a factory – or between social systems – such as international cooperation or conflict. Hence, social scientists are expected to study social bonds in addition to social contexts, for bonds are what hold systems together, and their weakening is what dismantles them. Parallel: Cognitive neuroscientists study neuron systems rather than individual neurons, just as linguists study sign systems (languages) rather than vocabularies, let alone isolated (and therefore meaningless) signs.

True, some of those who dislike the word 'system' prefer the term 'structure.' There has also been some vague talk of 'structuration' (Giddens 1984), apparently meaning the emergence of social systems. But structures are properties, not things, whereas social systems are concrete things. For example, a corporation is a system with a definite (though changing) structure, which is the set of bonds among its components as well as with its environment. The industrial sociologists and management scientists who study the social structure of a corporation do not investigate the structure of a structure – a meaningless expression; instead, they study the structure of a thing that happens to be a social artefact, such as a corporation.

The replacement of an individualist approach with a systemic one is bound to have important theoretical, empirical, and practical consequences, if only because it stimulates the shift from persons to social systems, and from inscrutable intentions to testable mechanismic explanations. For example, research and analysis in political science are bound to profit from such a 'paradigm shift' on the following counts: '(a) simultaneous downgrading and upgrading of contentious episodes as objects of study, (b) reorientation of explanations from episodes to processes, (c) comparative examination of mechanisms and processes as such, and (d) integration of cognitive, relational, and environmental mechanisms' (Tilly 2001: 36). The corresponding shift in the design of political policies would be equally beneficial.

In sum, individualism fails in social studies because it overlooks social structure; and holism fails even more spectacularly because it underrates individual action. Only systemism joins agency with structure, and urges the search for the mechanisms that make social systems and their constituents tick.

3 Emergence by Design

Emergence and convergence can occur either spontaneously or by design. The latter is of course the case with invention and manufacture, whether of ideas or of things – physical, biological, or social. For example, the invention of the electric motor required the combination of electromagnetic theory with mechanics; that of the nuclear bomb, the joining of nuclear physics with engineering; and that of effective psychotropic drugs, the convergence of biochemistry, cell biology, neuroscience, endocrinology, psychiatry, and pharmacology.

Another case in point is that of materials science – a typical multidisciplinary applied science. Consider for instance a chunk of steel used to fashion a sword. The swordsman takes a holistic view; the corpuscularian philosopher adopts a reductionist stand; and the solid-state physicist and the modern metallurgist take a systemic view. In fact, the latter is the view explicitly adopted by modern materials scientists, whether they deal with metals, ceramics, polymers, or resins. They have to if they wish to design new materials by controlling processes at various microphysical, mesophysical and macrophysical levels, from impurity dosage and deoxidation to hot working and tempering.

Unsurprisingly, one and the same invention is bound to be appreciated differently by different experts. Where the engineer sees a new calculator as an ingenious invention, the user sees it as a labour-saving tool, the investor as a financial opportunity, the merchant as a promising merchandise, the patent lawyer as something to be protected from imitators, and the office worker as either a labour-saving device or a job killer. The total value of the new gadget results from combining all these partial values. (Caution: We still do not know how to add heterogeneous values.)

Inventing is only the first step in a long and complex process. If promising, an invention must first be patented, and a financial or industrial patron must be found, for the patent to be finally 'embodied' into a material artefact ready to be marketed. The actual occurrence of this process depends not only upon the merits of the invention, but also on the outcome of the patent application, the current business climate, the vision of the prospective sponsor, the price, the ability of the business manager, the work conditions, the demand for the gadget, and so on. These and other factors constitute a package or system initially so vulnerable that, should any of them fail, the whole venture might founder (see Gilfillan 1970: 43–4).

Two types of invention must be distinguished: radical and improvement of an existing device. In the first case, the inventor (individual or team) triggers a chain reaction that may end up in the emergence of a new market – as in the cases of the printing press, the steam engine, the electric motor, the light bulb, the refrigerator, the radio, the antibiotic, the credit card, and the computer. But once a new product has carved out its own niche, its customers are likely to demand improvements in the original design, which, like every new and untried item, was bound to be defective in some respect. In this case, then, the arrow originating in the individual is reversed. These intertwined processes are sketched in the following Boudon-Coleman diagram.

Macro-level   Merchandise → Market
                       ↑   ↓
Micro-level   Invention   → Improvement

Obviously, without an original idea no such process could even start. But an invention (a brain process) would not be implemented, let alone socialized, without the concourse of a host of social circumstances. These range from the availability of venture capital to such social processes as patenting and litigating, manufacturing and marketing, population growth and colonization. Likewise, the widespread use of any gadget is bound to reveal imperfections calling for a succession of small improvements, some of which are tried out in the market and are either rewarded or punished by it (see, e.g., Petroski 1992).

4 Social Invention

What holds for engineering inventions also holds, *mutatis mutandis*, for social inventions, except that these are not (yet) patentable. For a social invention to succeed, it must be perceived as promoting the interests of a large number of people – even if, in fact, it ultimately hurts them. For example, the mutual-help societies and cooperatives that mushroomed in Europe during the nineteenth century met some of the needs of the industrial wage-earners uprooted from their countryside. And the Grameen Bank in Bangladesh, though invented by an economist (Muhammad Yunus), has been sensationally successful because it meets the needs and aspirations of many villagers; and because every borrower, far from dealing individually with the bank, is required to band together with four neighbours willing to see that her debt is repaid. The novelty resides in the fact that the efficiency of a financial product, credit, is tied to the cohesion of a social network, which in turn is due to reciprocal altruism.

Artefacts, then, should not be viewed in themselves, separate from the people who invent, produce, and utilize them – or are harmed by their use. After all, artefacts are only tools, and hence they cannot be understood in purely technological terms. For instance, it is mistaken to affirm that certain computers, such as IBM's famous Deep Blue, can beat chess masters. What is true is that computer-aided chess players can beat grand masters who do not use computers. The same holds for applied mathematicians who make computations using computers, *vis-à-vis* their colleagues who only use pencil and paper. What is superior to the individual artisan or player is the user-computer system, not the machine by itself. If the psychologists and philosophers who worship AI (Artificial Intelligence) knew this, they would exclude AI from the cognitive sciences, and would pay some attention to thinking brains instead. (More on this in chapter 11.)

This is not all: machines should be seen as components of sociotechnical systems, such as factories, ports, transport networks, shops, communications networks, hospitals, schools, and armies. Consequently, engineering projects should be placed in their wider economic, political, and cultural contexts. Exclusive attention to either equipment or profitability will not increase productivity unless the workers are educated and motivated by good working and living conditions, as well as given a say on the details of the production process. And the project will not be morally justified if the social and environmental costs of its operation exceed its benefits. In other words, humanist engineering – contrary to Taylorism, Fordism, and Stakhanovism – is systemic, not sectoral.

All this should be obvious to anyone with a systemic perspective and social sensibility. However, taking such a stand can be risky. For instance, the innovative and courageous Russian engineer Peter Palchinsky was jailed by the tsarist government, and executed by the Soviets, for stating precisely those views (see Graham 1993). This suggests that systemism can be subversive merely by insisting that, because every thing is a system or part of one, some social and interdisciplinary borders are artificial or even harmful, and hence they should be trespassed by the technologists and consumer advocates.

5 Philosophical Dividends of the Systemic Approach

The systemic approach we adopted from the start of this book has required the elucidation of several concepts and hypotheses that are philosophical because they occur in a large number of fields of inquiry. Let us only mention a few of them.

The very first philosophical concept that we came across is of course that of a concrete system, which involves the concepts of composition, environment, structure, and mechanism. The concept of a system is so pervasive and suggestive that it deserves being treated as a philosophical category on a par with the concepts of matter, space, time, law, mind, society, value, and norm.

Another philosophical category highlighted by the systemic approach is that of emergence. Emergence concerns both systems and their components. We say that a property of a system is *emergent on level L* if it is not possessed by any of the *A*-parts (or things belonging to level *L*) of the system. For example, ideation is a property of certain systems of neurons, not of individual neurons or of the entire nervous system. When two or more individuals (atoms, persons, or what have you) get together to form a system, each of them acquires at least one property that it lacked before – starting with that of being bound to other system components.

A third philosophical category we encountered when thinking about systems is that of level of organization, or integrative level, first used explicitly by biologists about half a century ago. We define a *level* as a collection of things characterized by a cluster of properties and relations among them. We distinguish five main such levels: physical, chemical, biological, social, and technological. In turn, every level may be split into as many sublevels as needed. (For example, the biological level may be subdivided into at least seven sublevels: cell, organ, organ system, multicellular organism, biopopulation, ecosystem, and biosphere.) And we stipulate that one level *precedes* another if all the things in the latter are combinations of things in (some or all of) the former. For example, the cell level precedes the organ level because organs are composed by cells. The level structure (or 'hierarchy') is thus the set of levels together with the relation of level precedence.

Further philosophical categories that were elucidated in the course of studying systems of various types are those of bond or link, assembly (in particular self-assembly), breakdown, stability, selection, adaptation, evolution, life, mind, person, sociosystem, culture, and history.

For example, systemism sheds light on the contemporary debates over abortion and stem-cell research and its applications (e.g., to the development of live prostheses). Are embryos persons? No, because a person is characterized by a brain capable of having mental experiences, and the brains of embryos do not have them. A fortiori, the cells of embryos do not have a higher moral status than those that we kill every time we scratch or pick at our body. Indeed, almost any cell can be used as a 'seed' to construct an entire body, because usually all the cells of a multicellular organism have the same genetic composition. Hence, to claim that embryos, or even single gametes, should be regarded as sacred, is like confusing the system with its components. (Ironically, this confusion is a point of contact between fundamentalist theology and radical reductionism.)

As for the philosophical hypotheses suggested by the systemic approach, let the following suffice. First, all systems but the universe receive inputs and are selective. That is, they react only to a subset of the totality of environmental actions impinging on them. Second, all systems but the universe react upon their environment: that is, their output is never nil. Third, all systems but the universe originate by assembly, in most cases spontaneously (self-assembly). Fourth, every assembly process is accompanied by the emergence of some properties and the loss of others (submergence). Fifth, all systems but the universe as a whole are bound to break down sooner or later. Sixth, all systems are subject to environmental selection. Seventh, every system belongs to some evolutionary lineage. Eight, all the processes of assembly of lower-level things result in systems belonging to levels above the physical. Ninth, the world is the system of systems. Tenth, the fruitful exploration of the world assumes real systems and produces epistemic systems.

The upshot is an ontology or world view that endorses the following philosophical doctrines:

1 *systemism*, for it holds that every thing is either a system or a component of one – but not *holism*, which maintains that wholes are prior to their components and incomprehensible by analysis;
2 *materialism*, for it countenances only material existence and discounts autonomous ideas, spirits, and the like – but not *physicalism*, which rejects emergence and thus ignores all the supraphysical levels;
3 *pluralism* with regard to the variety of things and processes, and hence the plurality of properties and laws – but not *dualism* with regard to the mind-body problem;
4 *emergentism* with regard to novelty, for it holds that, while some bulk properties are resultant or aggregate, others are emergent – but not *irrationalism* with regard to the possibility of explaining and predicting emergence;
5 *dynamicism*, for it assumes that every thing is in a state of flux in some respect or other – but not *dialectics* – for it rejects the tenets that every thing is a unity of opposites, and that every change consists in, or is caused by, some strife or ontic contradiction;
6 *evolutionism* with regard to the formation and breakdown of systems, for it maintains that systems of new kinds emerge from time to time and are selected by their environment – but neither *gradualism* nor *saltationism* – for it recognizes both smooth changes and leaps;
7 *determinism* as concerns events and processes, for holding them to be lawful and denying that any of them may come out of the blue or disappear

without leaving traces – but not *causalism* – for it recognizes both randomness and goal-striving as types of process alongside with causal ones;
8 *biosystemism* (or organicism) with regard to life, for it looks upon organisms as material systems that, though composed of chemicals, have properties that do not occur on other levels – but neither *vitalism* nor *mechanism* (or physicalism) nor *machinism* (or artificialism) – for it maintains the ontological irreducibility of living systems;
9 *psychosystemism* as concerns mind, for it holds that mental functions are emergent activities of systems of neurons – but neither *eliminative* nor *reductive* materialism – for it affirms that the mental, though explainable with the help of physical, chemical, biological, and social premises, is emergent relative to the physical and the chemical; and
10 *sociosystemism* with regard to society, for it claims that society is a system composed of subsystems and possessing properties (such as social stratification and a political power structure) that no individual has; hence neither *individualism* nor *collectivism*, and neither *spiritualism* nor *vulgar materialism* are satisfactory.

I will not apologize for having combined ten ontological *isms*. A single *ism* is usually inadequate and it may even be dangerous in being one-sided, closed, and rigid. Many-sidedness, openness, and flexibility are only achieved by building a *system of matching isms* in harmony with contemporary science and technology, and formulated in as exact a manner as possible – and by updating it as needed. Which is yet another systemist maxim.

Concluding Remarks

Whereas in previous chapters we argued for the suitability of the systemic approach to the study of found things, in the present chapter we have argued for the adoption of the same approach to the study and control of made things, from social systems to machines. Such methodological monism rests on the ontological thesis that all real existents are material, though not all of them are physical; and on the epistemological thesis that all concrete things exist outside the knower's mind, even if they owe their existence to some individual or collective agent. That is, our basic philosophical theses are *emergentist materialism* and *scientific realism*.

The methodological monism advanced here is of course at variance with the nature / culture duality, and the corresponding natural science / cultural science dualism, preached by Kant and his successors, in particular the hermeneutic school. But since our approach involves the notion of emergence, it is equally at variance with radical reductionism, in particular physicalism (or eliminative materialism).

However, compatibility with one's philosophy is less important than fertility. And there can be little doubt that the system-cum-emergence viewpoint has triumphed in the natural sciences and the technologies, and that it is making progress in social science too. Moreover, there is reason to think that the backwardness of social studies is largely due to resistance to that approach, a thesis that will be developed in some of the subsequent chapters. However, before doing that let us examine the main rivals of systemism, namely individualism and holism.

6

Individualism and Holism: Theoretical

Ordinarily we deal now with wholes, now with their parts; and we analyse or synthesize according to need. It is only when we are in a metaphysical or ideological mood that we tend to believe that there are only wholes (holism) or else only parts (individualism), and correspondingly exalt either synthesis or analysis – as if they were mutually exclusive rather than complementary. Given that 'part' makes no sense apart from 'whole,' and conversely, both individualism and holism are logically untenable. Yet, this logical hole is rarely if ever noticed, partly because the part–whole relation is seldom analysed.

Here is another logical objection to both individualism and holism: there are neither unrelated individuals nor relations above individuals. Relatedness is basic in all domains, whether factual, conceptual, or semiotic. In other words, there are neither stray items nor relations without relata. If individuals a and b are R-related, we state that Rab, as in 'Adam and Eve are siblings.' It is only by abstraction that we may focus now on a relation, now on its relata. To assert the ontological primacy of either is to misconstrue the very notion of a relation.

The holism–individualism polarization goes back to antiquity and is still central to current controversies on many subjects. This polarization is not only philosophical: it is also partly temperamental and partly ideological. Thus, the great Romantic poet and naturalist Goethe scolded Newton for having decomposed the pure white light-beam into rays of different colours. And most of the totalitarian ideologists, whether on the left or on the right, have extolled the absolute primacy of the whole over the part.

However, there is no need to look up any famous thinkers to discover individualist and holist modes of thought: we are prey to them in everyday life. Consider, for instance, the holist fallacy that, because two items are strongly connected, they must be one and the same. Or recall its dual, the individualist fallacy that, because two things are different, they cannot be connected. Both reasonings are fallacious for the same reason: in order for a tie to exist or to be severed, there must be at least two different items. The correct methodological maxim is neither 'Fuse whatever items are related' nor 'Disintegrate all wholes,' but rather 'Distinguish without necessarily detaching.'

Things are never as simple as holists and individualists believe them to be. A whole is such because it is composed of interconnected parts, and no part exists in isolation from everything else. Hence, both individualism and holism each contains a grain of truth. In the preceding chapters it has been argued that the whole truth lies in a sort of fusion of individualism and holism, namely systemism. In this chapter and the next we will attempt to refute in detail both individualism and holism. To this end, we shall start by showing that they are anything but monolithic.

Because individualism is the more elaborate of the two views, it is more challenging and rewarding to argue with individualists than with holists. (Besides, holists are hostile to analysis and reluctant to engage in rational debate.) This is why most of this chapter and the next will be devoted to analysing individualism into its various components – an analysis that, ironically, no individualist has so far performed.

I have set myself three tasks in this chapter. The first is to characterize, evaluate, interrelate, and exemplify the various types or components of individualism and its opposite. In each case two strengths of the doctrine will be distinguished: radical and moderate. The reader should have no difficulty in supplying the names of outstanding scholars who have argued for or against any of the various modes of individualism since ancient times. The second task is to confront individualism with its opposite, namely holism (or organicism). The third is to see whether we are forced to choose between the two, or whether an alternative to both is viable and preferable.

Three warnings are in order. First, I submit that logical discussion, though necessary, is insufficient for determining whether any given philosophical doctrine works: its compatibility with the bulk of relevant antecedent knowledge must be examined too. Second, it is unlikely that anyone has been consistent (or foolhardy) enough to uphold all ten kinds of individualism at once. Third, although individualism is often associated with rationalism, the two are logically independent. After all, Parmenides, Plato, Aquinas, Ibn Khaldûn, Comte, Marx, Peirce, Durkheim, and Dewey, to mention but a few, were anti-individualists as well as rationalists of sorts. And most shareholders, who are presumably individualists in more than one way, are swayed by greed and fear as well as by more or less rational argument.

1 Individual and Individualism, Whole and Holism

An individual is, of course, an object, whether concrete like an atom or abstract like a concept, that is undivided or is treated as a unit in some context or on some level. For instance, persons are individuals in social science but not in biology, which regards them as highly complex systems. Again, chemical and biological species are taxonomic units but not ontological individuals (recall chapter 2, section 4).

As for individualism, it is the view that, in the last analysis, everything is either an individual or a collection of individuals. According to Hobbes, Leibniz, Weber, William James, and Whitehead, to cite but a few, all existents are individuals, whereas all wholes are conceptual. This is a strong and pervasive ontological thesis, which underlies and often motivates another nine modes or facets of individualism: logical, semantical, epistemological, methodological, axiological, praxiological, ethical, historical, and political. We will examine all ten modes of individualism.

Oddly, though pervasive, individualism is usually noted only in relation to human affairs, particularly in the guises of methodological individualism and egoism. This may be due to the fact that, despite its pervasiveness, individualism – as will be argued below – does not constitute a viable world view. I hasten to warn, though, that its opposite, namely holism, is even less adequate if only because, in opposing analysis and favouring intuition over reason, it promotes intellectual sloth.

The multiplicity and interdependence of the components of individualism are seldom if ever acknowledged. If ignored, however, none of the individual components of individualism can be correctly understood and evaluated. By contrast, when the multiplicity of individualism is acknowledged, it is seen that its various components hang together both conceptually and practically. That is, they form a system or whole made up of interconnected parts – which of course goes against the very grain of individualism.

So much for introductory remarks on individuals and individualisms. As for wholes, they are, of course, complex objects that behave as units in some regards. They differ from aggregates or assemblages in that they possess emergent or global properties, that is, features that their constituents lack. Holists emphasize wholeness and emergence to the point of underrating the constituents and precursors of wholes. This prevents them from understanding the emergence mechanisms, and leads them to proclaim that wholes must be accepted reverently rather than analysed. In summary, holism is anti-analytic and therefore irrationalist.

2 Ontological

Ontological individualism is the thesis that every thing, indeed every possible object, is either an individual or a collection of individuals. Put negatively: There would be no wholes with properties of their own, that is, systemic or emergent properties. Ancient and early modern atomism, medieval nominalism, mereology (Lesnieswski's calculus of individuals), rational-choice theory, sociological and legal individualism, and libertarianism either exemplify or presuppose ontological individualism.

The doctrine comes in two strengths: radical and moderate. Radical individualists (nominalists) claim that individuals have no properties of their own other than that of being able to associate with other individuals to constitute further (complex) individuals. All attribution and all classing would be strictly conventional. As a consequence, there would be no natural kinds, such as chemical and biological species: all kinds would be conventional. Yet of course there are natural kinds in addition to whatever arbitrary collections one may imagine.

However, the main objection to radical individualism is that 'bare' individuals, that is, unaccompanied units, are hard to find. True, there are plenty of lone particles moving for a while in outer space and embedded in weak fields. But such loners are eventually captured by atoms or affected by strong fields. And an electron moving in a wire is said to be 'dressed,' because it is accompanied by other particles, which is why it is called a 'quasi-particle.' As for social individuals, the very notion is an oxymoron, since people are eminently sociable: indeed, every person, however withdrawn, belongs to several social networks at once.

A world of individuals would be deprived of universals, in particular laws. Hence, it would be lawless or chaotic in the original sense of the word. If – defying the laws of biology – there were humans in such a world, they would be unable to think in rational and general terms. Furthermore, they would be incapable of acting on the strength of rules grounded on laws, since these – the ontic universals par excellence – would not exist.

By contrast, moderate ontological individualism, exemplified by ancient atomism and seventeenth-century mechanism, admits properties and possibly natural kinds as well. But it still regards individuals as primary in every sense, and it overlooks or even denies the existence of systems. Undoubtedly, this view contains an important grain of truth: that all the known complex things result from the aggregation, assembly, or combination of simpler ones. For example, light beams are packages of photons, molecules emerge as combinations of atoms, multicellular organisms by either the combination or division of single cells, and social systems from the association of individuals.

However, none of these assembly processes occurs in a vacuum. Thus, every atom is embedded in fields of various kinds, and every human being is born into a social group, and is partly shaped by his or her natural and social environment. No man is an island – nor is an atom.

Moreover, some assembly processes result in systems, and every system has not only a composition but also a structure – the set of ties among its components. According to individualism, however, composition is everything, whereas structure is nothing. Hence, a consistent individualist will be unable to distinguish a snowflake from a water droplet, or a business firm from a club constituted by the same individuals. Likewise, the very existence of organisms is paradoxical to the upholders of the 'selfish gene' fantasy, since they claim that all the action occurs at the molecular level.

The oversight or misunderstanding of systems may be called the *individualist fallacy*. This consists in identifying a system with its composition, as exemplified by the assertions that a snowflake is 'nothing but' a bunch of H_2O molecules, an organism 'nothing but' a bag of atoms and molecules, and a society 'nothing but' a collection of individuals. All of this is mistaken because it involves overlooking structure, environment, and process. Shorter: Individualist nothing-but-ism is wrong because it denies the very existence of systems and their emergent properties.

Both in logic and in science, individuals and properties – whether intrinsic or relational – come together on the same footing: neither is prior to the other. In particular, there are no relations without relata – by definition of 'relation.' Moreover, every entity emerges and develops in interaction with other entities. This holds for persons and corporations as well as for molecules, cells, and other concrete entities.

Furthermore, any given individual, whether electron or person, is likely to behave differently in different contexts – for example, in a dyad, a triad, or a crowd. In sum, every thing in the world is connected, directly or indirectly, with other things. Except for the universe as a whole, the total loner, be it atom, person, or what have you, is a fiction. These are systemist theses.

Finally, individualism is bound to lead to inconsistency because it fails to analyse the environment (or 'situation') of an individual: it treats it as a whole, and rightly so. On this point individualism conforms to the scientific practice. For instance, the physicist or chemist who studies a process occurring in a container describes the latter in a global manner: he notes such properties as shape and temperature; and the biologist notes only some global features of the environment, such as geographical accidents and climate.

In short, ontological individualism does not work, except as a very crude approximation, namely, in the case of negligible interactions (as in a low-density gas or a sparsely populated region). However, it contains two important truths. These are the theses that only particulars (whether small or large) have real existence; and that there are no universals in themselves. Yet both truths are part of systemic ontology, to be sketched in section 10.

Still, ontological holism is not the viable alternative to individualism, for it leads to Parmenides' block universe or undifferentiated cosmic blob. In sociology it leads to eliminating the person and reifying such institutions as the law, science, and communication, which are said to be the behaving entities (Black 1976; Luhmann 1987). Thus, paradoxically, focusing exclusively on the whole leads to regarding it as an empty shell: radical holism equals nihilism.

3 Logical

Logical holism is the purely negative view that reason, in particular analysis of any kind, is at best limited, and at worse severely distorting. Logical holism is inherent in mysticism, intuitionism, and Hegelianism. Unsurprisingly, it has few adherents among contemporary philosophers, and none among mathematicians, scientists, or technologists.

The opposite of logical holism is logical individualism. This is the view that all constructs – concepts, propositions, and the like – are built out of conceptual or linguistic individuals, or zeroth-type items. It comes in two strengths: radical and moderate. Radical individualism denounces classes, or it tolerates them but regards them as virtual or fictitious – as if such individuals as points and numbers and functions and operations were any less fictitious.

Set theory treats sets as wholes with properties that their elements do not possess – for example, cardinality and inclusion in supersets. Since set theory is the foundation of mainstream mathematics, the adoption of radical logical individualism would cause the collapse of the entire mathematical building. (Substituting categories for sets does not improve things for the individualist, because the basic bricks of categories, namely arrows, or morphisms, are even more remote from individuals than are sets.)

Another consequence of radical individualism is that it cannot account for the unity of logical arguments and theories. Indeed, every argument is a whole, and more particularly a system, not a mere aggregate, of propositions. Think, for instance, of the basic rule of inference, the *modus ponens*: '$A, A \Rightarrow B \vdash B$.' The argument as a whole is consistent, and its conclusion emerges from its premises. Consistency is a global property, and emergence is the seal of systemicity.

The same holds, a fortiori, for theories. By definition, these are hypothetico-deductive systems of propositions, that is, potentially infinite systems of deductive arguments. The structure of any such system, that is, the relation that holds it together, is that of deducibility. And, *pace* extensionalism, this relation is not definable as an infinite set of ordered pairs of the form ⟨premise(s), conclusion(s)⟩. Indeed, in all logical calculi, the entailment relation is tacitly defined by a set of rules of inference.

Extensionalism is the moderate version of logical individualism. Extensionalism admits classes but holds that predicates should be defined as sets of individuals deemed to possess such attributes. In other words, logical extensionalism holds that predicates are identical with their extensions. Thus, 'is alive' would amount to the collection of all living things. In practice, however, one must use the predicate 'is alive,' or a conjunction of lower-level predicates, to characterize a class of living things. Moreover, different predicates may be coextensive, as is the case with 'is alive' and 'metabolizes.'

All non-arbitrary classes are generated by predicates. In the simplest case, that of a unary predicate P, the corresponding class is $C = \{x|Px\}$, read 'the set of all individuals with property P.' Something similar holds for predicates of higher degrees: binary, ternary, etc. Thus, we must have some concept of love before endeavouring to find its extension, that is, the class of ordered pairs of the form ⟨lover, loved⟩. In sum, sense precedes truth, and predicates precede (logically) kinds.

Extensionalism occurs in the standard characterization of a relation (in particular a function) as a set of ordered pairs or, in general, a set of ordered n-tuples. A first objection to this characterization is that it is only feasible for finite sets. And even in this case it only yields the extension of the relation, and is not always feasible. For example, the relation of predication is not definable as a set of subject-predicate couples. A second objection is that n-tuples have very different properties from their components – a simple case of emergence. For example, an ordered pair of even numbers involves an order relation and it is neither even nor odd. Furthermore, the standard set-theoretic definition of an ordered pair involves an order concept.

A third objection to extensionalism is that the most important of all relations in set theory, that of membership, or $\in$, is not definable as a set of ordered pairs of the forms ⟨individual, set⟩, or ⟨set, family of sets⟩. Instead, the $\in$ relation is defined implicitly, by the axioms in set theory where it occurs. If $\in$ were construed extensionally, it would have to be admitted that '$x \in y$' can be rewritten as '$\langle x, y \rangle \in \in$' – which is not precisely a well-formed formula.

Nor does one usually define functions as sets of ordered n-tuples, or tables. Again, this is possible only for finite sets such as a finite (hence miserly) sample of the non-denumerable set of ordered pairs $\langle x, \sin x \rangle$. Only the *graph* (extension) of a function is a set of ordered *n*-tuples – as Bourbaki notes. For example, the graph of a function $f: A \to B$ from set A into set B is $\Gamma(f) = \{\langle x, y \rangle | y = f(x)\}$. But the function f itself is defined otherwise, whether explicitly like the power function, or implicitly, for example, by a differential equation. (Moreover, the more interesting functions come in families of inter-definable functions, such as the trigonometric functions and the Legendre polynomials.)

We should then keep the difference between a predicate P defined on a domain D, and its extension $\mathcal{E}(P) = \{x \in D | Px\}$, read 'the collection of Ds that possess property P.' Moreover, we must distinguish this set from the collection $\mathcal{R}(P)$ of individuals that P refers to, that is, the reference class of P. One reason for this distinction is that it may well be that, whereas $\mathcal{E}(P)$ is empty, $\mathcal{R}(P)$ is non-empty. (Examples of predicates with a nonempty reference but an empty extension: 'the greatest number,' 'magnetic monopole,' 'perfectly competitive market.' Such unrealistic predicates are wrongly said to be non-referring.) Another reason is that, whereas the extension of an *n*-ary predicate is a set of *n*-tuples, the reference class of the same predicate is a set of individuals. For instance, the extension of the predicate 'helps' is the collection of ordered pairs ⟨helper, helped⟩, whereas the reference class of the same predicate is the logical union of the class of helpers with that of helped.

In sum, logical individualism is untenable. However, its failure makes no dent on logical analysis. It only shows that an analysed system is still a whole, or higher-order individual, with properties of its own, among them its structure. Moreover, only logical analysis can ascertain whether a given set is a system, that is, a collection every member of which is logically related to some other members of the same set. Hence, the demise of logical individualism poses no threat to rationalism.

The upshot for mathematics, science, and technology is that they would gain nothing, and lose much, if they were to eliminate predicates in favour of individuals or *n*-tuples of individuals. The reason is that there are no real bare individuals, devoid of properties: these are fictitious.

4 Semantic

Semantic holism is the view that the meaning of any single conceptual or linguistic item is determined by the entire body of intellectual culture. Since no single person can be expected to know everything, semantic holism is not viable. Worse, it implies that no one can ever grasp the meaning of anything – a defeatist and moreover obscurantist stand.

The opposite of semantic holism is semantic individualism. According to the latter, the meaning of a conceptual or linguistic whole, such as a sentence or the proposition it designates, is a function of the meanings of its parts. However, the function in question has never been specified. Moreover, it cannot be defined because the thesis is false, as shown by the following counterexamples.

Heidegger's pseudodefinition of time as 'the ripening of temporality' is meaningless even though its constituents make sense. Another example: the sentence 'That will do' gets its meaning from its context. A third one: the proverbial propositions 'Dog bites man' and 'Man bites dog' are not the same although they have the same constituents. A last example: the predicate 'good teacher' does not equal the conjunction of 'good' and 'teacher.' Instead, 'good teacher' is definable as the conjunction of 'teacher,' 'knows his subject,' 'loves his subject,' 'likes to teach,' 'clear,' 'inspiring,' 'dedicated,' 'patient,' 'considerate,' etc. In short, contrary to individualism, the units of meaning – concepts and their symbols – are not assembled like Lego pieces. Rather, they combine like atoms and molecules – or people, for that matter.

Linguists have known for nearly two centuries that every language is a system, and hence no linguistic expression is meaningful by itself, that is, separately from other expressions in the language (see chapter 4); so much so that a language may be analysed as a system with a definite composition (vocabulary), environment (the natural and social items referred to by expressions in the language), and structure (the syntax, semantics, phonology, and pragmatics of the language).

What holds for languages also holds, *mutatis mutandis*, for conceptual systems, in particular classifications and theories. Indeed, the sense or content of a part of such a system depends on the sense of other members of the whole: it is a contextual not an intrinsic property. For example, the meet ($\vee$) and join ($\wedge$) operators in a lattice intertwine so intimately, that they have no separate meanings. (Example: the absorptive laws $x \wedge (x \vee y) = x$ and $x \vee (x \wedge y) = x$.) And in classical particle mechanics, the sense of 'mass' depends on that of 'force' and vice versa, although both are undefined, and in particular not interdefinable. Their meanings are interdependent because they are related through Newton's second law of motion. Were it not for the latter, we would be unable to interpret mass as inertia, and force as a cause of acceleration.

What is true is that, contrary to semantic holism, in particular intuitionism, the linguistic and conceptual wholes, such as texts and theories, must be analysed to be correctly understood. And analysis is of course the breaking down of a whole into its constituents – without however severing the relations that hold them together. Moreover, conceptual analysis is best performed in the context of a conceptual system, preferably a hypothetico-deductive system or theory. For instance, to grasp the meaning of the technical concept of spin in microphysics, it is necessary to place this concept in some theory of elementary spinning particles – according to which spin is anything but a rotation. Incidentally, this example shows that ordinary-language analysis cannot ferret out the meaning of theoretical terms.

Semantic individualism also holds that truth values can be assigned or estimated one at a time. This presupposes that truth values inhere in propositions. But this is only true for logical truths and falsities – and even so only within a given logical calculus. The truth value of extra-logical propositions depends upon the truth value of others: axioms in the case of theorems, and empirical evidence in the case of low-level factual statements. In other words, the truth value of any proposition other than a logical formula depends upon other statements in the given context. In these cases one should not write 'p is true' but rather 'p is true in (or relative to) context C.'

In short, semantic individualism does not work, because it overlooks the web in which every construct and every sign is embedded. Still, its thesis that analysis is necessary stands and is important.

5 Epistemological

Epistemological holism is the view that, since everything hangs together, to get to know any particular we must first know the whole universe. Blaise Pascal rightly stigmatized this view as unviable four centuries ago.

The dual of epistemological holism is epistemological individualism. This is the thesis that, to get to know the world, it is necessary and sufficient to know the elementary or atomic facts – whence the name 'logical atomism' that Russell and Wittgenstein gave this doctrine. Any complex epistemic item would then be just a conjunction or disjunction of two or more atomic propositions, each describing (or even identical to) an atomic fact.

This view may hold for the knowledge of everyday facts recorded in such sentences as 'The cat is on the mat' – a favourite with linguistic philosophers. But it fails for the most interesting scientific statements. These are universal generalizations that cannot be reduced to conjunctions, because they involve quantification over infinite or even non-denumerable sets. (Example: 'For all t in $T \subseteq \mathbb{R} : f(t) = 0$,' where t designates the time variable, whose values lie on the real line $\mathbb{R}$, and '$f(t) = 0$' is a possible form of a law statement such as a rate equation or a dynamical law.)

A norm of epistemological individualism is that all problems should be tackled one at a time. But this is not how one actually proceeds in research. Indeed, posing any problem presupposes knowing the solution to logically previous problems. In turn, the solution to any interesting problem raises further problems. In short, problems come in packages or systems. The same holds for issues, or practical problems. For example, drug addiction is not successfully fought by just punishing drug pushers, let alone drug addicts. It might only be solved by attacking the economic and cultural roots of substance abuse, such as poverty, the competitive drug market, anomie, and ignorance. Thus, practical problems too assemble into systems; hence, the maxim 'One thing at a time' is a recipe for failure or even disaster (Hirschman 1990). Systemists should prefer the rule 'All things at the same time, though gradually.'

Epistemological individualism, like its ontological mate, may have been suggested by ancient atomism, but it fails in modern atomic physics. The reason is that a quantum-theoretical problem is not well posed unless a boundary condition is stated – and the boundary in question happens to be an idealized representation of the environment of the object under study. And an ill-posed problem has either no solution or no unique solution.

(More precisely, any problem in quantum physics boils down to stating both the state equation and the boundary condition. The latter specifies that the state function vanishes at the boundary. Now, a change in boundary may be accompanied by a qualitative change in the solution. For example, the state of a free electron confined within a box is represented by a standing wave; by contrast, if the box expands to infinity, the electron is represented by a propagating wave. Moreover, the form of the solution depends critically upon the shape of the box: the 'wave' may be plane, spherical, cylindrical, and so forth. In sum, there will be as many solutions to the problem as stylized environments.)

The point in recalling this example is that, far from being analysed, the environment idealized by the boundary condition (box) is taken as an unanalyzed macrophysical whole. The social analogue is the (macrosocial) situation or institution, something that is not describable in microsociological terms. This social context – particularly the economic, political, and ideological constraints and stimuli, as well as the mores and ethos of the epistemic community – is all too often overlooked by the individualist epistemologist, just as it is wildly exaggerated by his collectivist counterpart. Yet if cognition is detached from its social womb, it becomes impossible to understand how inquirers get to learn anything; why peer recognition is such a powerful motivation of research; or why other members of the scientific community, rather than the researcher himself, are eager to falsify his hypotheses.

Finally, epistemological individualism is defective also in focusing on the individual inquirer isolated from her epistemic community. It is not that the latter does the knowing, as the social constructivist-relativists claim: after all, social groups are brainless. Cognition is a brain process, but individuals learn not only through hard thinking and doing but also from one another. In particular, scientists belongs to scientific communities. And, as Robert K. Merton (1973) put it long ago, they are motivated by two mutually reinforcing reward mechanisms: intrinsic (the search for knowledge) and extrinsic (peer recognition). Further, these mechanisms are coupled, now constructively, now destructively.

Moreover, the members of every scientific community are expected to abide by such social rules as the open sharing and discussion of problems, methods, and findings; so much so that, to qualify for peer recognition, researchers pay a heavy peer-evaluation tax. In short, cognition is personal, but knowledge is social. 'I know X' is not the same as 'X is known [by the members of a given social group].'

6 Methodological

Methodological individualism is of course the normative counterpart of epistemological individualism. It holds that, since everything is either an individual or a collection of individuals, 'in the last instance' the study of anything is the study of individuals. In other words, the proper scientific procedure would be of the bottom-up kind: from part to whole. This micro-reductionist strategy is best known in social studies, but actually it has been attempted – as well as vehemently denounced as 'Cartesian' – in all fields.

For example, the properties of a solid would be known by analysing it into its constituent atoms or molecules, and those of a multicellular organism by reducing it to its cells. But solid-state physicists know that the first conjunct of the previous sentence is false. Indeed, the properties of a solid are not understood by modelling it as an aggregate of atoms, but by analysing it into three components: the ionized atoms, the electrons roaming among the latter, and the electromagnetic fields that accompany the ions and the electrons and that glue these constituents together. Hence atomic physics, though necessary, is not enough for understanding extended bodies. The disastrous consequence for radical reductionism should be obvious.

Likewise, biologists know that the second conjunct of the above claim is false as well, since cells can associate into organs, and the latter into larger systems whose biological roles are quite different from those of their constituents. Hence cellular biology is necessary but insufficient for understanding organs and, a fortiori, the organism as a whole: one must also investigate how cells connect to one another, for instance, through ions and hormones.

Methodological individualism works only for simple problems of the form, Given an individual, together with its law(s) and circumstance(s), figure out its

Macro-level		A → B	A → B
	Top-down ↓ ↑	Bottom-up ↑ ↓	
Micro-level		a → b	a ← b

Figure 6.1 Whenever a multi-level system is to be either studied or handled, two mutually complementary strategies should be tried: top-down (analysis) and bottom-up (synthesis)

behaviour. For instance, find the trajectory of a ball rolling down a ramp under the action of gravity – or the behaviour of a maximizing consumer in a given market. But the method fails whenever interaction is of the essence. For instance, it fails for a binary star and, a fortiori, for a system of a large number of bodies (or persons). Actually, even in the case of the single body the method gives only an approximate solution, for it neglects the reaction of the body upon both the constraint (e.g., the ramp) and the force field. Likewise, people are not passive agents either: they react upon the very networks in which they are embedded.

If methodological individualism were adequate, to know a triangle it should suffice to know its sides regardless of its relations, namely the inner angles – which is not even true in the exceptional case of equilateral triangles. Likewise, to know a human family it does not suffice to know its members: some knowledge of the relations among them and with other people is necessary as well, if only because they are bound to change in the course of time. In general, social facts can only be understood by embedding individual behaviour in its social matrix and by studying interactions among individuals. The composition and the structure of a system are every bit as inseparable in social matters as in natural ones. Detachment entails distinction but not conversely.

The dual of methodological individualism is methodological holism, or the view that all analysis is wrong because it allegedly destroys wholeness. This doctrine is closely allied to intuitionism, the enemy of both rationalism and empiricism. And, just as individualists are micro-reductionists, so holists are macro-reductionists. I submit that the right research strategy is to combine the bottom-up (synthetic) with the top-down (analytic) approaches, both of which relate the micro-level to the macro-level, instead of attempting to reduce the one to the other. Such combination, characteristic of the systemic approach in all research fields, retains the sound parts of individualism and holism. Systemism yields explanatory schemata like in figure 6.1, according as one starts with macro-facts (top-down analysis) or with micro-facts (bottom-up synthesis).

The partial failure of methodological individualism has an important consequence for the theory of scientific and technological explanation. According to the so-called covering-law 'model' of scientific explanation, to explain a fact is to show that it fits a pattern: that is, to subsume it under a law-like statement. But this is not what scientists or technologists call an explanation: they want to know how things work, that is, what makes them tick. This accounts for their preference for laws that sketch some mechanism or other – causal, random, or mixed – for the occurrence of the fact to be explained.

For example, Newton and his followers were not satisfied with the kinematical laws found by Galileo, Kepler, and Huygens: they also wished to know the causes of motion. Nor were Maxwell and Boltzmann satisfied with thermodynamics: they endeavoured to unveil the underlying mechanism – which turned out to be a combination of causation and chance. Again, it is not enough to state that remembered episodes are first 'stored' in short-term memory, then transferred to long-term memory. Cognitive neuroscientists want to find out how such memories emerge, work, connect, and deteriorate: they are after the neural mechanisms of learning, memory, and forgetting. In particular, they wish to know whether or not learning is the same as the strengthening of synaptic efficacy leading to the formation of new systems of neurons. They are not satisfied by being told that mental processes are cases of 'information processing' – whatever this may be.

Now every mechanism is a process in some concrete system, such as an atomic nucleus, crystal, cell, brain, ecosystem, or business. And the very concept of a system is alien to individualism, which recognizes only the components of systems – such as the trees in a forest and the individual members of an organization. Hence explanation proper, which invokes mechanisms, is beyond the ken of individualism. Consequently, methodological individualism erects an intolerable barrier to scientific understanding.

In short, methodological individualism does not work. Moreover, it cannot work, because the universe is not a mere aggregate of atomic facts but a system of systems, and because agents – in particular inquirers – are not self-reliant individuals, but nodes in social networks. Methodological holism fares even worse because, as will be shown in the next chapter, it involves replacing reason with intuition and giving up liberty – in particular the liberty to challenge the given social context. However, the practical aspect of the question deserves a separate chapter.

Concluding Remarks

There are three main world views concerning the structure of the universe and our knowledge of it. One is individualism, according to which everything is

96 Emergence

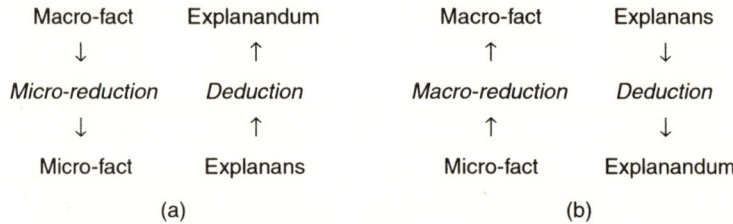

Figure 6.2 Two research strategies and modes of explanation involving a micro-level and a macro-level: (a) micro-reduction; (b) macro-reduction. They are mutually complementary rather than mutually exclusive.

either an individual or a collection of individuals. In negative terms: There would be no wholes except in fiction. The polar opposite of individualism is holism, according to which the universe is an undifferentiated blob, so that every part of it influences every other part. In negative terms: There would be no individuals except as fictions. The epistemogical concomitants of individualism and holism are rationalism and intuitionism respectively.

The alternative to both individualism and holism is systemism, a sort of synthesis of its rivals, according to which the universe is the maximal system, and everything in it is either a system or a component of a system. Moreover, systems are characterized by emergent properties. In negative terms: There are neither isolated individuals nor indecomposable wholes. Nor are all things at the same level of organization: they are distributed among a number of levels – physical, chemical, biological, social, and technical.

Some of the most awesome processes in both nature and society start on a micro-level and end up on a macro-level. For example, a nuclear chain reaction, unless moderated, ends up in an explosion; and an executive order of an all-powerful president may trigger a war. Conversely, a macro-process, such as a nuclear explosion or a war, may have lethal effects, with socially disruptive results, that in turn affect life, and so on. A single-level ontology cannot be expected to account for such zig-zaggings among levels.

A systemic ontology, particularly one involving the concepts of emergence and level, suggests a systemic epistemology involving different levels of analysis along with different ways of moving between them. In particular, systemism suggests combining the top-down and bottom-up research strategies, as shown in figure 6.2.

7

Individualism and Holism: Practical

The world can be seen as either a pile of things (individualism), a solid bloc (holism), or a system of systems (systemism). Consequently, our actions upon our surroundings may also be seen in either of three ways: the individual against the world, the world against the individual, or the individual interacting with his surroundings. The first view emphasizes autonomy, the second heteronomy, and the third interdependence. Equivalently: Individualism promotes self-reliance, independence, rights, and selfishness; holism emphasizes dependence, duties, conformity, and altruism; and systemism favours a combination of autonomy with heteronomy, independence with cooperation, selfishness with altruism, and rights with duties.

The axiological and moral concomitants of the three views in question can be summed up in so many maxims: 'What is good for the individual is good for society' (individualism); 'What is good for society is good for the individual' (holism); and 'What is good for the socially concerned individual is good for the just society' (systemism).

The counterparts of these doctrines in political philosophy should be obvious. They are libertarianism, totalitarianism, and democracy respectively. Tell me what ontology you hold, and I will guess what you think of government, civil liberties, social justice, and perhaps even the party you will vote for – provided your world view is consistent. (However, for better or for worse, few of us are fully consistent in these matters.)

1 Value Theory, Action Theory, and Ethics

Individualism and holism occur not only in theoretical philosophy but also in what may be called the philosophical technologies: value theory, action theory, and ethics (see Bunge 1989, 1998). Axiological (or value-theoretical) individualism holds that only individuals can evaluate; that there are only individual values; and that the part is more valuable than the whole – which is likely to be fictitious anyway. Praxiological (or action-theoretical) individualism focuses on individual action, and accordingly it overlooks both the social embeddedness of action and the interactions among individual actions. The ethical consequence is obvious: a moral or legal norm is morally justified only insofar as it benefits the individual.

I submit that just one of the three claims of axiological individualism is true, namely, that only individuals can perform valuations. Even so, we often evaluate under the influence of others, and sometimes under duress. Moreover, values are adopted or rejected by social groups, to the point that an individual's standing in a group depends upon his acceptance of the group's values. In short, valuation, though individual, is socially conditioned. (Methodological consequence: Axiology must stay close to sociology.)

The second thesis, that there are only individual values – such as well-being and liberty – makes no room for social values, such as security, peace, social cohesion, and justice. Yet most of us are attached to such values, not least because their realization is a necessary condition for that of a number of individual values. And no social value is an aggregate or combination of individual values. For example, individual goodwill does not suffice to build a good society.

The third thesis, that the person is more valuable than any of the social networks she belongs to, rests on the ontological presupposition that individuals are detachable from the systems they are embedded in. This thesis is every bit as wrong as the holistic view that individuals are expendable and must be subservient to the whole – state, church, party, corporation, or what have you.

One should not be forced to choose between the isolated individual and the supra-individual whole, because both are fictions. In reality, there are only interconnected individuals and the systems they constitute. Hence, when evaluating an individual action we should ask whether it is valuable to the social whole in question; and when evaluating the latter we should ask whether it promotes personal welfare.

Accordingly, free-riders and nihilists are just as reprehensible as exploiters and tyrants. It also follows that we should strive to minimize anomie – the discrepancy between personal achievement and social value or norm – by reforming both individual conduct and social structure. I submit that this systemic approach to axiology is free from the defects of its individualist and holist rivals. And it is the one that should help manage the unavoidable conflicts between individual and social values instead of suppressing either.

What holds for axiology also holds, *mutatis mutandis*, for praxiology and ethics. In all three fields, individualism overlooks the problems that originate in such macrosocial issues as overpopulation, poverty, sex discrimination, exploitation, tyranny, and war. And yet the victims of any of the latter by far outnumber the cases of suicide, abortion, prostitution, euthanasia, theft, and small-scale murder – the specialties of the individualist moral philosopher. In short, individualist moral philosophers focus on micro-moral problems, and thus overlook the macro-moral ones, which are far harder to tackle because they call for sociology, political science, and social policy-making.

By contrast, the systemist recommends focusing on the individual-in-society rather than on either individual or society, which is just an instance of the logical thesis that relations come together with their relata. There is a further reason: the practical and moral agent is neither the isolated individual nor society as a whole, but the person-in-society, at once constrained by some norms and stimulated and empowered by others.

An example should clarify the preceding. The practice of harvesting organs from executed prisoners is expanding. Utilitarians, who are individualists, are bound to approve of it: why let go to waste organs that could help the living? Others oppose this practice on religious grounds. A systemist opposes it for a different reason: because it condones the death penalty and promotes the organ-harvesting industry – which presumably lobbies for the preservation of capital punishment.

Utilitarianism is wrong in using a fuzzy notion of happiness (or utility) and in ignoring the social context of moral problems and individual action. It is also mistaken in seeking the greatest happiness of the greatest number, because these desiderata are mutually incompatible. (Call H the total amount of happiness to be divided equitably among n individuals, and h the slice of happiness allotted to each of them. Since $H = nh$, maximizing n implies minimizing h and conversely.) This is why utilitarianism is at best ineffectual. And negative utilitarianism ('Do no harm') is insufficient, for one ought to try and help others, defying custom or challenging the law if need be.

Since the sources of, and solutions to, so many practical and moral problems are partially social, a practical philosophy is impractical or worse unless it balances private and public concerns, thus tackling macro-ethical issues as well as micro-ethical ones. Such balancing involves acknowledging that any viable system of values and norms is ambivalent and therefore potentially contradictory rather than univalent and thus perfectly consistent (Merton 1976). Thus, we normally admit that liberty should be tempered by responsibility, competition in some regards by cooperation in others, and courage by prudence. We also supplement norms with counter-norms: self-preservation with concern for others, rights with duties, democracy with competence, and so on. Thus, even individualist or holist theorists are closet systemists in practice.

2 Historical and Political Individualism

Historical individualism is a philosophy of history, namely, the tenet that history is made by individuals. It comes in two versions. According to one of them, the main actors are heroes and villains; according to the other, every rational decision-maker is a historical agent. The obvious merit of historical individualism in either version is that it rejects inaccessible superhuman agencies such as fate, the general will, and the *Volk* – the Romantic idea of people-race-language-nation. The obvious flaw of the doctrine is that it overlooks the natural environment, tradition, and social networks, none of which is reducible to individuals.

Political individualism is the thesis that individual liberty is the maximal value. It is the same as libertarianism – rather than classical liberalism, which is consistent with democratic socialism. When joined with a solidarist morality, libertarianism entails that all political institutions should be suppressed: this is classical left-wing anarchism. And when joined with egoism, libertarianism entails that government should be minimal and should act exclusively in the service of those who have the wherewithal to enjoy liberty: this is contemporary right-wing libertarianism (or neo-liberalism). In other words, political individualism preaches either the elimination of all government or its shrinking to law-and-order forces.

Classical anarchism presupposes, as did Rousseau, that people are basically good and solidary, whence they need no external constraints. By contrast, contemporary libertarianism assumes, like Hobbes, that we are all evil and selfish, and hence in need of protection from ourselves. Neither presupposition is supported by social psychology. The latter confirms Robert Louis Stevenson's suggestion of a century ago, that we are a mixture of good and evil.

This may sound trite because it is, whereas individualism is off the mark if only because in real life every one of us needs help and seeks it, is willing to cooperate in some respects – and feels good if he does (Rilling et al. 2002). Political holism is no better for, whether in its mild communitarian version or in its ferocious totalitarian one, it drowns individuality. True, by comparison with either, political individualism looks attractive because of its praise of liberty. Still, just as holism justifies political oppression, individualism is socially dissolving – as Tocqueville noted long ago. Hence, neither is consistent with democracy.

However, because few people are consistent, one should not exaggerate the practical impact of either individualism or holism. Thus, the Nazi leaders preached holism to the masses while practising the most rapacious individualism. Conversely, the right-wing libertarians preach individualism for themselves, along with serfdom or worse for liberals, non-whites, and others. If one aims to judge political movements, their deeds are far more important than their speeches, which are often masking rather than revealing.

Fortunately, there is an alternative to both extremes. This is the systemic view that since the individual strives to survive, but cannot succeed without help, he must learn to combine competition with cooperation. The political corollary is that we need institutions, both governmental and non-governmental, to channel our prosocial impulses and hold the antisocial ones in check. Participatory democracy and polyarchy might fit this bill. But that is another story.

3 First Alternative to Individualism: Holism

Since individualism is deeply flawed in all its modes, an alternative to it is required. The first one to come to mind is, of course, holism (or collectivism). This is the amorphous-blob conception of the world: the view that the whole precedes and dominates the part, as a consequence of which the former is more valuable than the latter. The philosophies of Plato, Hegel, and Bergson are typically holistic.

Ontological holism asserts the priority and indecomposability of the whole. But of course a whole is not such unless it comprises parts. Hence, part and whole go hand in hand; so much so that a change in a part may cause a qualitative change in the whole and conversely, as when an individual initiates a social movement, and when the latter drags along an individual. Holism also claims that every thing interacts with everything else. But this is not true, because the intensity of most interactions decreases with distance or, in the case of society, with time. This makes it possible to isolate almost anything, at least in some regards, to some extent, and for a while.

According to logical holism, relations precede their relata. For example, Marx attempted to characterize the person as the set of her social relations. But this is of course logically incorrect, for relations come with their relata, and these with the former. Thus, the marriage relation does not exist without spouses, which in turn are not such unless related by marriage. (In abstract terms: A relation is not properly defined unless accompanied by the domain in which it holds, just as the latter is not fully characterized unless one specifies its structure, that is, the relations that hold among its members.) When proceeding rigorously, one always defines the total system, such as the relational system 'D together with $<$', or $S = <D, <>$. In short, logical holism is every bit as untenable as its dual, logical individualism.

Next comes semantic holism. This is the view that the meaning of any construct (or the signification of any sign) depends upon the entire body of knowledge (or text). This thesis has not been formalized, and it is hard to see how it could be. In any case, the thesis in question is false, as shown by the following counter-examples. The meaning of the implication relation is exhaustively determined by the predicate calculus, and that of 'photosynthesis' by biochemistry. In neither case do we need to rope in further fields of knowledge. In sum, semantic holism is false. Its merit is to stress that there are no stray constructs: that meaning is contextual. Yet, to be manageable, the context must be restricted. This limits the possibilities of convergence.

The next of kin is epistemological holism. This view may be compressed into three theses: on the source and subject of knowledge, and on the part–whole relation. The first is the claim that the highest, or perhaps even the sole, source of knowledge is the pristine, total, and instant intuitive apprehension of the whole, untainted by either experience or reason. Moreover, intuition would be infallible. This view – held by Bergson and Husserl among others – is so dogmatic, and so obviously at variance with all we know about cognition, that it is hardly worth being discussed. On the other hand, the problems of the kinds and roles of intuition, as well as of their relation with both experience and reason, are genuinely interesting questions, at once empirical and philosophical (see Bunge 1962a). But their discussion requires analytical tools that the holist rejects.

The holistic view on the source of knowledge is that the inquirer is society as a whole. This is the currently fashionable social-constructivist thesis, first advanced by Marx. Taken literally, this opinion is grotesque, since the organ of knowledge is the brain, whereas society has no brain. Moreover, social constructivism makes no room for original and particularly non-conformist thinking. The only merit of this view is that it corrects individualist epistemology, by reminding us that every inquirer is a member of one or more information networks. Still, by anchoring the individual too firmly to his community, it fails to explain creation and rebellion. After all, the fisherman goes out to catch fish, not nets. (See detailed criticisms of constructivism-relativism in Bunge 1999.)

Besides, holism fosters cultural relativism, that is, the view that each community has its own set of beliefs and values, which are neither better nor worse than those of other tribes. Needless to say, relativism is incompatible with the search for objective truth, which is cross-cultural: it leads to epistemological anarchism. Furthermore, because it denies the universal canons of valid argument, relativism does not even make rational debate possible among people from different cultures or even subcultures. Relativism is also inconsistent with the very idea of moral and political progress. And because it is localist rather than universalist, it does not even need the concept of humanity.

As for the holistic view on the epistemic part–whole relation, it comes in two strengths: radical and moderate. According to the former, to know the part we must know the whole. Since this is impossible, we are doomed to ignorance. By contrast, moderate holism holds only that human knowledge is a totality. This is true up to a point. In fact, human knowledge is indeed a system, but one whose components are not bound equally strongly. For example, geologists and mathematicians can work side by side without ever interacting significantly; and whereas biologists use some mathematics, mathematical research makes no use of biological findings.

Methodological holism holds that the whole needs no explanation – except perhaps in terms of its history – and that it explains the part. Thus, every particular biological process would be accounted for by a single overpowering vital force; mental processes would be explained as movements of the soul, or of its 'faculties' or 'modules'; and particular social facts would be accounted for by society-wide social forces, such as *Zeitgeist* and social learning – which are just as undefined as 'vital force' and 'soul.' Needless to say, all these are remnants of pre-scientific thinking.

Holism also claims that the confirmation or refutation of any thesis in any field of knowledge is bound to alter the entire system of human knowledge. For example, if quantum mechanics were to use a logic of its own (as some have claimed), then logic and physics should co-evolve. For better or for worse, this particular example is false. In fact, no one has ever found any new physical result with the help of quantum logic – which is not surprising, since quantum physics presupposes classical mathematics, whose underlying logic is classical. In sum, the various fields of research are indeed mutually related, but some ties are weaker than others. And empirical work cannot alter formal truths, for these do not represent any particular matters of fact: if they did, mathematical theorems would be tested in the lab. In conclusion, methodological holism does not work.

Axiological holism holds that any whole is more valuable than its parts, and that these are valuable only insofar as they serve the whole. The praxiological consequence is obvious: individual action should only be judged in terms of its contribution to the good of the whole. In turn, this entails that a norm is morally justified only if it guides actions that favour the whole: it inspires a duties-only morality. Political holism preaches the enslavement of the person to the powers that be – state, church, party, or corporation; all of which fits in with totalitarian ideologies, neither of which makes room for the private sphere. Maximal social cohesion and consensus are as bad as their polar opposites. Let us see why.

Emile Durkheim struck a deep chord in all of us when he claimed that social cohesion is the maximal social value. Indeed, we are all interested in avoiding strife and in preserving or even strenghtening the social cohesion of the social systems we belong to. The main reasons are personal and social. In personal terms, we all need and want to get on with our lives. The social reason is that, for a system to work efficiently, its parts must work together with minimal friction. If, by contrast, every member of the system attempts to 'do his own thing,' the bonds that hold the system together weaken to the point where the very existence of the system is endangered. This is one of the reasons that anarchism, whether left-wing or right-wing, is not viable.

However, cohesiveness must not be overdone to the point of hurting the system's members without compensation. When a marriage ceases to work, divorce is likely to be best for all the parties concerned. The same holds, *mutatis mutandis*, for business partnerships, political parties, and international alliances. All this has been known for centuries. Only the apostles of globalization have yet to learn this lesson: they keep repeating the mantra that free trade, privatization, financial liberalization, and the allegedly concomitant strenghtening of the world system are beneficial to all and, in particular, that they promote the equalization and democratization of nations and individuals as well as economic growth. There are two main arguments against this opinion.

The first is that economic statistics fail to support the contention about the growth-openness connection. Rodríguez and Rodrik (2001) conclude their methodological criticism of this hypothesis by stating that no significant association between economic growth and trade policies appears. Consequently, they 'dispute ... the view, increasingly common, that integration in the world economy is such a potent force for economic growth that it can effectively substitute for a development strategy.' As for income distribution, the recent annual reports of the UN Conference on Trade and Development show that in many countries, both developing and developed, globalization has been associated with increasing income inequality (Ocampo and Taylor 1998; Galbraith and Berner 2001).

A plausible mechanism to explain the perverse effects of globalization is this: in any dealings, of any kind, among unequally empowered parties, the more powerful are likely to take advantage of the less powerful – unless the handicaps are taken into account (as in golf), and a neutral umpire (like the Brussels Eurocracy) is charged with monitoring the score and correcting for inequalities. Shorter and more generally: Freedom of any kind only works among equals, unless a neutral power (as the liberal state was supposed to be before it chained itself to neo-liberal policies) can effectively guarantee it (see Bunge 1998).

A second argument has been provided by no less than the International Monetary Fund, a champion of free trade, in its *World Economic Outlook* for 2001. To begin with, it admits frankly that there is no significant relationship between capital liberalization and growth. Furthermore, the Fund confirms that all the leading economies are currently doing badly, so that none of them is in a position to bail out any of the others. Moreover, it admits that enhanced trade and financial linkages have increased the vulnerability of the world economy to shocks and heightened the risks of a self-reinforcing downturn. This is why India, China, and Russia – neither of which is fully integrated with the world economy – avoided the recession that struck the West in 2001.

Should it not have been obvious that the self-repairing ability of a system increases initially up to a maximum, and then decreases, as its cohesiveness, and hence rigidity, increases? And is it not equally clear that mainstream economists, who are methodological individualists and therefore sectoralists, are ill equipped to manage such a highly complex socio-economic system as the world economy? After all, their theory was crafted in the 1870s, and inspired by the retail shop.

In sum, holism is not a viable alternative to individualism.

4 Hybrids

The shortcomings of individualism and holism have suggested crossing them. There are two hybrids of individualism and holism. One may be called *individholism*, or individualism with a hidden holistic component; the other is *holindividualism*, or holism with a tacit individualist component. Both are conspicuous in social studies and their philosophy. Let us exemplify them.

The neo-classical microeconomists and other rational-choice theorists, as well as most hermeneuticists (or interpretivists), call themselves individualists. And so they are, though not consistently, because they often start their analyses with macro-situations that they do not analyse in terms of individual actions. The so-called 'logic of the situation' is a case in point, for it takes 'the situation' (the state of society) as an unanalysed whole. The same holds for free-market worship, in particular for the collective 'invisible hand,' which is no more real than such holistic fictions as collective memory, national destiny, and the will of the international community. Ditto for cultural relativism, since it regards culture as a whole; and for left-wing anarchism, because it espouses a communitarian morality.

In other words, the individholist smuggles in items that a consistent individualist should reject. Likewise holindividualism, as exemplified by Marxism, is inconsistent because it correctly admits the role of leaders who take initiatives and attempt to mobilize the masses or at least influence public opinion. A consistent holist places all the burden on such supreme but anonymous wholes as Nation, People, or History.

Institutional (or contextual) individualism, as sketched by Popper (1945), Bourricaud (1977), and Agassi (1987), is a rather widespread variety of individholism. This view is inadequate because it underrates interaction. As the individualist George Homans put it, 'Repeated interactions between particular persons are the very guts and marrow of social life' (1974: 57). In particular, institutional individualism does not account for team-playing, because it denies the very existence of teams as real entities distinct from their individual constituents. Yet teams do have (emergent) properties that their constituents lack – for instance, organization, coordination, and efficiency. This is why we bother to set them up or join them: with the expectation that they may accomplish what no individual could possibly accomplish by herself.

Since society is a system of systems, social scientists should be expected to study social systems, not individuals. Most macroeconomists, from François Quesnay (of *tableau économique* fame) to John Maynard Keynes (the father of modern macroeconomics) to Wassily Leontief (the inventor of input-output matrices) have regarded the economy as a system. Keynes said it explicitly: 'I am chiefly concerned with the behaviour of the economic system, as a whole' (1973: xxxii). Even Thomas C. Schelling, though a well-known enthusiast of rational-choice theory, notes that '[w]hat we typically deal with in economics, as in much of the social sciences, is a feedback system. And the feedback system "loop" is typically one of these relations that hold no matter how people behave' (1978: 50). Likewise, the general-equilibrium theorists, such as Gerard Debreu, regard the market as a whole that, unlike the households and firms that compose it, is said to be in equilibrium and to correct and run itself – surely two emergent if elusive properties.

What is wrong with individholism and holindividualism? Not much, since both can provide correct if incomplete analyses of some social facts – which is not surprising because they are cryptosystemist to the extent that they admit wholes with emergent properties. But they are inconsistent with their own declared intentions. Besides, although they may not sin by commission, they do sin by omission. Indeed, a deep bottom-up synthesis of a social fact, from a mere exchange of goods to a social revolution, will be correct only if supplemented with a top-down analysis of the same fact: recall chapter 5. Such dual study is typically systemic rather than either individualist or holist.

5 The Systemic Alternative

I submit that no scientist can live up to the extreme philosophies of individualism and holism. This is particularly obvious in the case of social studies, where the conflict between those philosophies was first noticed. In fact, even the most radical individualists admit, at least tacitly, that individuals, though assumed to be the source and origin of everything social, are likely to act differently in different situations or circumstances. Besides, they do not even attempt to analyse such situations in terms of individual actions. Likewise, even the most radical holist must acknowledge that individual actions, particularly when concerted, can generate, maintain, reform, or dismantle social systems. No wonder then that productive social scientists combine the bottom-up and top-down strategies (see figure 6.1).

The point then is not to attempt an impossible *reduction* of systems to mere aggregates of individuals, or the converse *reduction* of individual beliefs and actions to systemic properties. Rather, the point is to *relate* the microsocial and the macrosocial levels, showing how individuals *combine* (in particular compete and cooperate) with one another; and how, in turn, individual behaviour is *determined* (inhibited or stimulated) by a person's social environment rather than being totally free. A familiar example should serve to clarify this point.

Consider a girl faced with the problem of choosing a career. Obviously she is not totally free to choose. Even if she has a definite preference for a given career, and feels competent to embark on it, she may lack family support, not have access to a good school, or be deterred by a current low demand for graduates in her favourite field. Hence, her final choice will not be a faithful indicator of her preference, but rather of the latter conjoined with her circumstances. Something similar holds for all choice situations. It is not that we never have any choices, but that our preferences are distorted and constrained by circumstances beyond our control. Hence all the rational-choice theories, which are typically individualistic and take freedom of choice for granted, are every bit as flawed as the holistic views that ignore individual attitudes, preferences, and actions altogether. Once again, the systemic approach is the ticket.

To go back to the predicament of our young person. Suppose she chooses a given profession only because there is currently much demand for it. Such an individual decision is then determined by a macrosocial factor. Yet if many other students of our friend's generation make the same choice, then upon graduating they will face such stiff competition that many of them may not find employment. Shortage will have turned into glut. That is, a host of mutually independent, parallel, individual rational decisions will have an unintended and often perverse macrosocial effect (see Merton 1976).

The mechanism underlying the macro-to-macro Demand → Glut connection is this:

Demand for X at time t → Popular choice of X at time t.   *Macro-to-micro*
Popular choice of X at t → Glut of X at time $t + n$.   *Micro-to-macro*
Glut at time $t + n$ → Unemployment at time $t + n$.   *Macro-to-micro*

The moral of this story is that, because every individual belongs to some social system and behaves at least partly qua member of the system, it is mistaken to disregard the macrosocial level(s). Microsociological considerations must always intertwine with macrosociological ones, and conversely, because there are no fully autonomous individuals, families, firms, political, or cultural organizations. Every social unit belongs to at least one social system, and nowadays we are all components of the world system. Consequently, the task of social scientists is neither to study totally free persons nor to study social blocs as if they were opaque to analysis. Their task is to study the way individuals combine to produce social facts, how these in turn stimulate or inhibit individual agency, and how the various social systems interact.

I submit that systemism combines the sound components of individualism and holism: the former's thesis that there are only particulars, with the holistic emphasis on the peculiarities of wholes. Systemism holds that everything, whether concrete or abstract, is a system or a component of one or more systems, and that all of these have systemic or emergent properties. And it analyses a system into its composition, environment, and structure. If concrete, a system has also a mechanism or modus operandi : the processes that keep the system going – or end up undoing it.

Hence, the simplest model of a concrete system, such as a cell or a busines firm, is the composition-environment-structure-mechanism quadruple (see chapter 2, section 5). Individualists project this quadruple onto its first component, and holists onto the third. But bare individuals are fictitious: they are imagined by ripping open networks or dismantling systems. Whether in the external world or in the conceptual and semiotic realms, there are only interrelated individuals, that is, systems.

The systemic approach gives rise to an entire ontology (see Bunge 1979a). The popular confusion between systemism and holism has hampered the recognition and development of this new ontology. One of its distinctive features is that it connects items that individualists treat as mutually independent, without however making the holistic mistake of refusing to analyse such wholes and studying the mechanisms of their emergence and breakdown. An upshot of this approach is the thesis that society is a system of interconnected systems – the biological, economic, political, and cultural ones. A practical consequence of this thesis is that, to be successful, a national-development program must be at once biological, economic, political, and cultural: piecemeal reforms have at best short-lived results, at worst perverse effects.

On the other hand, there is no need to insist on logical systemism, because logic and mathematics are automatically systemic in dealing, not with either stray individuals or solid blocs, but with conceptual systems: arguments, algebraic systems, number systems, spaces, manifolds, function families, and so on. Nor need we dwell on semantic systemism, because it is generally understood that constructs and signs make sense only as components of systems, and that a proposition is assigned a truth value only on the strength of other propositions.

Since epistemological and methodological problems come in packages, they should be tackled as such, that is, systemically. This requires combining analysis with synthesis, reduction with fusion. The coalescence of different disciplines to form interdisciplines, such as biochemistry, cognitive neuroscience, social psychology, socioeconomics, and political sociology, is a triumph of the systemic approach – which is often adopted tacitly, though.

The systemic approach to axiology, or value theory, shows valuation to be a process occurring in an individual brain controlled by biological drives and social stimuli and constraints. Praxiology, or action theory, is similar, and so is ethics, or moral philosophy. In all three cases the systemic approach admits both the individual source and the social context of valuations, decisions, plans, actions, and norms of conduct. The individual, partly self-made and partly shaped by the environment, proposes and interacts with other people, but the environment disposes. A biological parallel is this: Emergence + Selection = Evolution.

Finally, the systemic approach to politics, the law, political science, and political and legal philosophy rests on an analysis of society into the different but interconnected subsystems within which the individual evaluates, decides, acts, and is acted upon. Political systemism overcomes the limitations of individualism (which focuses on the mythical independent and free citizen) and of holism (which focuses on the mythical overwhelming and allegedly unanalysable power).

The moral for the so-called policy sciences, or socio-technologies, is this. The systemic issues, such as those of poverty, war, and national debt, call for a systemic approach, because every one of them is a whole package of interrelated social ailments. For example, the individualist attempts to alleviate poverty (and assuage his conscience pangs) by giving alms to his favourite beggars; the holist favours social programs to alter the environment; and the systemist combines the latter with local organizations where the poor can help one another not only to survive but also to improve their lifestyles.

There is more to be said about systemism, but it will be said in the next chapter.

Concluding Remarks

A first upshot of our study in this chapter and the preceding is that individualism is not one but many-sided. Moreover, far from being mutually independent, these sides form a decagon. But this exemplifies the epistemological thesis of systemism, and thereby raises the Individualist's Dilemma: if thorough, he is inconsistent – and hence, if consistent, he is not thorough. In other words, individualism is self-destructive. This is why there is no individualist system, but only an individualist hydra that will grow a new head every time it loses one.

A second result is that individualism fails in all its modes, theoretical and practical. This result has been reached by checking the individualist theses against the relevant evidence. However, individualism never fails completely, for it focuses on an essential aspect of every system, namely its composition. Moreover, individualism often serves as a sound corrective to holism – which in turn is right in emphasizing the reality of certain wholes and their emergent properties.

A third result is that, since individualism fails, so does radical reduction, or top-down analysis with neglect of structure. By contrast, moderate reduction succeeds in some cases, whereas bridge-building – in particular the fusion of disciplines – succeeds in others. For example, chemistry uncovers the composition and structure of genes, but only cell biology exhibits their role or function in living beings. (Hence it is not true that genetics has been reduced to chemistry.) Likewise, physiology and biochemistry investigate digestion, but only ecology and ethology can tell us what and how much food an animal gets in a given environment.

A fourth result is that we are not necessarily impaled on the horns of the individualism-holism dilemma. Indeed, systemism is the correct alternative to any form of individualism, as well as of holism and their hybrids. After all, the world is a system, and so is human knowledge. Ignore the main associates of an individual – be it thing or construct – , and you will not know the individual. And ignore the individual, and you will not know the system.

A fifth upshot is, I submit, that there is a morally right and practically viable alternative to both political individualism and political holism. This is the systemist view that we should care for personal welfare and advancement as much as for the institutions that favour them – surely a platitude. In addition to this truism, however, systemism includes the controversial yet testable hypothesis that the best way to design, construct, maintain, or reform institutions is through a combination of social technology with participative and integral democracy – biological, economic, political, and cultural. However, this claim has yet to be empirically validated.

In sum, individualism works only marginally better than holism, because it refuses to admit emergence. Yet it makes an important conceptual and practical contribution, namely concern for the individual components of systems, which holism disdains. Systemism retains and works out this contribution, as well as the holistic emphasis on the peculiarities of wholes. Being a synthesis of sorts, it is bound to be rejected by the radical individualists as well as by the holists, though practised by moderate individualists and moderate holists alike.

Interestingly, both the inability to see patterns while being able to perceive the underlying components, and the inability to perceive the latter while being capable of forming the corresponding 'large pictures,' are well-studied cases of visual-form agnosia caused by certain neurological deficits. Individualism and holism are even more serious disorders, for they are bound to affect all our judgments, evaluations, and decisions. Fortunately, unlike their pathological counterparts, these epistemic deficits can be corrected painlessly and in the 'dry,' through mere philosophical discussion.

8

Three Views of Society

The natural scientists tackled and solved their main philosophical problems in the seventeenth century. They did so when Galileo, Kepler, Gilbert, Huygens, Boyle, Harvey, Torricelli, and their followers jettisoned both supernaturalism and apriorism, criticized both conventionalism and phenomenalism, and adopted more or less explicitly a radically new philosophy. This consisted of a naturalistic ontology together with a realist epistemology that combines the experimental method with mathematical modelling. Whereas naturalism encourages the study of reality and discourages myth-making, experiment tests guesses, and mathematical models unify and sometimes explain.

By contrast, the social sciences are still in the grips of ontological and methodological controversies (see, e.g., Boudon 1980; Bunge 1996, 1998; Weissman 2000; Trigger 2003a). Indeed, there is no consensus on either the nature of society or the best way to investigate or redesign it. No wonder then that the social sciences still lag behind the natural ones. In this chapter we shall examine the ontological side of this debate. The methodological side will be tackled in chapter 10.

1 The Two Classical Views of Society

It is usually taken for granted that there are only two possible views of society: that it is just a collection of individuals (individualism) or that it is a whole to which individuals are subjected (holism). Whereas individualism stresses action and underrates or even ignores social ties, holism underrates agency and overrates bonds. In other words, while individualists take individual behavior for granted and regard it as the source of social pattern, holists take the latter for granted and regard it as the source of individual behaviour. Neither regards the output – behaviour or structure, as the case may be – as problematic, let alone as only one of the sides of one and the same social coin.

Yet every case study calls for an investigation on both the micro-level and the macro-level. For instance, what drove that individual to become a suicide bomber? Presumably, his commitment to a political or religious cause in response to the continued occupation of his homeland by foreigners or infidels. And why does this occupation persist? Because a bunch of politicians and religious zealots want to increase their personal political power, others to carry out what they see as a sacred mission or even a divine commandment – or to favour the oil companies. Remove individual motives and actions, and nothing social remains; set structural constraints and stimuli aside, and no individual interests and possibilities remain. In other words, individual action is embedded in one or more social systems; and in turn, no such systems emerge, subsist, alter, or dismantle without individual action.

However, it is seldom recognized that society might be a system, that is, a complex thing constituted by interacting individuals. In fact, when rigorous contemporary social scientists hear the word 'system,' they are likely to draw their intellectual guns. They seem to feel threatened by a return to the holism of Ibn Khaldûn, Burke, Müller, Hegel, Comte, the mature Marx, Durkheim, Malinowski, Gini, or Parsons. Social scientists are rightly diffident of such imaginary wholes as collective memory, national spirit, manifest destiny, general will, and the responsibility of the international community. Thus, they take refuge in the equally doubtful individualism of Hobbes, Locke, Smith, Dilthey, Weber, Pareto, Popper, and the neo-classical microeconomists.

To be sure, the social individualists – pardon the oxymoron – do not deny that individual action is now constrained, now stimulated by the social context or situation. But they do not and cannot analyse a situation in individualist terms. Thus, for all their talk of the supremacy of the market and 'situational logic,' individualists leave markets and situations as unanalysed wholes. And they resist the very idea that individuals flock together – or are thrown together – into social systems such as families, gangs, tribes, villages, business firms, armies, schools, religious congregations, NGOs, political parties, or informal networks, all of which are every bit as real and concrete as their individual constituents. Individualists insist that all these are just collections of individuals, on a par with the throng of people in a busy avenue. Consequently, they ignore the emergent (or global) properties that characterize systems; in particular, they underrate social structure or even overlook it altogether.

Individualists are thus bound to miss one of the most important and intriguing of all kinds of events in society as well as in nature, namely the emergence of novelty. More precisely, they miss the emergence of things with systemic properties, that is, properties that their components or their precursors lack – such as cohesiveness, stability, income distribution, division of labour, social

114 Emergence

stratification, and social order. By the same token, individualists fail to realize the existence of systemic social issues, such as those of poverty, overpopulation, wealth concentration, political oppression, superstition, and underdevelopment. None of these issues can be solved by doing one thing at a time, because they affect several systems at once – the biological, economic, cultural, and political ones.

2 The Systemic Approach

Given the ubiquity of systems, it is advisable to adopt an entire systemic world view, as advanced in the previous chapters. Let us recapitulate the postulates of systemism:

1 Everything, whether concrete or abstract, is a system or an actual or potential component of a system;
2 systems have systemic (emergent) features that their components lack, whence
3 all problems should be approached in a systemic rather than in a sectoral fashion;
4 all ideas should be put together into systems (preferably theories); and
5 the testing of anything, whether idea, method, or artefact, assumes the validity of other items, which are taken as benchmarks until further notice.

Yet, social individualists resist the systemic approach. They insist on studying only the components of social systems, that is, individuals, while overlooking their structure or set of connections (internal and external). Perhaps theirs is a defensive strategy: they do not wish to be taken for holists, and they rightly mistrust the writers who call themselves system theorists, although actually they are holists. The opaque and long-winded utterances of these writers has given systemism a bad name. This may be why most social scientists nowadays shun the word 'system' even while studying social systems.

Fortunately, few authentic social scientists practise the philosophy they preach. For example, Karl Marx was a holist in epistemology: in fact, he was the grandfather of the now fashionable social constructivism, according to which ideas are social constructions rather than brain processes. Yet when it came to economic and political matters, Marx insisted that individual action is the source of all social change – though it happens in an inherited social milieu. Likewise, Max Weber popularized Dilthey's individualism, subjectivism, and anti-scientism. But he did not practise these philosophical views: instead, he proceeded scientifically when he studied such systems as the slave society, the caste system, feudalism, organized religion, bureaucracy, industrial capitalism, and legal codes – and the 'steel cage' in which the individual is locked up in modern times (Von Schelting 1934).

Closer to us, the self-styled individualist James S. Coleman stated that, according to his own variant of methodological individualism, '[t]he interaction among individuals is seen to result in emergent phenomena at the system level' (1990: 5). Moreover, he criticized the 'fiction that society consists of a set of independent individuals, each of whom acts to achieve goals that are independently arrived at, and that the functioning of society consists of the combination of these actions of independent individuals' (300). Moreover, he stated, 'the correct path for social theory is ... to maintain a single conception of what individuals are like and to generate the varying systemic functioning not from different kinds of creatures, but from different structures of relations within which these creatures find themselves' (197). In other words, once a social system is in place, individuals are partially shaped by it, and they become replaceable to some extent: their roles can be enacted by different persons.

3 From Statistics to Theoretical Models

The holist looks upon society as a whole that must not be analysed. In this way he may capture its systemic properties, but he fails to understand how it emerges. Suppose, for instance, that statistics show that the gross domestic product of a certain nation is rather high and has remained fairly constant over a certain period. In other words, the proverbial 'average person' has remained rather well off. However, this not a measure of general prosperity in that nation, because the average person is a statistical artefact. It may happen, as it often does these days, that, although the total wealth has increased, the rich have been getting richer, and the poor, poorer. Thus, the upward trend of one subpopulation has been balanced by the downward trend of the rest. To understand what is going on we must disaggregate some statistical data besides aggregating others. This is why mainstream social scientists are not holists.

Consider next the raw material of social research: statistics in the case of macrosocial research, and narratives (surveys, interviews, birth and death registers, ledgers, property deeds, letters, etc.) in that of microsocial studies. For instance, the student interested in the doubling of the lifespan of most human populations that occurred over the twentieth century will either get hold of documents concerning the members of a representative sample of the population of interest, or read the relevant tables and graphs. However, the microdemographer interested in explaining that rise will also ask questions involv-

Table 8.1 Longevity as a function of disease environment and income level in seventeenth-century England. The rich Londoners did not live longer than the rural poor in remote, unpolluted regions.

Income	Environment		
High	Urban elites 30–5 years	Small-town elites 35–40 years	Rural elites 40–50 years
Average	Urban areas 25–30 years	Typical rural 30–5 years	Remote 35–40 years
Low	Urban poor 20–5 years	Rural poor 25–30 years	Remote 30–5 years

ing macrosocial variables, such as where the subjects learned about healthier lifestyles, whether they had access to public sanitation and health-care services, whether their income was above the poverty level, whether their working hours and conditions fit the country's labour legislation, whether they practised family planning, and so on. That is, the investigator will place the individual in her social context: his unit of study is not the individual but the individual-in-her-society. This will allow him to aggregate individual data in various ways, thus climbing from the micro- to the macro-level.

If, on the other hand, the student resorts to social surveys (gathered from individuals or households), he will attempt to disaggregate the data into micro-level subpopulations, such as cohorts, income brackets, schooling levels, and so on. His unit of study is not the entire society but as small a segment of it as possible. However, he can also opt for a third strategy, that of combining personal data with statistics: this will enable him to explain life histories with the help of macrosocial data, and the latter in terms of the former.

This mixed strategy is likely to yield interesting results that neither of its rivals can produce. One such result is the analysis of mortality data for seventeenth-century England as a function of both disease environment and income level performed by Johansson (2000), and condensed in table 8.1.

The point then is not to opt for either the micro- or the macrosocial approaches, but to combine bottom-up with top-down research. As a matter of fact, social scientists routinely perform plenty of micro-macro analyses. These can be condensed in what I have called Boudon-Coleman diagrams (Bunge 1996, 1998). Here is a self-explanatory example:

Macro-level Economic growth → Population stagnation
  ↓   ↑
Micro-level Old-age security → Decline in fertility

A holist is likely to be baffled by the information that, in our time, economic growth is bound to be accompanied by population stagnation or even decline; and the individualist must take old-age security as a given – even if it depends partly upon a macro-system, namely government. The systemist dissolves the paradox by linking the system and individual levels. In so doing he unveils the mechanism that mediates two macro-variables.

Here is another example. There are two main currents in the study of ideas, the individualist or internalist, and the holist or externalist. Internalists focus on conceptual problems and solutions, whereas holists focus on networks, formal organizations, and business and political connections. The internalists tend to deal in disembodied ideas: they see the fish but miss the net. By contrast, the externalists tend to study groups while minimizing the importance of ideas: they see the net but miss the fish. The two parties do not talk to each other. And yet it should be obvious that each of them sheds light on only one of the sides of the coin, and that neither captures the complexity of the micro-macro links. For example, internalists cannot explain why science was born only once, in ancient Greece; why it withered a couple of centuries later; and why it was reborn at the beginning of the modern era, along with a revival of religious faith, a ferocious witch-hunting campaign, and popular enthusiasm for astrology, alchemy, and other superstitions. Externalists are equally at a loss to explain these processes, because they do not even distinguish science from technology, magic, religion, or philosophy, all of which are allegedly 'social constructions' contrived by groups or networks (see, e.g., Collins 1998).

By contrast, the systemist is likely to look at the problem this way. Every thinker is born into a pre-existing system loaded with a tradition that he either enriches or rebels against. He inherits some findings and problems and, if original, invents or discovers new ones. The solutions he proposes to the problems he tackles are born in his brain, not in society: social groups are brainless, hence they cannot think. Of course, social groups and circumstances stimulate or inhibit creativity. But their influence is not such that every idea has a social content, let alone a political purpose – notwithstanding Michel Foucault's *dicta*. For example, neither mathematics nor physics has a social content.

Obviously, the ideas in social studies have a social content. Hence, they must be judged by their adequate reflection of social facts or their efficiency in promoting social change. By contrast, the validity of mathematical proofs, and the truth of physical or biological theories, have nothing to do with social class, political power, or economic growth. These social factors are relevant only to the ability of individual researchers to conduct their work without distorting political or ideological pressures. For example, the cultural policy of classical liberalism, which is based on individualism, is one of benign neglect.

118 Emergence

By contrast, the totalitarian cultural policy, which is based on holism, is one of censorship. However, totalitarianism combined holism for the masses with individualism for the elite. (For the holism-Nazism connection see Kolnai 1938, Popper 1945, and Harrington 1996.)

Another good example is Tocqueville's (1998 [1856]) explanation of the backwardness of French, compared with English, agriculture in the eighteenth century. The proposed mechanism was landlord absenteeism, far more common in France than in England at that time. Whereas the typical French aristocrat left the administration of his land in the hands of a steward, and took up a position as a civil servant or a courtier, his typical English counterpart lived on his estate and saw personally to it that his land was well cultivated, his tenants paid their rent punctually, and his neighbours observed both law and custom. (Actually the rough squire, rather than the refined aristocrat, ruled the English countryside.) In sum, whereas the typical English landlord remained at the centre of his rural network, his French counterpart marginalized himself.

In turn, according to Tocqueville, the root of the said difference is macrosocial, namely the political organization, which was centralist in France and decentralized in England. A French aristocrat gained more political clout and prestige from shuffling papers, socializing, and scheming in Paris or Versailles than from puttering in his grounds, learning new cultivation methods, and acting as the local magistrate and tyrant. In this case, individual choice, and its consequence for rural life, were ultimately determined by the political system. As Tocqueville put it, 'The chief and permanent cause of this fact was ... the slow and constant actions of institutions' (ibid.: 181).

Boudon (1998) regards this case as confirming what he calls contextual individualism and cognitive rationality. I prefer to think of Tocqueville as a systemist *avant la lettre*, particularly since he noted the social and cultural aspects of the process in addition to its economic and political ones. Indeed, his main point is that landlord absenteeism destroyed the rural network traditionally centered in the landlord, in addition to impoverishing landlord and peasant alike. He was thus a socio-econo-politologist. Indeed, his explanation fits the following Boudon-Coleman diagram:

Macro-level   Political centralization   →   Impoverishment & alienation
                ↓   ↑
Micro-level   Landlord absenteeism   →   Agricultural stagnation & weakening of social ties

In this case, as in all other social processes, there are uncounted individual choices, decisions, and actions. Yet all of them are constrained by social structure, occur within or between social systems, and reinforce or weaken the bonds that keep these systems together. Action, bond, and context go together. Eliminate either of them, and no social fact remains.

4 The Science-Technology-Market Supersystem

Our fourth example will be the science-technology-market system. There are two main socio-economic views on the relation of technological innovation and the market. Individualists claim that the inventor proposes and the market disposes. Holists, by contrast, hold that all invention is market-driven: that the market demands and the inventor supplies. (Yet, ironically, all market worshippers espouse individualism.) Each party parades a large collection of examples, without bothering about counter-examples. I submit that only a systemic view of the matter attains the whole truth. (See also Gilfillan 1970 and Wiener 1993.)

The first thing to note is that there are big inventions and small ones: radical novelties and improvements. Whereas the former are motivated mainly by sheer curiosity and the love of tinkering, improvements may also be motivated by profit: these are often commissioned by the technologist's employer with a view to marketing the corresponding products. By contrast, some radically new inventions have created whole new markets. For example, the electrical industry was made possible by electrical engineering, which in turn depended upon experiments and theories on electricity and magnetism. In particular, Michael Faraday discovered the principle of electromagnetic induction, which Joseph Henry used to design the electric motor, and Nikola Tesla to design the dynamo. Industry transmuted these and many other bits of scientific and technological ingenuity into welfare and wealth. This is only one of many science-technology-industry systems. The market does not create: it only demands and rejects – that is, rewards or punishes. Moreover, it usually rewards followers rather than pioneers. It would be as silly to underrate the power of the market as to regard it as the fountain of technological ingenuity.

Fifth example: the combination of competition with cooperation. Whereas individualists stress competition or conflict, holists emphasize cooperation or solidarity. (Marxism is a special case: it stresses interclass warfare but intraclass solidarity.) Actually, competition and cooperation coexist in all social systems, though not in the same respects. Indeed, a social system cannot emerge, let alone persist, without a modicum of (spontaneous or coordinated) cooperation in some regard. And once a system is in place, competition in some respect is bound to arise in its midst precisely because of a common interest in some scarce resource, such as attention, affection, time, space, food, money, jobs, or what have you.

Think, for instance, of a scientific community. Some post-Mertonian sociologists of science, notably Latour and Woolgar (1986), claim that there is no such community: that individual scientists engage in a selfish and unscrupulous struggle for power. These writers are poorly informed in this regard as in others. Indeed, as Merton (1968) noted, scientists, far from being dispassionate observers, are motivated by curiosity and peer recognition. And although researchers compete for peer recognition, they also learn from one another: scientific research is a social endeavour even when it has no social content and no practical value. As Wolpert states, 'In order to promote the success of their ideas, scientists must thus adopt a strategy of competition and cooperation, of altruism and selfishness' (1992: 88). In any event, Latour's claim that 'science is politics by other means' has recently been falsified by an empirical study of citations (Baldi 1998).

Another clear example of the need to combine the grains of truth contained in individualism with those inherent in holism is the emergence of social norms. While holists take them for granted, and hold that the individual simply bows to them, individualists claim that norms are invented, and are universally adopted if perceived to enhance collective welfare – or, as Hume put it, out of 'a general sense of common interest.' There is some truth in each of these views; but not the whole truth, if only because, as Coleman noted (1990: 243–63), there are two kinds of norm: conjoint, which favours all the people who adopt it; and disjoint, which favours the interests of some (usually the powerful) at the expense of the others (usually the weak). For instance, while it is in everyone's interest to abstain from unnecessarily polluting the environment, only some people benefited from slavery, compulsory religious worship, and corporal punishment. But it so happens that neither individualists nor holists pay any attention to social stratification. Hence, neither school explains the emergence of social norms or their submergence in times of deep social crisis – let alone the existence of counter-norms such as 'Promote universal free trade – except of course when it hurts the sectors deserving of government subsidies.'

Here is my last example. Boudon (1974), who calls himself a contextual individualist – an individholist in my jargon – has shown that the proliferation of universities after the Second World War has had an unintended perverse effect. This is the emergence of a sizeable intellectual proletariat, and a concomitant increase in social inequality. The mechanism is as follows: as the number of university graduates increases, the queues of candidates waiting for qualified jobs lengthen. The practical moral is obvious: There are two ways to check the massive unemployment of university graduates. One is to impose admission quotas in the professional faculties; the other is to raise the level of job specification in industry and government – that is, to influence choice from above in order to minimize failure at the bottom. However, social policy deserves another section.

5 Implications for Social-policy Design

The preceding suggests two important points, one theoretical and the other practical. The first is that any deep explanation of social change calls for the unveiling of social mechanisms, which in turn involves micro-macro analyses. This is so because every individual action is partially constrained or stimulated by macrosocial circumstances, which may in turn be affected to some extent by individual actions.

The second point is that effective social policies should be based on correct hypotheses concerning the social mechanisms of interest. The reason is that a social policy is expected to design or redesign some social mechanism – for instance, of health care, wealth redistribution, or conflict resolution. By contrast, the intuitive and the empirical approaches to social policy-making are wasteful and often counterproductive. For example, contrary to popular wisdom, a raise in the minimum wage does not increase unemployment, but benefits the economy as a whole because it increases purchasing power, as thus demand (Card 1995).

There is an additional reason for favouring systemic social policies, namely that serious social issues, such as poverty, political marginality, and illiteracy, are systemic by definition. That is, they involve many interrelated features and even several social systems at a time. For example, an effective policy of national development must involve factors of various kinds: environmental (e.g., protection of forests and fisheries), biological (e.g., health care and planned parenthood), economic (e.g., industrialization and improved infrastructure), political (in particular, political liberty and participation), and cultural (in particular, education and encouragement of the arts and sciences).

The reason for the need for such a multifactorial approach is that all those factors are interrelated. For example, there is no modern industry without educated manpower, and no education on an empty stomach, let alone on a gut full of parasites. For this reason, the sectoral and one-problem-at-a-time approach is bound to fail. Even such a staunch individualist as the financial wizard George Soros has concluded (1998: 226) that, contrary to the opinion of his erstwhile teacher Karl R. Popper, piecemeal social engineering does not solve systemic problems. Soros suggests that these should be tackled radically and in all their complexity.

122 Emergence

Contrast the systemic approach to social issues with its rivals. The radical individualists oppose all social planning in the name of individual liberties (a.k.a. privileges). Hence, they leave individuals to their own resources – which, in an inegalitarian society, are meagre for the vast majority. By contrast, holists swear by top-down planning. Consequently, even when they address the basic needs of the common people, the holists are likely to ignore the rights and aspirations of individuals. In either case, the powerless individual, whether forsaken or corralled, has nothing to gain. The systemic approach to social policy-design is quite different from both libertarianism and totalitarianism: it attempts to involve the interested parties in the planning process, and designs social systems and processes likely to improve individual well-being, revising the plans as often as required by the changing circumstances (see Schönwandt 2002).

6 Social Studies are About Social Systems

Consider again the 1789 French Revolution, that inexhaustible source of social thought. Despite its shattering worldwide consequences, it was a cakewalk. Indeed, the central government fell without bloodshed in the course of an evening. Tocqueville (1998 [1856]) explained this process clearly and in systemic terms, namely, as a result of the replacement of the feudal social networks with four closed and mutually hostile castes: those constituted by the peasants, the bourgeoisie, the aristocrats, and the Crown. The traditional networks were ripped apart when, in the previous century, the landowners had abandoned their land and left their tenants to their own devices as a consequence of the concentration of both government and the nobility in Paris and Versailles. This is how it happened that 'the ties of patronage and dependence which formerly bound the great rural landowners to the peasants had been relaxed or broken' (ibid.: 188). The king was thus a victim of his own art of 'dividing people in order to govern them more absolutely' (191).

There was more: The centralization of political power left a political vacuum that was filled by the intellectuals, most of whom criticized the unjust social order. This explains the disproportionate influence of the *philosophes*, in particular the Encyclopedists: they occupied the place that the aristocrats occupied in England and elsewhere at the time. 'An aristocracy in its vigor not only runs affairs, it still directs opinion, sets the tone for writers, and lends authority to ideas. In the eighteenth century, the French nobility had entirely lost this part of its empire; its moral authority had followed the fortunes of its power: the place that it had occupied in the government was empty, and writers could occupy it at their leisure and fill it completely' (ibid.: 198). Yet a century and a half later, the author of a huge treatise on the sociology of philosophy (Collins 1998) devotes a single page to the Encyclopedists, and fails to explain their remarkable influence, but devotes many laudatory pages to the Counter-Enlightenment, from Hegel and Herder to Nietzsche, Husserl, and Heidegger.

Let us now jump forward two centuries and face the Middle East hornet's nest. There can be little doubt that this, together with the United States, is a system, and moreover a very unstable one. Indeed, a rash action of any of the actors could destabilize or even destroy the whole region. Imagine this scenario. Either of the powers constituting the system provokes another, which retaliates by sending missiles over its neighbour, which responds by massive air attacks or even nuclear bombs. As a result, at least two Middle Eastern nations lay in ruins – and this because of the sectoral view, ambition, and recklessness of the politicians who believe themselves to be in charge.

The point of these stories is to remind ourselves that, contrary to the radical-individualist tenet, society is not an unstructured collection of independent individuals. It is, instead, a system of interacting individuals organized into systems or networks of various kinds. In fact, it is well known that every one of us belongs at once to several 'circles' (systems): kinship, friendship, and collegial networks, business firms, schools, labour unions, clubs, religious congregations, and so forth. This explains our plural 'identity.'

To be sure, the emergence, maintenance, repair, or dismantling of any social system can ultimately be explained only in terms of individual preferences, decisions, and actions. Still, these individual events are, in turn, largely determined by interaction and context. People cultivate the relations, and support the systems, that benefit them, while they neglect or sabotage those that hurt them. In sum, agency and structure are only two sides of the same coin.

Now, individuals are studied by natural science and psychology, which – along with anthropology, linguistics, demography, and epidemiology – is one of the biosocial intersciences. The social sciences proper, such as sociology and economics, do not study individuals except as components of social systems. Thus, anthropologists study entire communities such as villages and tribes. Sociologists study social systems, all the way from the childless couple to the world system. Economists study the social systems engaged in production, services, or trade. Politologists study power relations in all systems, particularly the political ones. And historians – unlike biographers – study social (structural) changes on all scales.

It is not enough for a social scientist to point out the social context or circumstance of a fact. He is expected to study social facts, and these happen to occur in social systems – as is the case with a strike in a business concern – or between social systems – as in an international conflict. Hence, he must study social *bonds* in addition to social contexts, for bonds are what hold systems together.

In short, the social sciences study social systems. True, some students, such as Anthony Giddens, prefer 'structure' to 'system.' But structures are properties of things, not things, whereas social systems are concrete things. For example, a corporation is a system with a definite (though perhaps changing) structure, or set of bonds among its components and its environment. The socio-economist who studies the social structure of a corporation does not investigate the structure of a structure – a meaningless expression – but the structure of a thing.

Furthermore, the interconnectedness of social facts ought to be reflected in the social sciences studies. That is, the frontiers among them ought to be trespassed, as Hirschman (1981) has insisted, because they are artificial. The reason is that all the social sciences refer to a single entity: society. In other words, we should promote mongrel disciplines, or interdisciplines, such as socio-economics, political sociology, economic anthropology, sociolinguistics, and biological sociology (not to be mistaken for sociobiology).

7 The Competitive Advantage of Systemism

To appreciate the advantages of systemism over its two rivals, it may help to consider briefly three examples, one in sociology, another in management, and the third in political science. A family sociologist interested in understanding why so many families are breaking up these days is unlikely to be satisfied by lamentations over the decline of family values that followed on the wake of the decline of religion. Nor will he be convinced by the rational-choice theorist who regards the family as a production unit that may cease to produce a certain number of goods, from meals to children; or because, after cold calculation, one of the spouses realizes that his or her marriage had been a mistake from the start (Becker 1976: 244).

Neither of these views focuses on the weakening of the interpersonal bonds that gave rise to the family to begin with. In modern society, most people marry or divorce neither because of social pressures nor because of economic calculations. People marry primarily because they feel affinity, fall in love, and share interests to the extent that they wish to live together; and they divorce when they fall out of love, their interests diverge, they suffer from work-related stress or do not earn enough, or some other cause. In these matters, interaction or its weakening is all-important, whereas rational calculation and the institutional context are secondary.

Our second example is that of a management consultant, operations research expert, or industrial sociologist intent on understanding how a given business firm works, or has ceased to work efficiently. Presumably, he will not be satisfied with holistic considerations about the firm's goal or the business environment. Nor will he try to guess subjective utilities and probabilities, to check whether the managers have succeeded or failed in maximizing their expected utilities. The competent consultant knows that those numbers are inaccessible and, like everything subjective, are at most an object of study, never a tool of scientific analysis. He will focus instead on the social structure of the firm, and on various mechanisms that keep the firm going, or that have been allowed to deteriorate or become obsolete. Indeed, he will study the three main mechanisms: work, management, and firm–environment interactions. Unless corrected, a dysfunction in any of the three will jeopardize the firm's survival, for it will result in a decrease in efficiency, as measured by the output-input ratio – not the inaccessible expected utility. (For the centrality of mechanisms in social studies, see Hedström and Swedberg 1998; Bunge 1999; Pickel 2001; and Tilly 2001.)

Finally, a political example: Why did the Soviet empire collapse? Recall that nobody – in particular, neither the Cold War think-tanks nor the Marxist-Leninist schoolmen – predicted this momentous event. It especially took by surprise the holistic futurologists, the game-theory modellers, and the data-gatherers-and-hunters of the various American 'intelligence' agencies. I submit that these political prophets, analysts and spies failed because they did not study seriously the various subsystems of the Soviet society and their interrelations.

In particular, those experts failed to realize that the so-called dictatorship of the proletariat had cut the non-coercive bonds that hold a civil society together; that the top-down planned economy did not deliver enough consumer goods, and functioned at a low technological level in everything but space exploration and arms manufacture; and that the official Marxist-Leninist ideology had stunted cultural development and ceased to command popular allegiance because it was finally perceived as having become an obsolete dogma only good as a tool of social control. The result of the malfunction of all three subsystems – the economy, the polity, and the culture – was a rather backward and rigid society made up of poorly motivated, mutually suspicious, and grumbling individuals. Gorbachev's reforms came too late, were not radical enough, were not implemented, and had the perverse effects of disappointing everyone, relaxing discipline all around, and eroding the state authority (see details in Bunge 1998: 205–11).

In sum, the systemic approach is superior to both individualism and holism because, instead of studying either empty wholes or individuals that only share a context, it focuses on social systems and the mechanisms that make them tick, namely interpersonal ties. Interaction – especially participation and cooperation – is the mortar of society. The context or institutional framework

Table 8.2 Three ontologies and their concomitant epistemologies and methodologies

Ontology	Epistemology	Methodology
Individualism	Rationalism or empiricism	Analysis: micro-reduction
Holism	Intuitionism	Synthesis: macro-reduction
Systemism	Scientific realism	Analysis and synthesis

is nothing but the social system (or supersystem) within which individuals and groups act. And the situation that methodological individualists invoke is nothing but the momentary state of such a system.

Concluding Remarks

To conclude. The slogan *Divide et impera*, in its methodological sense, suggests identifying the components of a system, whether concrete like a business firm or conceptual like a valid argument. This is fine but insufficient, because a system is not an unstructured aggregate but a structured thing that results from a combination of its components, and one that exhibits emergent properties. Therefore, we also need to practise the dual slogan, *Coniuga et impera*, if we are to uncover the structure and mechanism of any systems and hope to control them. In short, the correct strategy for tackling systems of any sort is *Divide et coniuga*.

The preceding is particularly important in the study of society and the design of social policies. Neither of the two most influential approaches to the study and management of social affairs is completely adequate, let alone practically efficient, for the following reasons. Individualism is flawed because it underrates or even overlooks bonds; and holism is inadequate because it underrates individuals. By contrast, systemism makes room for both. Moreover, it emphasizes the role of the environment, and suggests studying or altering the mechanisms of both stasis and change. The consequence for political philosophy and social-policy design is that systemism takes into account social values (ignored by individualists) as well as individual values (held in contempt by holists). Hence, it is more likely than its rivals to inspire and defend policies that combine competition with cooperation, and enhance individual welfare and liberty while strengthening or reforming the requisite institutions.

Finally, note the epistemological and methodological partners of the three ontological doctrines examined above: see table 8.2. We shall take a closer look at them in the following chapter.

PART II

Convergence

9

Reduction and Reductionism

The convergence of disciplines can be either horizontal or vertical. The former occurs when two or more disciplines merge on an equal footing, as in the cases of cognitive neuroscience and socio-economics. In contradistinction, vertical emergence is the subordination or reduction of one discipline to another, as in the case of the reduction of thermodynamics to statistical mechanics.

In turn, there are two kinds of reduction: downwards and upwards, or micro-reduction and macroreduction respectively. Whereas micro-reduction is analysis or decomposition of wholes into their parts, macro-reduction is synthesis or aggregation of individuals into wholes. And reductionism is of course the methodological doctrine that recommends reduction as the only way to understanding.

Micro-reductionism is the methodological partner of individualism, while macroreductionism is that of holism. In the following we shall concentrate on the former, because it is the most popular strategy of the two, despite which it is often misunderstood: its practitioners exaggerate its power, whereas its opponents denounce it angrily. As usual, we shall choose a third way.

If everything is either an individual or a mere collection of individuals, then the understanding of a whole is only brought about by diving down to the very bottom of things – that is, by identifying the (putative) ultimate constituents. Thus, light beams will be understood in terms of photons; atoms in terms of elementary particles; cells in terms of organelles and their components; multicellular organisms in terms of cells; social groups in terms of persons; propositions in terms of concepts; texts in terms of sentences – and so on. In short, micro would explain macro without further ado.

The sensational success of micro-reduction in modern science has given the impression that the concepts of scientific method and reduction are coextensive: that to conduct scientific research is basically to try and reduce wholes to their parts. Not surprisingly, the enemies of science – such as the New Age followers and the postmodernists – are vehement anti-reductionists.

The success of micro-reduction has obscured the fact that in most cases it has been partial rather than total. There are two main reasons for such limitation. The first is that a system, such as an atom, a cell, or a family, has a structure as well as a composition. In other words, an integrated whole is not just a collection of basic entities: it is a new entity with (emergent) properties of its own (recall chapter 1).

The second reason for the limitation of micro-reduction is that reference to the environment of the thing of interest is unavoidable, and the environment belongs to a higher-order level than the thing in question. This holds for physical atoms as well as for social atoms. Indeed, a well-posed problem in atomic physics or in field physics includes the boundary conditions, which constitute an abbreviated description of the macrophysical environment. Likewise, a well-posed problem in psychology or in social science includes explicit reference to the macrosocial environment, in particular the embedding system or supersystem. In other words, what passes for reduction is often a far more complex operation. Let us therefore analyse it.

1 Reduction Operations

Reduction is a kind of analysis. It can be ontological or epistemological: that is, it can bear on things or on ideas. In either case, to reduce A to B is either to *identify* A with B, or to *include* A in B, or to assert that every A is either an *aggregate*, a *combination*, or an *average* of B's, or else a *manifestation* or an *image* of B. It is to assert that, although A and B may appear to be very different from one another, they are actually the same, or that A is a species of the genus B, or that every A results somehow from B's – or, put more vaguely, that A 'boils down' to B or that, 'in the last analysis,' all A's are B's.

The following well-known hypotheses are instances of reduction, whether genuine or illegitimate: The heavenly bodies are ordinary bodies satisfying the laws of mechanics; heat is random molecular motion; light beams are electromagnetic wave-packets; chemical reactions are atomic or molecular combinations, dissociations, or substitutions; life processes are combinations of chemical processes; humans are animals; mental processes are brain processes; and social facts result from individual actions – or conversely.

At least four of the above hypotheses exemplify a special kind of reduction, namely micro-reduction. *Micro-reduction* is the operation whereby things on a macro-level are assumed to be either aggregates or combinations of micro-entities; macro-properties are assumed to result either from the mere aggregation of micro-properties, or from a combination of the latter; and macro-processes are shown to be effects of micro-processes. In short, micro-reduction is the accounting for wholes by their parts.

The converse operation, whereby the behaviour of individuals is explained by their place or function in a whole, may be called *macro-reduction*. A classical example is the attempt, by Marxists, behaviourists, and ecological psychologists, to explain human behaviour exclusively in terms of environmental or structural features. For instance, workers are assumed to vote for left-wing parties, and all new ideas are said to reflect, in the last analysis, changes in the mode of production.

Though never mentioned as a case of macro-reduction, this is what the so-called Copenhagen interpretation of quantum mechanics is. Indeed, it postulates that every microphysical process is produced by an experimenter handling a measuring device: the micro would depend on the macro, and there would be no micro-processes except in the laboratory. Despite its consecration in uncounted textbooks, this assumption is inconsistent with the central axiom of quantum mechanics, namely the state equation, which involves no variables describing measuring devices, let alone experimenters (Bunge 1967b, 1973a).

Whereas micro-reduction inheres in individualism, macro-reduction is typical of holism. Of the two kinds of reduction, the former is so much more common, successful, and prestigious than the latter that the term 'reduction' is usually taken to mean 'micro-reduction.'

The following are examples of micro-reduction: The magnetic field surrounding a magnet results from the alignment of the magnetic moments of the component atoms; water bodies are composed of molecules resulting from the combination of hydrogen and oxygen atoms; plants and animals are systems of cells; and all social facts are ultimately rooted in individual actions.

Whereas micro-reduction is an epistemic operation, *reductionism* is a research strategy, namely, the methodological principle according to which (micro) reduction is in all cases both necessary and sufficient to account for wholes and their properties. By contrast, macro-reductionism is usually called *anti-reductionism*. The ontological partner of (micro)reductionism is atomism, while that of antireductionism is holism. In recent years anti-reductionism has become one of the battle cries of the 'postmodernist' reaction against science and rationality generally. This movement rejects micro-reduction not because of its limitation but because it attempts to explain.

There is no gainsaying the sensational successes of reduction in all the sciences ever since Descartes formulated explicitly the project of reducing everything except the mind to mechanical entities and processes – or, as he put it, to *figures et mouvements*. (No wonder Descartes is the *bête noire* of the 'postmoderns,' both on the right and on the left.) One only needs to recall the colossal achievements of mechanics, and later on of nuclear, atomic, and molecular physics, as well as of molecular biology, to understand the popularity of reductionism among scientists, and their reluctance to acknowledge that reduction may be limited after all.

True, mechanism declined since the birth of field physics and thermodynamics in the mid-nineteenth century (see, e.g., D'Abro 1939). It is now generally understood that mechanics is only one chapter of physics, whence it is impossible to reduce everything to mechanics, even to quantum mechanics – which hardly qualifies as mechanics since it does not calculate trajectories. Moreover, not even the whole of physics suffices for biology – which is not to say that vitalism has been resurrected. Evolutionary biology, born in 1859, killed physicalism in showing that bio-evolution fits no physical laws – although it does not violate any either. The same holds for the social sciences. The only physical law that these need is that of energy conservation.

However, the sharp decline of physicalism has not been the end of reductionism. Quite the contrary, reduction is still extremely successful. Witness the near-completion of the Human Genome Project in 2000, which is neither more nor less than the analysis of the twenty-three human chromosomes into their component genes. The idea behind that project was that, once the genome was known, the rest would follow shortly.

Though nearly completed, this project turned out to be disappointing to the radical reductionists: it showed that we have only about 32,000 genes, only 30 per cent more than the nematode *Caenorhabditis elegans*, which is just one millimetre long and has 959 cells, whereas we have about 100 trillion; and whereas we have about 100 billion neurons, the tiny worm has only 302. By comparison with rice we are even worse off, since rice has many more genes than people do, and yet it has still to think the dumbest of ideas. The moral is clear: It is not so much how many genes you've got, but what they do – in particular, how they combine with one another and which proteins they help synthesize – and how the cells are organized.

Still, micro-reduction will continue to be successful in all of the sciences because all real things happen to be either systems or components of such (recall chapter 2). To reject micro-reduction altogether is to deprive oneself of the joy of understanding many things and processes, and of the power that this knowledge confers.

Yet, as will be seen below, micro-reduction is not omnipotent: it has limitations. In general, analysis is not enough; eventually it must be supplemented with synthesis. This is because the world and our knowledge of it happen to be systems rather than mere aggregates of independent units. For example, to know what a particular gene does is to know how it interacts with other genes and which proteins it helps synthesize or what function it regulates.

However, knowledge of the genome does not entail knowledge of the proteome. The reason is that the genes 'specify' the composition of the proteins but not their configuration or shape. This gap suffices to dash geneticism, which is the project of reducing all of the sciences of man to genetics via the chain advanced by Wilson (1975) and Dawkins (1976):

Genome → Proteome → Cell → Multicellular organism → Society.

If the very first arrow is so far fictitious, why trust that the succeeding arrows in the sequence will prove to be unproblematic?

Micro-reduction, even when feasible, is seldom sufficient for explanation, let alone control. To advance or apply knowledge, it is often necessary to combine two or more theories or even entire research fields rather than reducing one to the other. Witness the very existence of physical chemistry, biochemistry, physiological and social psychology, bio-economics, economic sociology, and hundreds of other interdisciplines.

In the following we shall begin by attempting to identify the ontological roots of the limitations of the micro-reduction operation. Subsequently we shall examine micro-reduction at work in physics, chemistry, biology, psychology, and social science. This study should exhibit the limits of micro-reduction as well as its power.

2 Micro-levels, Macro-levels, and Their Relations

We may distinguish at least two levels in systems of any kind: the macro-level and the micro-level. The *macro-level* is the kind itself, that is, the collection of all the systems sharing certain peculiar properties. The corresponding *micro-level* is the collection of all the components of the systems in question. (In a moment it will be seen that there may be more than one micro-level.) For example, the atomic level is the set of atoms, while the molecular level is the set of molecules. (The fact that the molecular level is composed of a number of sublevels is beside the point.) In general, an n-th level system is composed of things on level $n-1$.

A classical example of the micro-macro distinction is that of the treatment of a macrophysical system by statistical mechanics and thermodynamics. The former explains or reduces the latter. For example, temperature is analysed as the average kinetic energy of the components. An example from social science is this: A sharp increase or decline in a stock market may be caused by a mere rumour about certain unusual earnings, mergers, bankruptcies, accounting malpractices, or macroeconomic forecasts. Here the fear or greed of individuals in great numbers triggers a macrosocial event.

The distinction between levels need not be arbitrary or even a matter of mere detail in description: it often has a real counterpart in the qualitative differences between systems and their components. For example, although a hurricane is made up of molecules, its spiral shape is not contained in its molecular components. Still, levels are collections of things, and hence are concepts, not concrete things. Therefore, levels cannot act upon one another. In particular, the expression 'micro-macro interaction' must be taken as an ellipsis. Indeed, it does not denote an interaction between micro and macro levels but an interaction between entities belonging to a micro-level and things belonging to a macro-level.

Actually, only particle physics can get by with a single micro-level – as long as the particle's environment is treated as an unanalysed whole. All the other sciences study systems or even supersystems composed by systems, so that they involve the distinction between a number of micro-levels. In other words, most sciences tackle nested systems ('hierarchies'). Think, for example, of the human brain, with its many subsystems such as the thalamus, the hippocampus, and the primary visual cortex – every one of which is composed of further systems, namely neurons and glial cells. The complexity of the real systems studied in most sciences forces us to analyse the concept of composition into as many levels as needed (recall chapter 2, section 5).

All of the factual sciences are faced with micro/macro gaps, because all of them study systems of some kind or other, and all systems have components (the micro-aspect) as well as macro-properties of their own (the macro-aspect). In many cases one knows how to solve problems concerning the micro-level or the macro-level in question, but one does not know how to relate them. In particular, one seldom knows how to account for macro-features in terms of micro-entities and their properties and changes thereof. Consequently the micro-specialists (e.g., atomic physicists and microeconomists) and macro-specialists (e.g., experts in fluid dynamics and macroeconomists) outnumber the experts in gap bridging.

Every problem concerning a micro-macro relation is intrinsically difficult. This difficulty is compounded by the dearth of careful philosophical analyses of micro–macro relations. We shall proceed to sketching such an analysis. The very first task we must perform is that of distinguishing two basic types of micro–macro relations: *de re*, or ontological, and *de dicto*, or epistemological. Ontological micro–macro relations are a particular case of the part–whole relation, whereas epistemological micro–macro relations conceptualize relations between micro-levels and macro-levels. Let me explain.

The assembly of two or more atoms (or molecules, or cells, or animals) into a higher-level entity is a case of ontological micro-to-macro relation. Likewise, the converse process of system dismantling exemplifies an ontological macro-to-micro relation. The water-condensation effect of a table-salt molecule, the effect of the pacemaker cells on the heart, and of a leader on an organization are further examples of the ontological micro-to-macro relation. When a limb is cut off, its cells die; when 'the sun goes down,' the average kinetic energy of the air molecules decreases; and when an organization is banned, all of its members are affected. These are instances of the ontological macro-to-micro relation. In all of these cases links (or bonds or couplings) between micro and macro things or processes are set up, maintained, altered, or severed.

No such bonding relations are involved in the relations among levels of organization because – as noted above – levels are sets, hence concepts, not concrete things or processes. An example is the famous formula for the entropy S of a thermodynamic system in a given macro-state in terms of the number W of atomic or molecular states or configurations compatible with the given macro-state. Indeed, this formula, '$S = k\ ln\ W$,' is a micro-to-macro relation of the epistemological type. So are the formulas for the specific heat, the conductivity, and the refractive index of a body in terms of the properties of its atomic constituents. Ditto Hebb's theory of learning in terms of the reinforcement of inter-neuronal connections: here the macro-level is composed by brain subsystems capable of performing mental functions, and the micro-level by neurons. There is no action of neurons on the brain or on the mind: There is only a conceptual relation between two levels of organization.

3 Same-level and Inter-level Relations

Combining the micro-macro distinction with the ontologico-epistemological one, we get a total of eight interlevel relations:

1 *Micro–micro (mm)*
   (a) *Ontological*, e.g., atomic collisions; the love bond.
   (b) *Epistemological*, e.g., quantum theories of atoms; psychological theories of interpersonal relations.
2 *Micro–Macro (mM)* or *Bottom-up*
   (a) *Ontological*, e.g., the interaction between an electron and an atom as a whole; a social movement triggered by a charismatic leader.
   (b) *Epistemological*, e.g., statistical mechanics; a theory of animal behaviour triggered by micro-stimuli, such as a handful of photons striking a retina.

3 *Macro–micro (Mm)* or *Top-down*
   (a) *Ontological*, e.g., the action of a flood or an earthquake on an animal; the effect of governments on individuals.
   (b) *Epistemological*, e.g., a theory of an intrusive measurement on a microphysical entity; a model of the course of a ship adrift in an oceanic current.
4 *Macro–macro* (MM)
   (a) *Ontological*, e.g., Sun–Earth interaction; a theory of rivalry between groups of animals (e.g., families).
   (b) *Epistemological*, e.g., plate-tectonics theory; international–relations models.

These distinctions are relevant to the theories of definition and of explanation. They are also pertinent to the old dispute between reductionists and anti-reductionists, a philosophical controversy that occurs in every science. For instance, molecular biologists fight with organismic biologists, and rational-choice theorists fight with collectivists. Whereas reductionists claim that only *mm* and *mM* relations have explanatory power, their opponents write these off and hold that only *Mm* and *MM* relations can explain.

In our view each of the two contenders is partially right, hence partially wrong: since all four relations exist, all of them pose problems. In particular, we need to investigate how individuals of all kinds interact (*mm*), and how they assemble forming systems (*mM*). We also need to know how being a part of a system affects the individual (*Mm*), and how systems affect one another (*MM*). The need for such broader research projects shows that the radical reductionists are as wrong as the radical anti-reductionists. Hence, we had better adopt the systemic approach, which embraces all four relations – and more when required.

Evidently, whenever we distinguish more than two levels we are faced with several further relations. For example, interpolating a meso-level between a micro- and a macro-level, and adding a mega-level on top of the latter, we get four same-level relations plus three inter-level relations, thus a total of seven without skipping levels – actually fourteen when the ontological-epistemological distinction is drawn. This happens, for instance, when relations between genome, cell, whole organism, and ecosystem are investigated. Another well-known finer-level distinction is that between individual actor (nano), firm (micro), cartel (meso), national economy (macro), and world economy (mega).

4 Inter-level Hypotheses and Explanations

Let us reword the above in terms of the kinds of proposition that can be constructed when distinguishing just two levels, a micro-level (*m*) and a macro-level (*M*). By combining the corresponding concepts we may form propositions of four different kinds, two same-level and two inter-level ones:

1 *mm*, e.g., hypotheses concerning specific nuclear forces, inter-neuronal connections, and face-to-face personal relations.
2 *mM*, e.g., the formulas of statistical mechanics, genotype–phenotype relations, and the macrosocial outcomes of individual actions such as voting.
3 *Mm*, e.g., Lorentz's formula for the force exerted by an external magnetic field on an electron, and the hypotheses on the influence of social structure upon individual behaviour.
4 *MM*, e.g., Newton's law of gravity, the rate equations of chemical kinetics, the ecological relations between biodiversity and biomass productivity, and data on international conflicts.

As with propositions, so with explanations. That is, in principle, an explanation may contain same-level or inter-level explanatory premises or conclusions (Bunge 1967a). Moreover, explanations of all four kinds are needed because ours is a world of systems, and systems must be understood on their own level as well as resulting from the assembly of smaller units and constraining the behaviour of their components.

In other words, ontology must be realistic and it must guide epistemology if the latter is to be of any use in research. (This view of the dependence of epistemology upon ontology is in sharp contrast with the positivist thesis that the 'logic of science,' and in particular the analysis of reduction, must be free from ontological assumptions: see, e.g., Carnap 1938.) One-sided ontologies, such as individualism (atomism) and collectivism (holism), ban each other's explanation types and thus curtail the power of science and technology. Only a systemic ontology encourages the search for explanations of all four kinds.

In the following we shall examine a few typical examples drawn from five branches of contemporary science: physics, chemistry, biology, psychology, and sociology.

5 From Physics to Chemistry

The earliest cases of reduction in physics were those of astronomy to physics, and of statics to dynamics. The former is philosophically the more interesting, because it eliminated the theological distinction between celestial and terrestrial bodies. We won't dwell on it, because it is well known. We shall peek instead at the relation between dynamics and statics, because it is rather puzzling and never correctly analysed in the philosophical literature.

At first blush, statics is the particular (or degenerate) case of dynamics, when all the velocities vanish because the forces at play balance one another. Yet, the converse does not hold: the components of a mechanical system may be in motion, yet in internal equilibrium with one another, hence in equilibrium as a whole. This is why dynamics can be formally reduced to statics.

D'Alembert achieved this formal reduction upon reinterpreting Newton's second law of motion, '$F = ma$.' He viewed '$I = -ma$' as the (inertial) force created by the motion, and balancing the applied force F. Thus, Newton's law of motion may be rewritten as the equilibrium condition $F + I = 0$. Of course, the reduction thus achieved is merely formal (mathematical), since actually the system may be in motion relative to some frame of reference. Still, the moral of this story is that reduction is not the same as deduction. (Reduction implies deduction but, since the converse implication does not hold, reduction is not equivalent to deduction.)

Our second example is provided by quantum mechanics. This theory is usually regarded as the nucleus of microphysics, and moreover as the key to the reduction of macrophysical to microphysical things and processes. This view is roughly correct, though only with qualifications. The first qualification is that, so far, only classical particle mechanics has been reduced to quantum mechanics: the dynamical theory of continuous media, in particular liquids, is still in force despite decades of efforts to explain liquids in quantum-mechanical terms. However, this research project too is in force.

Second, and more fundamentally, the quantum theory contains a number of concepts borrowed from macrophysics, such as those of space, time, mass, electric charge, classical linear momentum, and classical energy. (Thus, four of these concepts, namely x, t, p, and E, occur in the most elementary of all state functions, namely the plane 'wave' $\Psi = A\, exp\, [i(px - Et)/h]$.) Besides, the boundary conditions, which are part of the statement of every problem in quantum theory, constitute schematic macrophysical representations of the environment. (Example: The condition that the state function vanishes on the surface of a container – provided the latter is so thick, that it won't allow for tunnelling.)

A third example is this: In every measurement, the measuring apparatus is treated as a macrophysical system described in classical or at most semi-classical terms. A reason for this is that, as Bohr (1958) emphasized, a nuclear, atomic, or molecular measuring instrument must contain an amplifier, since only macrophysical events are observable.

A fourth and even more dramatic and popular case is that of inseparability: A quantum system remains a system even if its components become spatially separated. The system components, and their corresponding states, are said to be *entangled*. (The constituents of the system are so strongly coupled that the system state cannot be written as the product of the states of the components.)

What holds for quantum physics holds, a fortiori, for quantum chemistry. This discipline contains not only the above-mentioned macrophysical concepts but also certain macrochemical ones. Thus, one of the accomplishments of quantum chemistry is the *ab initio* calculation of the equilibrium constants of chemical reactions. In classical chemistry, these constants are treated as empirical parameters. In quantum chemistry, they are part of the theory of chemical reactions construed as inelastic scattering (collision) processes. However, this theory presupposes the rate equations of chemical kinetics, a part of classical chemistry.

Indeed, consider the problem of calculating the rate (or equilibrium) constant of a chemical reaction of the type '$A + B \to C$.' The phenomenological (macrochemical) equation for the rate of formation of the reaction product C is not deduced but is postulated when the rate constant is calculated in quantum-theoretical terms. Hence, quantum chemistry does not follow from quantum mechanics without further ado. In other words, chemistry has not been fully reduced to physics. The epistemological reduction is only partial even though the ontological one is total. (For details see Lévy 1979, Bunge 1982.)

6 Biology, Ecology, and Psychology

Undoubtedly, the most sensational advances in contemporary biology have been inspired by the thesis that organisms are nothing but bags of chemicals, whence biology is just extremely complex chemistry (see, e.g., Bernard 1865). Yet this thesis, though heuristically enormously powerful, is only partially true. We shall show this to be true in the cases of genetics and of the very definition of the concept of life.

At first sight, the discovery that the genetic material is composed of DNA molecules proves that genetics has been reduced to chemistry (e.g., Schaffner 1969). However, chemistry only accounts for DNA chemistry: it tells us nothing about the biological functions of DNA – for instance, that it controls morphogenesis and protein synthesis. In other words, DNA does not perform any such functions when outside a cell, anymore than a stray screw holds anything together. (Besides, DNA does nothing by itself: it is at the mercy of the enzymes and RNAs that determine which genes are to be expressed or silenced. In other words, the genetic code is not the prime motor it was once believed to be. This is what epigenesis is all about.)

The reason for the insufficiency of biochemistry to explain life is of course that the very concept of a living cell is alien to chemistry. True, the cell components are physical or chemical entities, but these are organized in a cell in characteristically biological ways (recall chapter 3, section 4). It is also true that every known property of a cell, except for that of being alive, is shared by some physical or chemical systems. Yet only living cells possess jointly the dozen or so properties that characterize organisms, among them metabolism and self-repair. Consequently biology, though certainly based on physics and chemistry, is not fully reducible to the latter. For example, any system of biological classification based exclusively on the degree of similarity in DNA – as, for example, between us and chimps – is bound to fail, for missing the supramolecular features of organisms.

What holds for cell biology holds a fortiori for organismic biology. Indeed, all the organs of a multicellular organism must work individually as well as in concert for it to be fit. And a healthy organism engages in whole-organism processes, such as metabolism, motion, interaction with other organisms, and reproduction. The radical reductionist cannot explain such higher-level processes, and the holist denies the need for such explanation. Only the systemist will endeavour to explain the whole by its parts and their interactions. This is what the physician does, for instance, when diagnosing anemia of a certain type as a result of iron deficiency, and hemochromatosis as a consequence of iron excess.

Reductionists are stubbornly opposed to emergence and levels. To them, every property is a property of a basic constituent or, at most, a statistical average; and only the bottom level matters. Yet, even reductionists value organisms more than their elementary constituents. And everyone knows that ordinarily parts come cheaper than wholes. For example, the total worth of the atoms in a human body is about one dollar; that of the usable tissues (excluding organs) is more than $200,000; and that of a young unskilled worker is more than $1,000,000.

Ecology too is the subject of spirited controversies between reductionists and holists (see, e.g., Saarinen 1980, Looijen 2000). The radical reductionists, such as Daniel Simberloff, claim that ecosystems are nothing but accidental collections of populations, and they concentrate on the binary competitions among them represented by the famous Lotka-Volterra equations. They refuse to acknowledge that communities and ecosystems are characterized by such emergents as niche, food web, species diversity, equilibrium, productivity, and sustainability.

The holists (or 'functional ecologists'), such as Eugene Odum, take advantage of this weakness of the reductionist camp, study whole-system properties and processes, and regard ecosystems as self-regulating systems aiming for stability. A critical result of this approach is the hypothesis that the more diverse ecosystems are the more stable. But a recent experiment has falsified this hypothesis, showing instead that optimal biospecies diversity is not maximal (Pfisterer and Schmid 2002). This and other results have vindicated the analytic method.

Mainstream ecology is systemist rather than either holist or individualist, for it adopts 'a moderate or mixed approach in which it is acknowledged that communities and ecosystems are discrete higher-level entities with their own emergent properties, but which seeks the explanation of these properties especially in interactions between the component species [populations]' (Looijen 2000: 153).

What about psychology: is it reducible to biology? Assume, for the sake of the argument, that all mental processes are brain processes (ontological reduction). Does this entail that psychology is a branch of biology and, in particular, of neuroscience (epistemological reduction)? Not quite, and this for the following reasons. First, because brain processes are influenced by social stimuli, such as words and encounters with friends or foes. Now, such psychosocial processes are studied by social psychology, which employs sociological categories, such as those of social group and occupation, which are not reducible to neuroscience. A second reason is that psychology employs concepts of its own, such as those of emotion, consciousness, and personality, as well as peculiar techniques, such as interrogation and suggestion, that go beyond biology.

We conclude then that, even though the psychoneural-identity hypothesis is a clear case of ontological reduction, and sensationally fertile to boot, psychology is not reducible to neuroscience even though it has a large overlap with it. (For details see Bunge 1990.) Shorter: Ontological reduction does not imply epistemological reduction. This problem will reappear in chapter 11, with reference to cognitive neuroscience.

7 From Biology to Social Science: Human Sociobiology and the IQ Debate

Human sociobiology is the attempt to 'biologize' social science and, in particular, to reduce it to genetics (Wilson 1975, Dawkins 1976). Its basic tenets are that (a) 'the organism is only DNA's way of making more DNA'; (b) we are genetically programmed to behave as we do; (c) all social items are designed by natural selection to enhance adaptation; and, consequently, (d) social science is to be reconstructed as a branch of biology.

The centrepiece of human sociobiology is its hypothesis of kin selection, according to which we are 'designed' by natural selection so as to behave altruistically towards those who share most of our genes with us – that is, our kin, in particular child, parent, or sibling. Paradoxically, such unselfish behaviour is regarded as selfish, for it tends to perpetuate our genes. So, we are unselfish to the extent that we are selfish. This thesis is not only paradoxical, not to say self-contradictory (Stove 1995). It is also at variance with the evidence. Indeed, young men tend to behave more generously towards their girlfriends, to whom ordinarily they are not genetically related, than to their parents; the present Queen of England is said to feel closer to her dogs and horses than to her family; and a pet may imprint on its owner.

Despite its popular success, sociobiology is not viable, mainly for three reasons. First, whereas biological adaptation, when it occurs, results from a very slow and rather erratic trial-and-error process, social adaptation – when beneficial – can be very quick thanks to the enormous plasticity of the human brain and the resulting behavioural plasticity. Second, we often engage in maladaptive activities, such as taking drugs, gaming, supporting tyrants, adopting ideologies that restrict individual or social development – or committing suicide. Third, most people engage in plenty of activities that have no primarily biological motivation or import, such as chatting, playing, watching TV, worshipping, reading poetry, listening to music, designing experiments, reconstructing the past, proving theorems, and philosophizing.

There is a further reason why social science cannot be 'biologicized': every human society, however pristine, is ruled not only by laws of nature but also by customs, norms, conventions, and institutions that, far from being natural, are social inventions, such as languages, monogamy, hospitals, schools, commodification, moral and legal codes, and 'surgical air strikes.' Although these social items are certainly compatible with the laws of nature, they are not always biologically beneficial. Above all, they are made and unmade, that is, unnatural. Just think of the deep and quick changes in industry, trade, warfare, the law, technology, science, and the arts since the time of the American and French Revolutions. Can any of these changes truly be said to have resulted from purely biological urges, let alone from gene mutations, or from a long process of natural selection?

Besides, most biological drives are met or frustrated through social mechanisms, such as communication, mutual help, work, trade, legal and moral coercion, and violence. It is equally true that, since all social facts are the work of living beings, social structure can either favour or stunt biological functions. For example, an economic order can either help or hinder the meeting of basic biological needs; and custom and the law can either control or tolerate antisocial behaviour elicited by testosterone excess.

Yet such social channelling of biological processes only speaks for biosociology – not to be mistaken for sociobiology. For example, testosterone excess might help explain why the overwhelming majority of murderers worldwide are males; but it does not explain why the American murder rates are many times their Swedish, Turkish, Indian, or Japanese counterparts. Surely such differences call for sociological and historical explanations – for instance, in terms of social inequalities, unemployment, anomie, solidarity, criminal law, and tradition. (More on this in chapter 13.)

In short, human sociobiology and its inheritors – evolutionary psychology, Darwinian medicine, and biological ethics – are sheer fantasies. One might as well argue for the survival of the rudest, smartest, most competitive, risk-loving, or prettiest, since rudeness gets rid of rivals, smartness allows for the choice of the most suitable means, competitiveness makes for grabbing the most first, gains involve risk-taking, and everyone would like to mate with the prettiest. However, one may also argue persuasively for the survival of the gentlest, dumbest, most accommodating, most prudent, and ugliest, since everyone protects the gentle, the dumb pose no threat to the strong, the cooperative get help in return, the prudent incur less risk, and the ugly are unlikely to exhibit themselves in public. That is, the rudest, smartest, most competitive, most valiant, and prettiest would win the battles in the struggle for life; but the gentlest, dumbest, most cooperative, most cautious, and ugliest would win the war for life, because they are spared the rigours of battle.

Though mutually incompatible, both stories will sound equally plausible to different people. But of course neither of them enjoys the support of empirical evidence: we simply do not know which of those features is positively correlated to progeny size in different natural and social environments. All we know is that social conventions and tastes have changed in the course of history. For example, nowadays the Venus of Willemsdorf would be treated for obesity; great fertility is frowned upon; peace activists are often beaten up by the police; the modern scholar treats the novelty-fearing schoolmen with scorn, and so on. (See Kitcher 1985, Stove 1995, Lewontin 2000, Dover 2000, and Dubrovsky 2002 for criticisms of the over-ambitious project of accounting for everything mental and social exclusively in biological terms and, in particular, in terms of genes and natural selection.)

In addition to being scientifically unsound, biological reductionism – particularly in its genetic-determinist version – has been misused to provide the semblance of a scientific justification for superstition, rape, war, and 'ethnic cleansing'; for treating women as mentally deficient, and criminals as irredeemable; and for neglecting public education on the assumption that IQ is mostly hereditary. Let us glimpse at the latter.

The claim that mental abilities are mostly inherited is based on four main assumptions: (a) genotype alone determines phenotype; (b) the concept of intelligence is well defined; (c) IQ is a reliable measure of intelligence; and (d) the statistical analysis of the variance (or square of the mean standard deviation) of the IQ distribution in a population can tell us how much IQ is inherited and how much learned. However, all four assumptions are false, as will be shown anon.

The first assumption is false, because (a) even the clones of fruit flies can exhibit some gross morphological differences; (b) the human genome, which consists of about 32,000 genes, is insufficient for determining ('specifying') the 10^{14} or so brain synapses; and (c) a gene may function differently in different environments, which is why identical twins behave differently if raised in different families (see, e.g., Collins et al. 2000).

The second hypothesis is false too, because no theory of general intelligence is available (see, e.g., Sternberg 1985). Hence, the third assumption is groundless too: if we don't know what X is, neither can we know whether some other variable is a reliable indicator of X. And the fourth assumption is false for the following reasons highlighted by Kempthorne in a classic paper (1978) that the innatists seem to ignore.

Kempthorne rightly stressed that variance measures diversity, not variation or change, much less a controlled one – which alone could establish causation. (Recall that x and y can be said to be causally related if and only if (a) there is a functional relation between x and y; and (b) there is a mechanism whereby an increment or decrement in x is followed by a change in y.) In particular, only genetic manipulation could establish the hypothesis that the replacement of one set of genes with another is followed by a change in certain (well-defined and reliably measured) mental abilities.

8 Kinds and Limits of Reduction

In section 1 we examined the ontological concepts of reduction. We shall now tackle their epistemological counterparts. An epistemological reduction may bear on concepts, propositions, explanations, or hypothetico-deductive systems. To reduce a concept A to a concept B is to define A in terms of B, where B refers to a thing, property, or process on either the same or a different (lower or higher) level than that of the referent(s) of A. Such a stipulation will be called a *reductive definition*. (In the philosophical literature reductive definitions are usually called 'bridge hypotheses,' presumably because they are often originally proposed as hypotheses. History without analysis can be misleading.)

A downward reductive (i.e., microreductive) definition may also be called a *top-down* definition. Example: 'Heat $=_{df}$ Random atomic or molecular motion.' On the other hand, the upward reductive (i.e., macro-reductive) definitions may be called *bottom-up* definitions. Example: 'Conformism $=_{df}$ An individual's bowing to the ruling customs or norms.' But there are also *same-level* definitions, such as 'Light is electromagnetic radiation.'

The reduction of a *proposition* results from replacing at least one of the predicates occurring in it with the definiens of a reductive definition. For example, the psychological proposition 'X was forming a linguistic expression' is reducible to the neurophysiological proposition 'X's Wernicke's area was active,' by virtue of the reductive definition 'Formation of linguistic expressions $=_{df}$ Specific activity of the Wernicke area.' (Notice that this statement was born as a hypothesis. It becomes a definition in a neurolinguistic theory: it was born as a bridge, and is now a glue. More on this in chapter 17)

An *explanation* can be said to be reductive if and only if at least one of its explanans premises is a reductive definition or a reduced proposition. For example, the explanation of the existence of a concrete system in terms of the links among its parts is of the micro-reductive (or bottom-up) kind. On the other hand, the explanation of the behaviour of a component in terms of the place or function it holds in a system is of the macroreductive (or top-down) type. Work on a car assembly line (or on the origin of life) induces explanations of the first kind, while the car mechanic (and the medic) typically resorts to explanations of the second type.

The analysis of theory reduction is somewhat more complex. Call T_1 and T_2 two theories (hypothetico-deductive systems) sharing some referents; further, call R a set of reductive definitions, and S a set of subsidiary hypotheses not contained in either T_1 or T_2. (However, these hypotheses must be couched in the language resulting from the union of the languages of T_1 and T_2 if they are to blend with the latter.) We stipulate that

(a) T_2 is *fully* (or *strongly*) reducible to $T_1 =_{df} T_2$ follows logically from the union of T_1 and R;
(b) T_2 is *partially* (or *weakly*) reducible to $T_1 =_{df} T_2$ follows logically from the union of T_1, R, and S.

Ray optics is strongly reducible to wave optics by way of the reductive definition 'Ray $=_{df}$ Normal to wave front.' In turn, wave optics is strongly reducible to electromagnetism by virtue of the reductive definition of 'light' as electromagnetic radiation of wavelengths lying within a certain interval. On the other hand, the kinetic theory of gases is only weakly reducible to particle mechanics because, in addition to the microreductive definitions of the concepts of pressure and temperature, this theory includes the subsidiary hypothesis of molecular chaos (or random initial distribution of positions and velocities).

Likewise, as we saw above, quantum chemistry, cell biology, psychology, and social science are only weakly (partially) reducible to the corresponding lower-level disciplines. We also saw that even the quantum theory contains some classical concepts as well as subsidiary hypotheses, for example, about macrophysical boundaries, so that it does not effect a complete reduction to microphysical concepts. (More on the various kinds and aspects of reduction in Bunge 1977b, 1983b, and 1989.)

9 Reductionism and Materialism

Every success of scientific micro-reduction can be made to count as a victory of materialism, whereas every limitation of that strategy is sometimes regarded as a defeat of it (see, e.g., Popper 1970). However, these rival ontologies are seldom carefully characterized. In particular, materialism is often confused with physicalism, or the thesis that everything is physical or reducible to physical elements; and many an appreciation of the role of ideas in social life is characterized as idealist.

These are confusions. The standard histories of philosophy are partly to blame for them because of their nearly uniform disregard for materialism, although it is as old as idealism, and certainly more influential than the latter in modern science. (The only well-known history of materialism is Friedrich Lange's [1905]. But this work is badly dated and severely biased, because Lange was a neo-Kantian and early fictionist hostile to materialism.) To avoid such confusions and oversights, let us propose the following distinctions between three broad kinds of materialism. Contemporary materialism is a family with three main members: physicalism, or vulgar materialism; dialectical materialism, the Marxist philosophy; and emergentist (or modern) materialism.

Physicalism (or vulgar materialism) is radically reductionist. Indeed, it states that everything is physical. Consequently it claims that, although there may be different levels of analysis or description, these have no counterparts in reality. Ancient Greek and Indian atomism, as well as the mechanistic world view that dominated natural science between about 1600 and 1900, have been the highlights of physicalism.

Dialectical materialism, crafted by Engels, Lenin, and a number of Soviet philosophers, is a sort of synthesis of eighteenth-century materialism (mostly physicalist) and Hegel's dialectics. It has therefore the merits of the former and the absurdities of the latter. The main false theses of dialectics are that every thing is a unity of opposites, and that all changes derive from such 'contradictions' or 'struggles of opposites.' The mere existence of elementary particles, such as electrons, and of cooperation on all levels – from self-assembly and cell clumping to social cooperation – falsifies these theses. These cases also indict dialectical materialism as an a priori philosophy eager to find examples but reluctant to admit counter-examples. (For detailed criticisms see Bunge 1981a.) However, dialectical materialism has the merit of emphasizing qualitative novelty, or emergence.

Emergentist (or modern) materialism avoids the oversimplifications of physicalism and the obscurities and sophistries of dialectics (see Novikoff 1945; Sellars, McGill, and Farber 1949; Warren, 1970; Bunge 1977a, 1977b, 1979a, 1980, 1981a; and Blitz 1992). It asserts that, although every real existent is material, material things fall into at least five qualitatively different integrative levels: physical, chemical, biological, social, and technical. The things in every level are composed of lower-level things, and possess emergent properties that their constituents lack. For example, a brain subsystem capable of having mental experiences of some kind is composed of neurons, glial cells, and cells of other kinds, none of which is capable of minding; likewise, a business firm, though composed of persons, offers products that no individual could produce.

Because it joins materialism and emergentism, emergentist materialism avoids such oversimplifications as eliminative materialism (the denial of the existence of mental processes) and sociobiology (the denial of specifically social categories irreducible to biological ones). On the positive side, emergentist materialism encourages the research of emergence mechanisms, and promotes the interdisciplinary mergers required to explain emergence.

Concluding Remarks

We shall now make bold and generalize the preceding conclusions by stating that, while partial micro-reduction is often successful, full micro-reduction seldom is. What often does work is relating two or more levels without attempting to reduce one to the other, as suggested by the following simple example.

It is well known that, when the interest rates rise above a certain level, the construction industry declines. In obvious symbols, $R \Rightarrow \neg C$. This relation between two macroeconomic variables can be explained as follows. If the interests rates rise beyond a certain level, poor people cannot afford to buy or build houses, as a consequence of which the construction industry declines. In obvious symbols,

$R \Rightarrow \neg B$ Macro–micro
$\neg B \Rightarrow \neg C$ Micro–macro
$\therefore R \Rightarrow \neg C$ Macro–macro

The failures of full micro-reduction may be explained by the hypothesis that every real thing, except for the universe as a whole, is embedded in some higher-level system or other. Hence, every *mM* relation is accompanied by some *Mm* relation, and both have often *mm* or *MM* concomitants. For this reason same-level definitions and explanations must be supplemented with inter-level (in particular bottom-up and top-down) definitions or explanations. Which goes to show that, to be of any use to science and technology, epistemology must match ontology. More precisely, a realistic epistemology must be paired off with a systemic ontology.

In short, although reduction ought to be pushed as far as possible, we should realize its limits: we should settle for partial (or weak) reduction whenever full (or strong) reduction is unattainable. This methodological maxim distinguishes moderate from radical reductionism. The former is a more realistic research strategy than the latter, and either is more powerful than anti-reductionism. Besides, reduction, which results in vertical or inter-level integration, should be supplemented with merger or horizontal (same-level) integration. This dual research strategy should work because, as argued earlier, the world happens to be a system of systems rather than either a pile of disjoint items (the individualist view) or an unanalysable bloc (the holist dogma). However, before studying integration it will be convenient to examine some famous cases of failed reduction.

10

A Pack of Failed Reductionist Projects

So far we have examined reduction and reductionism in general terms. We have found that reduction, though often successful, is necessarily limited by the occurrence of emergence along with the formation of systems, and submergence along with their dismantling. It was therefore suggested that the moderate version of reductionism is superior to the radical one. In the present chapter it will argued that some of the fashionable reductionist projects are not just limited but utter failures.

1 Physicalism

Reductionism has been rampant since the 1600s. Nowadays it is found among the believers in a 'theory of everything' – if not quantum theory then perhaps the next superstring theory; among those who claim that the sequenced genome is the Book of Life; in the sociobiologists who hold that human psychology and social science are reducible to evolutionary biology or even genetics; among the computer experts who declare that all processes, from planetary motion to metabolism to evolution to feeling and thought, are computations; and among the self-styled economic imperialists, who attempt to account for all social facts in terms of the hypothesis that all humans are always intent on maximizing their expected utilities at all costs and under all circumstances. Reductionism is popular because it is believed to be the royal road to the ultimate explanation of everything, on top of which it effects the unification of the sciences. I will challenge these beliefs, though without surrendering to holism or the attending intuitionism.

Physicalism is of course the oldest and most successful of all the reductionist projects. Undoubtedly it has been very fruitful, since it has spawned physical chemistry, biophysics, and bioengineering. But it is unworkable because such key biological concepts as sex, immunity, the food web, competition, health, and plague do not apply to physical things except metaphorically: they designate supraphysical emergents. Physicalism is even more impotent in social science: think, for instance, of the concept of cohesion of a system, which is measurable by the binding energy at the atomic level, and by the frequency of cooperative interpersonal interactions at the social level.

Though dead in biology and social science, physicalism is still favoured by many physicists, who regard quantum theory, or else string theory, as the 'theory of everything,' or universal science. However, this strategy has yet to conquer the whole of physics. For example, no one has yet derived the central law of fluid dynamics – the two-centuries old Navier-Stokes equation – from quantum mechanics. True, some macrophysical things are quantons: Bose-Einstein condensates, superconducting rings, bodies of liquid helium, and possibly black holes, quasars, and other macrophysical systems as well. However, quantum mechanics plays no role in accounting for such common macrophysical processes as evaporation or freezing, wind or rain. There is certainly talk of 'the wave function of the universe,' complete with measuring instruments and observers. But nobody has the foggiest idea what such a function looks like. All we are told is that its symbol is Ψ – a case of hand-waving in lieu of serious science.

Unsurprisingly, the universal quantization strategy has not worked in biology. Indeed, physics knows nothing about life, sickness, or death; so much so that nobody has got the slightest idea as to how to write a state (or Schrödinger) equation for the humblest of bacteria. If quantum physics cannot handle bacteria, why should it be able to account for persons and the social systems they constitute? And if the quantum theory cannot tackle metabolism, disease, perception, or self-deception, why hope that it will be able to account for human cooperation and strife, the adoption and rejection of social conventions, or the rise and decline of empires? Why should physical theories, which are systems of law statements, be expected to account for social conventions and historical accidents? Promissory notes are no substitutes for serious research projects – and nobody would hold any hopes for quantum sociology. So much for physicalism.

2 Computationism

Nowadays the most popular of all reductionist projects is informationism, a.k.a. computationism. According to it, all things are bundles of information, and all processes are computations. For instance, the biochemists Adam Arkin and John Ross (1994) stated that glycolysis *is* a computation – not just that it can be simulated on a computer. The philosopher Daniel Dennett (1995) equated evolution with a family of 'substrate-neutral' algorithms – which he did not care to define. The media expert Nicholas Negroponte (1996) assures us that atoms do not matter: that only their organization or information matters. And the physicist John A. Wheeler, who used to think that the building bricks of the universe are propositions, has lately claimed that there is no matter: that all the 'its' are bits or derive from them. For example, a hamburger would consist of so many megabytes. If only cows could be made to believe that the butcher is not after their meat but only after the information they carry!

Clearly, if stuff does not matter, if only form or structure does, then all things are bundles of bits, and all processes are computations in accordance with computer programs or algorithms. And if this is so, then all the sciences are ultimately reducible to computer science. This is indeed a tacit postulate of information-processing psychology, as well as of many a project in AI (Artificial Intelligence) and AL (Artificial Life).

Computationism has also been adopted enthusiastically by many philosophers of mind (see, e.g., Dennett 1991 and Churchland and Sejnowski 1993). Ironically, although some of these philosophers believe themselves to be materialists of the physicalist variety, actually they are dualists and even Platonists, because they write about brains storing and processing symbols, and computer programs being 'instantiated' (embodied) in brains or in robots.

Imagine, for a moment, as Kary and Mahner (2002) ask us to do in jest, that after some smart tinkering in his garage, someone were to come up with a silicon-based artefact that looked alive or even as if it were thinking. How could he ascertain that it is indeed an organism and moreover a minding one? The functionalist would say: 'Look at what it does. Things are just what they do, regardless of the stuff they are "instantiated" in.' But the aeronautical engineer would disagree: He might note that, although gliders, propeller planes, and jets fly, they do so according to different mechanisms. The glider is swept by air currents, the propeller aircraft calls for complex equations representing vortex motion, and the jet abides by momentum conservation (or equality of action and reaction).

Nor would biological psychologists be easily persuaded: they would demand to see living cells, or even large systems of interconnected neurons firing synchronically, not just chips and wires and things. Moreover, they might want to perform electrophysiological and biochemical tests, to find out more than mere shallow analogies between the said artefact and a really living and minding thing. Of course, the computer programmer has no use for blood and urine tests and the like. He may say that his device is alive, or thinks, if such and such figures or numerals appear on the screen of a computer connected to it. Fair enough: it is his prerogative to restrict his attention to whatever similarities he may discern. But then he cannot claim to have made a biological or psychological finding, because biologists happen to study organisms, and psychologists some of the processes that go on in the brains of some animals. In other words, things are not the same as their artificial simulates. In particular, a computer simulation of a physical, chemical, biological, or social process is not equivalent to the original process: at most, it is similar to some aspects of it. (For an analysis of analogy and simulation see Bunge 1973b; for the limitations of computers see Bunge 1956; Kary and Mahner 2002.)

No astronomer would claim that the planets compute their orbits as they revolve around the Sun. And yet something like this is what the computationalist asks us to believe: that chemicals compute as they react with one another; and that brains compute as they perceive, feel, or think – as a consequence of which similarly programmed robots must be attributed a consciousness, feelings, and perhaps even a moral conscience. The computationalist mistakes partial similarity for identity.

3 Linguistic Imperialism

Let us now examine briefly two older naive reductionist proposals to unify all of the sciences. One of them was proposed by the sociologist and philosopher Otto Neurath (1931, 1944). This was to adopt a single language in all the sciences, namely that of (macro)physics, in particular the one employed in writing laboratory protocols. This project, adopted at one time by the Vienna Circle, was prompted by both love of science and fear of metaphysics. Unsurprisingly, it went unheeded in biology, psychology, and social science. The closest these disciplines came to using the language of physics was in biophysics and psychophysics. However, these are interdisciplines, not cases of physicalist reduction. Besides, no theoretical physicist, biologist, or sociologist can restrict his vocabulary to the small subset used in writing laboratory protocols. The reason is that all scientific theories contain predicates that refer to unobservable entities and processes, such as 'binding energy,' 'anomie,' and 'national economy.'

Carnap (1938) was more explicit: he claimed that what he called 'reduction statements' posit unobservable-observable relations, such as 'If there is an electric current in this wire, then if a compass is placed in its vicinity, the compass needle will be seen to move, and conversely.' But of course such statements are what Bridgman called 'operational definitions' and I call 'indicator hypotheses.' These occur not in theories, but in the theory/laboratory (or field) interface. For example, the fact that inflation is a cause of anxiety and stress does not allow one to reduce psychiatry to economics or conversely; it only shows that psychiatrists should not be blind to the socio-economic environment of their patients. Philosophers would have paid no attention to Carnap's proposal if they had studied the nature of indicators and their crucial role in theory testing. (The earliest such study is found in Bunge 1967a.)

A recent, kindred, and more popular view is that scientific research boils down to making inscriptions, chatting, horse-trading, intriguing, and fighting (Latour and Woolgar 1986). This quaint opinion seems to have been motivated by both a passion for symbolism and a fear of the scientific method. Anyway, it denies the very rationale of scientific research, namely, the search for truth. Hence, it cannot even distinguish measuring or calculating, on the one hand, from exchanging gossip in the corridor or chatting over the net, on the other.

Both reductionisms are versions of what may be called *linguistic imperialism* – the former allied to positivism and the latter to hermeneutics. Neither of them will do, because what are at issue in science are facts and ideas, not just symbols – unless of course one perpetrates philosophical hermeneutics, according to which social facts are 'texts or like texts.' For example, some hermeneuticists view nations as narrations, and political science as discourse analysis (see Bhabha 1990). But of course nations, unlike narrations, have territories and material resources, they are peopled and have institutions, and so on. Besides, if politics were discourse, what would political discourse be, and how should it be studied: as literary criticism? (More on hermeneutics in chapter 13.)

As for the symbols that occur in science and technology, they make sense only in so far as they represent ideas, things, or processes. This is why mathematical formalisms acquire a non-mathematical content or meaning only when supplemented with semantic assumptions, such as '$P(t)$ represents the population of the given territory at time t.' The ultimate aim of all the basic sciences, except for linguistics and archaeology, is not to discover the meanings of any signs, but to find out how things work, that is, to unveil their mechanisms. In this endeavour, signs are replaceable auxiliaries, since they stand for non-symbolic items.

The universal languages at once viable and useful are those of logic and mathematics. But of course the interpretation of their symbols will necessarily vary with the research field. Thus, the linear function may occur in economics to describe the increase in profit with productivity; and it may occur in criminology to describe the rise in crime rate with the consumption of alcohol and crack cocaine.

So much for three ambitious yet hopelessly naive reductionist illusions. Let us next examine a few other popular reductionist projects that failed: biologism, psychologism, politicism, economism, and culturalism.

4 Biologism I: Sociobiology

Biologism is of course biological reductionism. The earliest examples of biologism were racism and the thesis that all human types, from slave to criminal to genius, are born as such. For example, Aristotle believed that there are 'natural slaves' and 'natural free men.' Social Darwinism, popular around 1900, is a modern heir to that doctrine. And the old but still popular phrase 'blood tells' speaks a library. A famous defender of this thesis was the Italian anthropologist Cesare Lombroso, who held that criminals are born, not made, and that criminality can be read in the face: narrow forehead, asymmetric face, salient ears, crooked nose, and so on. This is how murderers are still often depicted in cartoons. So ingrained was biologism in the culture of the late nineteenth century, that no one asked Lombroso for empirical evidence, and he was honoured by many a learned society.

The prime contemporary examples of biological reductionism are human sociobiology and its offshoot, evolutionary psychology. Sociobiology has made impressive inroads in social studies since its inception in the mid-1960s. The reason for this expansion is that it claims to explain everything social in terms of the most successful of all biological theories, namely the neo-Darwinian theory of evolution, a synthesis of Darwinism with genetics. Moreover, in Wilson's own words (1975), sociobiology was 'the new synthesis': that of biology and sociology. However, it actually attempted two reductions at once: those of social science to biology, and of biology to genetics – as a consequence of which all social facts would ultimately be explainable in terms of DNA (see Van der Dennen, Smillie, and Wilson 1999).

In particular, sociobiology endeavored to solve what is usually taken to be the main theoretical challenge to evolutionary biology: How could unselfishness, which 'evidently' reduces individual fitness, evolve by natural selection? There have been various sociobiological answers to this question, the most popular of which seems to be the hypothesis of kin selection (Hamilton 1964; Sober and Wilson 1998), which holds that helping others occurs mainly or even exclusively among relatives, because the helpers compromise their fertilizing ability and thus depress their Darwinian fitness (progeny size). This hypothesis is open to the following objections.

Sociobiologists assume that biology can explain altruism, real or apparent, in all animal species, from bee to man. This universality claim is at variance with the well-known fact that human evolution is social, hence partly artificial, as well as biological. In particular, customs and the legal and moral norms are made and remade, not inborn. Moreover, not all of them promote survival: think of the norms that have favoured warriors, parasites, tyrants, or obscurantists.

And how could any unfair norms ever be challenged if they were locked in our genomes? How could genetics explain the relatively recent abolition of slavery and the death penalty, or the shift from the idea of justice as revenge to that of justice as equity? Socio-genomics may work for eusocial insects, but not for humans, because human society, unlike the anthill, is artificial, not natural.

Besides, there is no empirical evidence for the hypothesis in question, advanced speculatively twenty-five years before DNA fingerprinting made it possible to ascertain kinship in non-human animals. It is only thanks to techniques introduced in 1989 that it has recently been ascertained that in several species – our own included – there is cooperation between non-relatives along with competition between relatives (see, e.g., Cockburn 1998; Clutton-Brock 2002; West, Pen, and Griffin 2002). Cooperativeness is particularly noticeable among the great apes, perhaps because these, unlike monkeys, are capable of empathy (Waal 1996).

Furthermore, recent brain imaging studies (Rilling et al. 2002) on people playing Prisoner's Dilemma games have shown that we feel good when behaving cooperatively toward strangers. Even the exact brain areas that carry this feeling have been pinpointed. This is not to say that reciprocal altruism is in our genes: social psychologists have found that moral feelings develop along with social life (Moessinger 1988). We also need company to learn to speak and work and play.

In all sustainable human societies, cooperation is rewarded in the long run – at least in the form of public recognition – whereas selfishness is not. Indeed, we despise free riders and punish them in some way or other, even if the person administering the punishment runs some risk – as recent experiments have shown (Fehr and Gächter 2000). One may behave aggressively towards people in out-groups but, to retain one's good standing in one's own social circles, a modicum of in-group tolerance and solidarity is required. Given that humans are helpless at birth, as well as sociable from birth, it should have come as no surprise to find that they are naturally cooperative in some respects while competitive in others. Hence, the real problem is not so much to explain altruism as to explain consistent selfishness. The only known tribe of utterly selfish people, the Ik of northern Uganda, became nearly extinct because they were unable to coexist (Turnbull 1972).

Third, human altruism is a subject for psychology and sociology, not armchair evolutionary biology. As indicated above, a good way to investigate it in an objective manner is to observe the brains of people engaged in transactions involving the possibility of either cooperation or defection. Another is to study how the destitute manage to survive. Lomnitz (1977) has done this, showing that the dwellers of Mexican shanty towns survive by practicing reciprocal altruism (quid pro quo). A third is to investigate the voluntary associations (NGOs) that people support mainly out of altruism – even if they claim to do it out of 'enlightened egoism.' Such socio-psychological study shows remarkable differences among societies, depending on the strength of non-biological factors, such as level of economic development, political organization, and religious heterogeneity (Curtis, Baer, and Grab 2001). Genetics cannot explain such regional differences in altruism. In particular, whereas distrust of otherness may be instinctive, racism is an ideological contraption with economic and political roots and uses (see Fredrickson 2002).

A fourth point: Because of their trust in adaptation, sociobiologists tend to exaggerate social harmony at the expense of social conflict. Indeed, they claim that social norms, far from being social inventions – and as such pernicious nearly as often as beneficial – have emerged spontaneously 'to make human groups function as adaptive units' (Sober and Wilson 1998: 173). This view, first proposed by Hume, and later endorsed by Hayek and other conservative ideologists, presupposes that all human groups (systems) grow naturally and are adaptive. But surely most human systems and institutions are social inventions, if not always carefully designed ones. Witness governments, armies, schools, hospitals, churches, clubs, professional associations, and corporations.

What holds for social systems holds also for social values and norms: these, too, are social constructions rather than biological goals or constraints. The proof is that in every social system, from family to business to scientific community to religious congregation to government, mutually contradictory values and norms develop (see Merton 1976). For example, honesty is the best business policy – as long as it does not cost too much. Voting should be free – but no effort or trick should be spared to win elections. Scientists should share their findings – but not before staking their priority claims. The flexibility of social norms is in sharp contrast with the inflexibility of biological laws.

Moreover, some social norms hinder rather than favour the satisfaction of biological needs. Just think of asceticism and the legal obstacles to family planning, or of gender and ethnic discriminations, social parasitism and the caste system, ruinous business competition and equally ruinous corporate mergers, corporate and political corruption, war and colonialism, and further examples of maladaptive or divisive social behaviour. Or should we believe that social injustice and the concomitant socio-economic inequalities, which are increasing in today's world, are the inevitable and even adaptive result of natural selection?

5 Biologism II: Evolutionary Psychology

Charles Darwin was the founder of evolutionary psychology. He realized that his own theory of evolution, jointly with the psychoneural identity hypothesis, implies the conjecture that behaviour, emotion, and ideation evolve alongside anatomical and physiological traits (see Bunge 1979b). This is why Karl Lashley asserted that '[t]he evolution of mind is the evolution of nervous mechanisms' (1949: 32). However, nervous tissue does not fosilize, so that animal psychologists are restricted to doing comparative psychology, a science pioneered by Darwin (1871) and founded formally in 1894 by Conwy Lloyd Morgan – who, incidentally, was one of the earliest emergentists (1923).

Comparative psychology has produced many anecdotes and speculations, and a small but important and growing collection of rather solid scientific findings. One of these is that emotion is likely to have emerged with reptiles about 200 million years ago (Cabanac 1999). This conjecture is based on the finding that the temperature of modern lizards and mammals, but not of frogs and fish, rises when they are handled gently. (Caution: This and other physiological indicators, like increased heartbeat, are coarse and ambiguous. Indeed, they may indicate pleasure, fear, or neutral arousal.)

In the case of our remote human ancestors, one can make educated guesses about their mental life on the basis of the utensils, tools, weapons, paintings, hearths, debris, and other artefacts they left. This is what cognitive archaeology is all about (see Donald 1991; Renfrew and Zubrow 1994; Trigger 2003).

All that is hard and tentative work. A much easier and therefore far more popular approach to evolutionary problems is to speculate freely in the manner of Leda Cosmides, John Tooby, David M. Buss, and others, with the applause of such media stars as Richard Dawkins, Steven Pinker, and Daniel Dennett (see, e.g., Cosmides and Tooby 1987; Barkow, Cosmides, and Tooby 1992; Buss et al. 1998). This brand of evolutionary psychology, an offspring of human sociobiology, combines six hypotheses: reproductionism, computationism, nativism, adaptationism, cognition-emotion disconnection, and the 'computational theory of social exchange.' I shall argue that all these assumptions are false.

I call 'reproductionism' the thesis that reproduction takes precedence over all other biological functions. This thesis is suggested by the definition of natural selection as differential reproductive success. (This fancy technical concept does not replace the classical notion of selection as preserving-and-culling: see Gould 2002: 659.) I submit that, although sex is a powerful drive, self-preservation is even stronger among vertebrates. Indeed, ethologists know that a hunted or hungry animal gives priority to security or food respectively over the search for a mate. And historians know that wars are motivated by the wish to keep or steal natural resources, markets, or trade routes rather than sex partners. This is why primatology is no substitute for social science.

Paradoxically, the second hypothesis, namely that the brain (or the mind) is a computer, is non-biological. Indeed, it denies the relevance of the material 'substrate' of the mind – hence that of neuroscience as well. Besides, this assumption ignores all of the non-algorithmic mental processes, such as altering mood, feeling and emoting, evaluating and criticizing, finding problems and forming new concepts, unearthing presuppositions and taking initiatives. None of these processes is programmable, because none of them is predictable.

The third hypothesis of evolutionary psychology is the innateness of the cognitive programs or algorithms that allegedly drive behaviour. No parent or teacher can accept this assumption. Indeed, it is common knowledge that sphincter control, walking, swimming, and talking are learned – not to mention tool-making, hunting, farming, getting along with other people, observing (and challenging) social norms, courting, worshipping, arguing, planning, politicking, counting, saving, and using algorithms. Besides, even if our entire subjective (mental) life were fully programmed, surely some programs would allow for internally generated processes, rather than conforming to the behaviourist stimulus-response dogma. Still, Cosmides and Tooby (1987) espouse precisely this dogma when asserting that their putative programs 'map informational input into behavioral output.'

The fourth assumption of evolutionary psychology is that such cognitive programs were 'designed' (preserved and improved on by natural selection) for survival: they would be adaptations. That is, they would be perfect or nearly so: functional, efficient, reliable, precise, and economic (Williams 1966). Furthermore, those programs were allegedly designed to cope with 'the Pleistocene environment.' All of this is neat but fantastic. To begin with, we do not know precisely what that environment was like – except that it cannot have been uniform throughout a period that extended from circa 1,600,000 to 10,000 years BP, and included huge climatic oscillations as well as the reversal of the magnetic poles. Moreover, it is not even known exactly where and when *Homo erectus* emerged; in particular, it is not known whether its members radiated from East Africa or, on the contrary, travelled from Eurasia to Africa (Asfaw et al. 2002). Besides, it is hard to understand how rapid progress could have occurred over the past 100,000 years or so, if humans were programmed to survive in the past rather than having the potential to cope with the present and craft their own future.

Darwin, for one, knew that the key to survival is not adaptation but adaptability, or the ability to change in response to environmental changes. Psychologists, anthropologists, and historians have confirmed this insight: they tell us that the trademark of humans is not programmed rigidity but versatility and creativity. We are encyclopedists rather than specialists, and adapters no less than adaptable. We are also often spontaneous, improvisers and creative rather than stimulus-bound patients. Moreover, improvisation – the ability to alter the genetic 'project' to face environmental emergencies, as in the case of acquired immunity – may be a feature of all organisms (Koshland 2002).

Besides, human creativity opposes both hard-wired behaviour and blind natural selection. Furthermore, some inventions, such as housing, sanitation, vaccination, life insurance, cooperatives, and the welfare state, have helped us opt out of natural selection. Social evolution can thus occasionally override biological evolution. Finally, as Stephen Jay Gould and Richard Lewontin (1979) noted in their famous puncturing of adaptationism, only some inheritable features are adaptations. Others are 'exaptations,' that is, features co-opted opportunistically to perform fitness-enhancing functions other than the original ones; and still others are 'spandrels,' that is, features with no original adaptive functions.

It is too hard and too early to tell which among the current features of the human mind are adaptations, which exaptations, which spandrels – and which shortcomings. For instance, are forgetfulness and absent-midedness liabilities, assets, or either in different circumstances? (See Schacter 2001). In any case, all this is a matter for scientific research, not journalistic fantasizing. And the research in question cannot be exclusively biological, because humans are subject to artificial (social) selection, which interferes (now destructively, now constructively) with natural selection. For example, olfactory acuity is adaptive in a rural environment but an encumbrance in an urban one; and meekness is advantageous under a dictatorship but not in a democratic society.

The fifth hypothesis, that cognition is detachable from emotion, is required to support the second one, that all mental processes are computations. Damasio (1994) has called it 'Descartes's error,' for it is well known that the organ of cognition (the neocortex) is anatomically connected to that of emotion (the limbic system), and that such connection is two-way (e.g., Barbas 1995). Moreover, the paths from the latter to the former are more numerous than those in the opposite direction. This explains why emotion can now energize cognition, now block it. The same anatomical fact also helps explain why social scientists must go beyond rational-choice theory if they wish to explain how social networks emerge, how marketing experts alter our natural preferences, how stock brokers take advantage of fear and greed, or how politicians transmute apprehensions and hopes into votes (see Massey 2002).

The sixth hypothesis of evolutionary psychology, namely the 'computational theory of social exchange,' states that we are born with an algorithm for assessing the costs and benefits of our actions. This would be a general-purpose (or item-independent) survival kit crafted by natural selection. More precisely, the costs and benefits would be perceived rather than objective. This qualification is reasonable in view of the so-called Thomas's 'theorem,' according to which we do not react so much to stimuli as to the way we perceive them (Merton 1968). Yet our social perceptions are often wrong, as when most Germans supported Nazism believing that it would bring them prosperity and glory rather than misery and shame. And wrong social perceptions are likely to lead to unsuccessful actions, even fitness-reducing ones. Only correct social perceptions, and the corresponding objective evaluations of costs and benefits, can enhance survivorship and Darwinian fitness. So, even if we were endowed from birth with a cost-benefit estimating algorithm (an implausible assumption), it would be unrelated to evolution because it would only supply subjective costs and benefits (Lloyd 1999: 227).

So much for the six general hypotheses of evolutionary psychology. What about the evidence for its specific conjectures? At first sight, the empirical evidence for these fantasies is overwhelming, since they appear to explain every possible bit of observed behaviour, from selfishness to altruism, from mate search to religious worship. However, the quality of such empirical evidence is dubious, if only because the alleged cognitive programs are not specified. To state that there is an algorithm for behaviour X is about as precise, informative, and testable as saying that X is destiny. Second, and equally important: algorithms proper are explicit and precise rules that can only be crafted and utilized by experts. To attribute to babies the possession of inherited algorithms is as plausible as attributing to them the ability to identify grammars or solve equations.

Evolutionary psychiatry does not fare better, for it postulates that all mental disorders are adaptations, that is, features that favour individual survival (Nesse and Williams 1994). Thus, depressives and schizophrenics, no less than paranoids and autists, should count themselves lucky, since their suffering is for their own good or that of their progeny. There are at least four objections to this postulate. One is that it contradicts all that is known in clinical psychiatry: mental disorders are severely maladaptive, since they prevent their victims from functioning normally, bring unhappiness to them and their entourage, and decrease their fertility. A second objection to the postulate in question is that, contrary to evolutionary biology, it assumes that adaptation is the driving force of evolution rather than a dicey outcome of it. The third is that 'its authors do not ask *whether* every disease has evolutionary causes, but *assume* this in order to explain all diseases in such terms' (Dubrovsky 2002: 9). If disease is every bit as adaptive as health, should we not jettison the Hippocratic oath altogether? Further, how do we account for the emergence of health care and medicine, and why do we care for them? And can such tacit rejection of evolutionary medicine be explained in evolutionary terms?

Unsurprisingly, the attempts of evolutionary psychologists to explain the emergence of behaviour patterns and social systems have been unsuccessful. For example, most of them claim – along with their sociobiological predecessors – that human males are naturally promiscuous (or even polygynous) because they wish to maximize the number of their children. While it is true that nearly all primate males are promiscuous, it is false that they are so for that reason. In fact, promiscuous men typically do not care for the welfare of their progeny: they just want to maximize their own sexual pleasure and social prestige without regard to the consequences for their progeny.

The failure of evolutionary psychology to explain the emergence of social systems and social mores and norms is even more dismal. For instance, it does not even attempt to explain the origin of private property, the state, feudalism, capitalism, organized religion, voluntary associations, or political movements. It does not account for the variability of institutions, or even for the variety of sexual mores and kinship systems. It could not do so because it ignores (a) the behavioural plasticity that results from the conjunction of neural plasticity with environmental changes, and (b) the traditions and social circumstances that prompt or prevent social changes.

True, not all evolutionary psychologists are equally doctrinaire. For example, a few of them own that social constraints and particular circumstances play a role in the emergence of social norms such as marriage rules. Thus, Kanazawa and Still (2001) admit that the marriage norms in a society depend critically upon the distribution of economic resources among men: whereas economic inequality favours polygyny, equality correlates positively with monogamy. Still, those authors also assume that women are free to choose their mates, and that their choice is only guided by economic considerations: they would always prefer the wealthiest men, for they are expected to maximize their offspring's welfare.

In short, according to Kanazawa and Still, marriage norms would be the spontaneous outcome of free choices made by women driven solely by their desire to secure good providers for their children. But this assumption flies in the face of the well-known facts that in most societies few women enjoy such freedom; that the life of a concubine slaving under a first-wife is far from enviable; and that a debt-ridden farmer cannot afford to pay dowries for his daughters, but may feel compelled to sell them to his landowner or to a brothel. And in more advanced societies, where women do have a say in mate choice, social equivalence, congeniality, sexual attraction, and love are likely to weigh more than concern for child welfare – especially in modern societies, where a large and increasing percentage of women remain single, childless, or both.

To conclude, evolutionary psychology in its present state is not a science but at best an emerging science and at worst a piece of science fiction. It is also a challenge to formulate an authentically scientific research project – next time one that is compatible with biology, cognitive neuroscience, anthropology, and sociology, as well as testable, at least in principle, against the archaeological record and sociological research. Future evolutionary psychologists will also have to learn from social psychologists the ways children acquire moral norms, such as that of reciprocity, in the home, on the playground, or at school. And they will have to remember that, as far back as in 1893, Thomas Henry Huxley wrote that 'the ethical progress of society depends, not on imitating the cosmic [evolutionary] process, still less in running away from it, but in combating it' (Huxley and Huxley 1947: 82).

So much for biologism. Let us now examine its next of kin.

6 Psychologism

Psychologism is of course the thesis that everything social is ultimately psychological, and hence all the social sciences are in principle reducible to psychology. This thesis was advanced by scholars otherwise as different as the empiricist John Stuart Mill (1952), the idealist Wilhelm Dilthey (1959), and the behaviourist George C. Homans (1974) among others. According to them, every social fact is the outcome of individual actions steered by the actor's beliefs, values, goals, and intentions. In this perspective neither nature nor the social environment would play any role except in constraining individual agency: all individuals would basically be free agents pursuing their private interests.

The simplicity and apparent unifying power of the psychologist project makes it very attractive at first sight. But the results of this project are meagre. For one thing, they all concern individual behaviour, which that variety of psychological reductionism purports to explain in terms of a single principle, that of maximizing expected utilities. None of them concerns such macrosocial facts as wealth concentration, unemployment, business cycles, environmental degradation, and international conflict, which affect us all and cannot be explained exclusively as outcomes of individual choices. On the contrary, these and other macrosocial features explain much individual behaviour. For instance, people tend to consume less during economic slumps and war; the unemployed are far more likely to break the law than the employed; and most people adopt the values and beliefs of the ruling class.

Take, for instance, the formation of attitudes towards others, such as trust, cooperativeness, conformism, and their duals. Although attitudes are psychological features, they do not emerge in a social vacuum, but are shaped by social structure. This could not be otherwise because, far from being intrinsic properties (conceptualized as unary attributes), social attitudes are relational. Indeed, a statement about trust is of the form 'A trusts B with respect to [or to do] C' (Coleman 1990; Cook and Hardin 2001). Moreover, trust emerges (and submerges) over time, in the course of repeated interactions. This development is part of the process of social learning. In short, unlike visual acuity, musical ability, and depression, trust is not an individual trait and thus a subject for individual psychology. Trust can only be studied adequately by social psychology, the fusion of two sciences.

A related fusion relevant to our discussion is that of social psychology with neuroscience, to constitute social cognitive neuroscience. This discipline, which bridges the brain to society, investigates in depth such common social-psychological phenomena as attitudes and the perception of other people's behaviour (see, e.g., Cacioppo and Petty 1983; Ochsner and Lieberman 2001). However, let us go back to psychologism.

Psychoanalysis is of course the most popular and amusing version of psychologism. Freud held, in particular, that social behaviour is determined early in life by toilet training and love of parent. However, in the course of one century, psychoanalysts have not produced a single piece of experimental evidence for their fantasies. They thrive in private practice and in the popular press, not in the laboratory. Indeed, only once in the course of a whole century have psychoanalysts attempted to perform an experiment (Vaughan et al. 2000). However, this was not an experiment proper, because it did not involve a control group. By contrast, scientific psychologists have produced plenty of evidence against those psychoanalytic fantasies that are testable at all (see, e.g., Crews 1998; Torrey 1992; Wolf 1995). One of the latest casualties is the catharsis hypothesis: recent experiments have confirmed, once again and to the chagrin of the television industry, that viewing TV violence induces aggressive behaviour (Johnson et al. 2002).

Psychologism was also the rationale for the name 'behavioural science' given psychology and the social sciences between about 1930 and 1970. This denomination has all but disappeared, along with the behaviourist psychology (or rather non-psychology) that inspired it. The current eclipse of the name is likely to be due to the barrenness of the psychologist project. For instance, hardly anyone believes nowadays that infancy is destiny, let alone that wars are caused by the mythical Oedipus complex or by the desire to seduce, rape, or steal as many women as possible. We have learned about class wars, ideological wars, and wars over land, oil, diamonds, water, markets, or political constituencies. Psychologism has only survived in the rational-choice school, of which more in the next section.

In short, to paraphrase what the biophysicist A.V. Hill (1956) wrote about the physicists of his day who jumped on the biology bandwagon without first bothering to learn about the ways of organisms: A student of society who regards social science as simply a branch of biology or of psychology has no important future in social science. Whenever a process crosses two or more levels of organization, its study must involve several levels of analysis, neither of which is necessarily more important than the others.

7 Sociologism, Economism, Politicism, and Culturalism

We may call 'sociologism' the thesis that everything human is to be explained in sociological terms. Vygotsky (1978) was an eminent member of this school. He deliberately set out to 'liberate psychology from its biological prison,' making symbols and actions the centres of psychological investigation for being 'the very essence of complex human behaviour.' In correcting one imbalance – the neglect of social stimuli – he reinforced another – the neglect of that which makes symbols and controls action, namely the brain. However, he came close to the mark when stating that 'the most significant moment in the course of intellectual development, which gives birth to the purely human forms of practical and abstract intelligence, occurs when speech and practical activity, two previously completely independent lines of development, converge' (1978: 24). Vygotsky's valid contributions to social psychology should therefore be distinguished from his metatheoretical manifesto.

Economism, self-styled 'economic imperialism,' is currently the most popular of the reductionist projects among the students of society, particularly those of a rationalist mindset. It comes in two varieties: collectivist (or holist) and individualist. The prime example of collectivist economism is Marxism, which focuses on impersonal economic forces that generate class struggles. Marxist economism is macro-reductionist, or of the top-down type: that is, it attempts to explain the part by the whole, which in turn remains unexplained. For example, it accounts for political sympathy in terms of one's position in the class structure; for incarceration as disposal of the labour surplus; and for technological innovation in terms of market demand.

By contrast, rational-choice theory, as practised for instance by Gary S. Becker, Mancur Olson, Thomas C. Schelling, and James S. Coleman, is micro-reductionist or of the bottom-up kind: it attempts to explain the whole by the parts. Indeed, it focuses on individual choice and action guided exclusively by self-interest, though constrained by institutions. In either version, economism holds that all social facts are ultimately economic – that people always act to advance their own interests. (See, e.g., Becker 1976; Hogarth and Reder 1987; Moser 1990; and the journal *Rationality and Society*.)

Individualist economism, that is, rational-choice theory, is currently in vogue, presumably because it looks scientific in addition to purporting to explain much with little, thus producing the illusion that it unifies all the social sciences around a single postulate. However, it can be shown that rational-choice theory is conceptually fuzzy, empirically groundless, or both (Bunge 1996, 1998, 1999, and chapter 13). Indeed, when the utility functions in a rational-choice model are not specified, as is generally the case, untestability is added to vagueness. And when they are mathematically well defined, the choice among them is arbitrary, since no experimental data support any of them. Besides, experimental economics has shown that we are risk-averse rather than utility-maximizers (Kahneman, Slovic, and Tversky 1982).

While it is true that it would be foolish to underrate economic interests or rational calculation, it is also true that economism embraces so much, that it explains nothing in particular. For instance, it does not explain why some people stay single whereas others marry; why most people prefer owning houses and cars instead of doing the rational thing, which is renting them; why the famous central banker Alan Greenspan said that the stock market around the year 2000 was characterized by 'irrational exuberance'; why globalization embraces only merchandises and benefits mainly the powerful; why income inequalities continue to grow world-wide despite spectacular gains in productivity; why economic prosperity is often accompanied by high unemployment; or why religious fundamentalism continues to grow nearly everywhere in what is supposed to be the age of science and technology. Surely more than rationality and economics, whether mainstream or heterodox, is needed to explain these and other economic pathologies, as Corrado Gini (1952) called them.

Let us turn next to a neighbour and rival of economism, namely politicism. This is the view, popular in the late 1960s, that all human endeavours are political or at least coloured by politics. An early proponent of this view was Antonio Gramsci, the neo-Marxist who reacted against Marx's economic determinism. He wrote about the hegemonic power of the state, denied the existence of politically neutral subjects, and emphasized the need for political action. (Ironically, his fascists jailers concurred.) According to the French structuralist Michel Foucault, even art would be cultivated for the sake of power. The sociologist of science Bruno Latour has claimed that scientific research is 'the continuation of politics by other means.' The clue to understanding any cultural activity would be, *Cherchez le pouvoir*. And, of course, politicism is at the core of academic feminism, in particular feminist philosophy. According to its radical wing, all sex would constitute rape, logic would be 'phallocentric,' and Newton's equations of motion would constitute a 'rape manual,' as Harding (1986) claimed.

There is no shred of evidence for this radical view. Moreover, it makes a mockery of all the non-political activities, from earning one's livelihood to the search for truth and the creation and enjoyment of beauty. And it involves an egregious ignorance of family life, farming, education, volunteer work, stargazing, theorem-proving, net surfing, sports, and other activities where the struggle for power plays at best a subordinate role. Which is not to deny that social life has a political side, since it involves competition for scarce resources, from good jobs and money to love and political allegiance. So much for politicism.

Let us finally examine culturalism, particularly as practised by the internalist historians of ideas, and as propounded by the idealist philosopher Wilhelm Dilthey (1959), the French semioticians, and the anthropologist Clifford Geertz (1973). According to culturalism (or idealism or hermeneuticism), all social facts, or at least their ultimate sources, are cultural or symbolic and, more particularly, spiritual. In other words, social facts would be 'texts or like texts' calling for 'interpretation' rather than explanation. Ideologies, rites, and ceremonies would be paramount, whereas natural resources, work, and strife would be incidental. It would not matter how people make a living: whether by farming or manufacturing, organizing or teaching, stealing or oppressing. Nor would love, hatred, and fear count. Only ideas – particularly myths – and the ways they are expressed and transmitted would matter. For example, a culturalist student of racism will focus on racist ideologies, eschewing any consideration of racist practices, such as the Jim Crow laws in the southern United States, and the Jewish pogroms. Consequently, he will fail to understand why racism has been so widely practised despite being blatantly false and immoral – namely, because it is as economically profitable as gender and class discrimination.

The methodological consequence of culturalism is obvious: The social studies would be *Geisteswissenschaften* (sciences of the spirit) or, in French, *sciences morales*. It would therefore be misguided to try and render them scientific – an attempt that Hilary Putnam once called 'barbarous.' I submit that this view, which stems from Kant, has stunted the development of the social studies as well as their use in tackling effectively social issues. This is not to deny that culturalism has a grain of truth. On the contrary, every important social fact has a cultural component – as will be argued in the next chapter. But because it ignores biological drives, as well as production and the struggle for power, culturalism cannot explain such conspicuous macrosocial facts as demographic shifts, mass migrations, economic cycles, social revolutions, or the use of morals and religions as tools of social control. (See Trigger 2003 for detailed criticisms.)

I submit that all of the above reductionist strategies have failed to solve any interesting social problems. Unsurprisingly, they have not helped design viable and morally admissible social policies. They have equally failed in unifying the social sciences. Indeed, every one of those imperialisms – biological, psychological, linguistic, economic, etc. – is without an empire. They are at best failed research projects, and at worst ideologies. (See further criticisms in Kincaid, 1997.) It is therefore necessary to look for a viable alternative, one free from the simple myth that social life and social change have a single source.

Concluding Remarks

Reduction is a strategy for coping with the bewildering diversity of reality and the concomitant diversity of the sciences of reality. Yet, for better or for worse, reduction has failed more often than it has triumphed, largely because it has denied emergence. Let us therefore get into the more modest but more rewarding business of finding out what is common to the various social sciences in addition to logic and the scientific method. It will turn out that their integration is more fertile than any attempted reduction.

11

Why Integration Succeeds in Social Studies

The social studies are notoriously fragmented. For example, the typical economist does not listen to demographers; political scientists are rarely interested in cultural studies; and most students in the field of cultural studies pay no attention to economics. Worse, every discipline is divided into equally isolated subdisciplines. For example, educational sociology is usually pursued independently from economics and politology; and the study of social inequality, gender discrimination, and racism are ordinarily disjunct from political science and the sociology of religion. I submit that such fragmentation is artificial and an obstacle to the advancement of knowledge.

Such fragmentation is artificial because all of the students of society are expected to describe and explain social facts, and every social fact is likely to have multiple aspects – biological, economic, political, and cultural. For example, where land is scarce, population growth worsens such scarcity; and this event is in turn likely to trigger violence, with its biological, political, and cultural concomitants. Given the multifaceted nature of social events, the interdisciplinary barriers would seem to stem at best from differences in emphasis, at worst from tunnel vision or turf protection.

The frontiers in question are not only artificial. They are also deplorable, because they split systemic problems, such as those of the excessive concentration of wealth and power; they also block the flow of ideas, data, and methods that could be used in more than one discipline. For example, they discourage the investigation of socio-economic features such as income distribution; of biosocial ties such as the association between morbidity and income; and of economico-politico-cultural ones such as the business-politics, and religion-politics connections. After all, as Braudel said (1969: 85), all of the social sciences are interested in the same subject: the past, present, and future actions of people.

The policy-makers, legislators, and public servants who overlook such ties among different aspects of social life are unlikely to help solve any sizeable social issues. For example, one of the main causes of underdevelopment is extreme concentration of economic and political power; a deficient healthcare system maintains high morbidity, which is detrimental to both learning and productivity; and both religious fundamentalism and terrorism are bound to flourish in economically depressed and politically oppressed regions. Given the many causes of underdevelopment, any sectoral approach to this problem is bound to fail. To generalize: Fragmentation leads to theoretical shallowness, which in turn hampers social progress.

If the fragmentation of the social sciences and technologies is both artificial and harmful, it should be overcome. But how? That is, how can the social sciences be unified without loss of depth, diversity, and rigour? Reduction cannot be the answer because it has been tried without success. Indeed, in the previous chapter we saw that the various reductionist strategies that have been tried in the social studies – in particular biologism, psychologism, economism, politicism, and culturalism – have failed. If this contention is true, we must find out why. And the answer to this question should suggest an alternative strategy, as indeed it does. The alternative in question is cross-disciplinarity.

True, cross-disciplinarity has been recommended for several decades (see, e.g., Handy and Kurtz 1964; Sherif and Sherif 1969; Bunge 1983b; and Boudon 2003); so much so that the grant application forms of many science-funding agencies have a slot for cross-disciplinary relevance. However, genuine cross-disciplinary research is still scarce in the social sciences. It is not even clear what 'cross-disciplinarity' means, or how it differs from dilettantism. Therefore, there may still be room for one more plea for the two variants of cross-disciplinarity – namely multidisciplinarity and interdisciplinarity – in social studies.

The thesis to be suggested in this chapter is, in a nutshell, that social studies ought to be cross-disciplinary because all social facts, particularly if macrosocial, are multidimensional. More precisely, these facts have at once biopsychological, economic, political, and cultural aspects as well as environmental causes and effects. If this is true, then the right research strategy is integration or cross-disciplinarity rather than reduction. To put it in metaphorical terms: To explain a social fact we must look not only underneath and above it, but also around it. And such contextualization requires the intervention of additional disciplines. Shorter: Emergence calls for convergence.

1 The B-E-P-C Square

I take it for granted that the social sciences proper study social facts rather than individual ones. However, I dispute Durkheim's thesis that every social fact originates in another social fact. And I object equally strongly to radical individualism, which either denies the existence of social facts, or claims that every social fact originates in some individual decision, rather than in interactions among individuals. Even such a radical individualist as Homans (1974) stressed that interaction is the hub of social life. Think of group-hunting or trading, building a public facility or chatting, healing or teaching.

I submit, moreover, that every social fact has five different but closely linked aspects: environmental (N), biopsychological (B), economic (E), political (P), and cultural (C). I also suggest that a social change may originate in any of these sources, so that there is no single prime social mover, not even 'in the last analysis.' Graphically, the conjunction of these two theses looks like this:

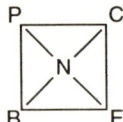

The edges in this diagram represent actions and fluxes of various kinds, from energy and information transfers to more or less subtle social actions and signals, from wink and poster to greeting and shoving. The same figure also suggests the multi-causation thesis that in society there is no first motor, since the N, B, E, P, and C factors can be ordered in $5! = 120$ different ways.

This view, then, is one of multiple and frequent reciprocal causation: at any one time, every person and every social system is the recipient and the effector of a large number of stimuli of many different types and intensities. Often, however, when studied synchronically, one of the aspects may be more salient than the others, in which case these can be legitimately neglected to a first approximation, so that unidisciplinarity may succeed. However, when attention shifts from point events to long-term processes, it is often seen that variables of all kinds are relevant and take turns (Braudel 1969). When this happens, unidisciplinarity fails, and cross-disciplinarity is called for.

This view is illustrated by the following sample out of the said 120 possible permutations.

Example of $N \rightarrow E \rightarrow B \rightarrow P \rightarrow C$. It is likely that the Sumerian and Maya civilizations decayed as a result of severe and prolonged droughts that desiccated the irrigation canals, which in turn ruined the agriculture, which caused famines, which may in turn have caused political unrest and cultural decadence.

Example of $B \rightarrow E \rightarrow P \rightarrow C$. A lethal plague (B) concentrates wealth among a few families (E), which enhances their political power (P) and puts them at the head of a new cultural movement (C). This was a major mechanism of the Florentine Renaissance – which is not to deny that it was the work of a few hundred exceptionally able intellectuals, artists, craftsmen, merchants, and politicians. The natural environment (N) was affected by the urbanization caused by the prosperity of artisans and merchants, as well as by the increase in agriculture stimulated by the population increase.

Example of $E \to B \to P \to C$. The first Industrial Revolution (E) increased the economic deprivation of craftsmen and labourers, which in turn decreased their biological well-being (B), which radicalized their politics (P), a fact that influenced the culture (C). The environment (N) was affected by both industrial pollution and the expansion of the regions devoted to grazing for the production of wool for the textile mills.

Example of $P \to B \to E \to C$. A government's neo-liberal policy (P) causes the deterioration of public health (B), which decreases the productivity of the workforce (E), which in turn depresses public school attendance (C). The environment (N) is affected by the decay of inner cities, the expansion of shanty towns, and the repeal of environment-protection legislation.

Example of $C \to B \to E \to P$. A vigorous medical campaign (C) improves the health of a large group (B), which has a favourable economic impact (E), and strengthens the group's political clout (P). The quality of air, water, and soil increase due to the improvements in sanitation and housing.

Obviously, in principle, cases of purely economic, purely political, or purely cultural facts might be cited. (I am presupposing that there are biological and psychological facts, such as cell division and smelling, lacking in any appreciable social inputs.) I invite the reader to try and find such counter-examples. I submit that my examples suggest that the view advanced above is at once plausible and suggestive, hence worthy of being investigated.

I also submit that the interdependence of the five aspects in question is the material basis of the cross-disciplinary approach to the study of social problems, as well as of the design of systemic (as opposed to sectoral) social policies. This approach comes in two variants: multidisciplinarity, or the addition of research fields, and interdisciplinarity, or their intersection. Let us look at a few examples of each. The logic, semantics, and methodology of cross-disciplinarity will be investigated in chapter 17.

2 Social Multidisciplinarity

An early example of multidisciplinary social science is Ibn Khaldûn's monumental *Muqaddimah* (1377?). This work was, in fact, a compendium of geography, sociology (rural and urban), politology, economics, and culturology. It seems that it had no parallel in the Western world until Alexis de Tocqueville's two main books (1835, 1856), Max Weber's *Wirtschaft und Gesellschaft* (1922), Fernand Braudel's *La Méditerranée* (1949), and Michael Mann's *Sources of Social Power* (1986, 1993). None of these authors emphasized the frontiers among the social sciences. However, they did not offer unifying principles either. (Weber proposed them in his methodological essays, only to forget them in his substantive work.)

As a matter of fact, all of the classical sociologists were trespassers, as Boudon (1998) has stressed. Not even Pareto, a champion of the strict separation of economics (allegedly the science of free rational behaviour) and sociology (the science of bound non-rational behaviour) stuck to this division. Indeed, he pioneered the mathematical modelling of the distribution of income, a typically socio-economic problem. The fact that his "law" is not true is beside the point. The point is that Pareto preached divergence while practising convergence. Moreover, the problem of income distribution is central to socio-economics, a merger of sociology and economics (see, e.g., Swedberg 1990 and the *Journal of Socio-Economics*).

Let us next consider briefly a few contemporary examples of multidisciplinarity: the studies of social norms, science, recent human evolution, national development, and the reconstruction of the ex-socialist nations. The emergence, maintenance, and violation of social norms has attracted the curiosity of social scientists, jurists, philosophers, and others. However, the editors of a recent anthology on the subject (Hechter and Opp 2001: xviii) admit that, '[f]or better or worse, the literature on social norms consists of a loose collection of mostly independent propositions that are scattered in the literatures of the various disciplines.' Not only is there no adequate theory of the emergence of such norms; there is not even a generally accepted definition of the concept in question. Indeed, whereas for some authors a social norm has a moral content, for others it only specifies what is customary or convenient to do. No wonder then that the field is multidisciplinary, hence potentially dispersive, rather than interdisciplinary, hence cohesive.

Our next example is the study of science, a subject for philosophy, anthropology, psychology, sociology, politology, and history. Although it is sometimes admitted that all of these studies complement one another, good multidisciplinary research in the sciences of science is rare. In particular, just as in the past internalism was the dominant approach, nowadays the social studies of science focus all too often on external circumstances and thus ignore the scientific problems, theories, and methods – that is, they overlook what motivates scientists (see Bunge 1999). Consequently, the externalists do not explain satisfactorily why scientists cooperate in some respects while competing in others.

Roughly, the philosophy of science is expected to tell us what science is in contradistinction to other branches of culture, such as technology, art, and ideology. (Incidentally, the postmodernist emphasis on text, discourse, metaphor, rhetoric, social convention, and power struggle is incapable of distinguishing science from non-science, in particular pseudo-science, religion, and politics.) The psychologists of science investigate what makes scientists run, whereas the anthropologists, sociologists, and politologists of science study scientific communities and exhibit the social stimuli and constraints on individual investigators. And of course the historians of science are expected to show how and why science has evolved the way it has and not otherwise – for instance, why logic and physics were born in ancient Greece rather than Sumer.

These various studies of science are not just complementary to one another: they feed and control one another. For example, the philosopher should regard the history of science not as a parallel discipline but as a purveyor of examples and counter-examples, and thus as a sort of laboratory to test philosophical hypotheses on science. Another good example is the study of the interplay between the two main motivations (or reward mechanisms) of basic research, namely disinterested curiosity and peer recognition. Robert K. Merton (1973) was the first to note that these two stimuli, the intrinsic and the extrinsic, are what make scientists tick: they investigate because they want to know, and they wish their work to be known and appreciated by their peers, or even by society as a whole. Merton was also the first scholar to cultivate at the same time the psychology, sociology, and history of science – and all three against a realist epistemological background.

Merton also noted that, although those two drives – curiosity and ambition – reinforce each other, occasionally one of them can get in the way of the other. For example, the pressure to 'publish or perish' leads untenured investigators to tackle only short-term projects. The practical reason is that, if they embark on ambitious long-term projects, they take large risks and may spend long periods without publishing, and therefore without receiving external support. And if they spend too much time procuring such support for their projects, they may end up as bureaucrats. In this case they may tend to do research by proxy rather than hands-on. As a consequence, they are likely to become rusty, and hence be forced to rely on graduate and postdoctoral students. Worse, excessive reliance on unsupervised graduate students and postdoctoral fellows may result in shoddy work or even fraud. (These perverse effects of the peer-recognition mechanism and the accompanying peer-review system have been the subject of Carl Djerassi's novels.) In short, to understand why scientists work, we must investigate their twin engines, the search for knowledge and the search for fame, and must show how these interfere with one another, now constructively, now destructively.

Another good instance of multidisciplinary research is the investigation of recent human evolution by Luca Cavalli-Sforza and his co-workers (1994). This team has tracked the migrations of groups possessing agricultural skills, from the Fertile Crescent to western Europe, over about ten millennia since the Neolithic revolution. To do this they combined genetics with archaeology, anthropology, and linguistics. Neither of these disciplines by itself would have succeeded in the attempt, precisely because biologism (in particular geneticism) is every bit as one-sided as culturalism (or sociologism). After all, before the invention of writing and the mail, skills and ideas could only travel in brains.

Our fourth case intended to show the need for multidisciplinarity is the problem of national development. Most governments of the underdeveloped ("developing") nations, as well as the International Monetary Fund, ask only for the opinion of economists in advanced countries about this problem. These alleged experts have elaborated the so-called Washington consensus, a single policy prescription for all countries regardless of their history, endowments, needs, aspirations, and possibilities. This alleged panacea consists in the following commandments: Privatize, embrace free trade, encourage foreign investment, tame inflation, and balance the budget by cutting social expenditures. Their only concerns are to come *in articulo mortis* to the rescue of inept governments, and to protect the interests of the lenders. Whether this recipe actually enhances welfare is not their concern: they are apriorists, not realists, hence dogmatic, not critical. Consequently, they do not take the blame for the failures of their pet projects, such as those in Indonesia and Argentina.

A number of studies have shown that the Washington-consensus policy tends to increase income inequality (e.g., Ocampo and Taylor 1998, Galbraith and Berner 2001) and poverty (Stiglitz 2001), as well as to weaken social cohesion (e.g., Deblock and Brunelle 2000). Further studies have shown that, in the long run, foreign investment has a negative effect on employment and other variables (e.g., Dixon and Boswell 1996). The recent cases of the late Soviet empire and Argentina (for a decade the star pupil of the IMF) are blatant counter-examples to the dogmas in question.

Moreover, that is not how the European countries, the USA, or Japan became modern nations. In fact, they developed at the same time their economies, polities, and cultures under protectionist governments. That is, they practised integral development – albeit in a tacit fashion. I suggest that the success of this development was due to the fact that, as noted elsewhere (Bunge 1979, 1997a [1980]), every moderately complex society may be regarded as a supersystem composed of one natural system, the biological one (which includes the kinship system), and three artificial systems: the economy, the polity, and the culture. These systems interact strongly with one another. For example, industrial development requires not only capital but also a healthy and literate workforce, ingenious technologists, as well as suitable and stable political and legal frameworks – a stability that is best attained through political democracy. An advancing society may thus be likened to a four-wheel-drive car.

In sum, sustained, though not necessarily quick, national development is integral rather than unilateral. Consequently, a suitable study of growth will recruit demographers, epidemiologists, sociologists, political scientists, and culturologists in addition to economists. Interestingly, Kuznets (1955) had advanced a similar thesis in his foundational if statistically baseless paper on the relation between economic growth and income inequality. Half a century later, Ignacy Sachs (2000) and James K. Galbraith (2001) took Kuznets's work seriously, and found that economic development depends not only on microeconomic (or market) variables, but also on such political variables as macroeconomic policies and the struggle for political power.

Nor are social scientists the only experts that the study and design of national development require. These problems should also be tackled by jurists and experts in public health and public education – as well as philosophers ready to remind the specialist that tunnel vision amounts to near-total blindness. Systemic social engineering may succeed where piecemeal social engineering has failed. Which should be obvious, since systemic problems – that is, issues where variables of many different kinds are entangled – call for systemic solutions. Even George Soros (1998), an individualist and erstwhile student of Popper's, concurs.

The theoretical and practical inadequacy of the unidisciplinary approach to development may be seen in such extreme late-twentieth-century tragedies as the Rwandan genocides in the 1990s and the birth pangs of the post-communist societies. The former may be understood as an instance of a C → B → E → P process strongly conditioned by a severe natural constraint (N), namely a small and mountainous territory. The C trigger was the idea that people should procreate freely – a gospel preached by the Catholic missionaries. This practice (B) helped overpopulation, which led to a severe shortage of land (E), which in turn caused a murderous political strife (P).

Our fifth and last example will be the transformation of the pseudo-socialist societies, in particular the former Soviet ones, into pseudo-capitalist ones. This transformation may be explained schematically as a process of the P → E → B → C kind. Indeed, the change of *nomenklatura* (P) led to an economic debacle (E), which depressed overnight the standard of living (B), which in turn led to a precipitous cultural decline (C). The environment (N) helped along, with abnormally cold weather; and the collapse of the manufactures (E) worsened the reckless exploitation of natural resources (N). Given the many-sidedness of this transformation, no unidisciplinary study of it could possibly result in either insight or useful advice (see Pickel 2001). In particular, economism fails because real transactions are socially embedded rather than being conducted in an abstract market.

In sum, multidisciplinarity is particularly indicated in the field of social studies, because these happen to study organisms belonging in social networks and engaged in economic, political, and cultural activities, all of which interact with nature.

3 Social Interdisciplinarity

Let us now take a look at the important interdisciplinary research reported on by Robert Fogel in his 1994 Nobel lecture, 'Economic Growth, Population Theory, and Physiology: The Bearing of Long-term Processes on the Making of Economic Policy.' Notice the five disciplines involved in this work: physiology, demography, history, descriptive (or positive) economics, and economic policy. Fogel exhibits, among other things, the dependence of morbidity and longevity upon the body-mass index (weight/height squared), and of productivity upon health, two strong correlations that he followed over two centuries.

A purely economic or purely demographic approach would have yielded neither Fogel's findings nor his forecasts and recommendations concerning demands for public health care or for pension policies. Much the same holds for the economico-political approach to the study of public health adopted by Vicente Navarro (2002) and his co-authors in the *International Journal of Health Services*. Only this interdisciplinary approach can grapple at once with the effects of social inequalities on health and quality of life, and with the economic interests behind the public health policies pursued by certain governments.

Social facts are multifaceted because, contrary to the radical individualist tenet, society is not an unstructured collection of independent individuals, but a supersystem of interacting individuals organized into systems or networks of various kinds and of different cohesion strengths, from the family, the school, and the religious congregation, to the supermarket chain and the nation.

Moreover, because social facts are multifaceted, social issues are best tackled in either a multidisciplinary or an interdisciplinary manner. This is why the US National Science Foundation set up in 1996 the multidisciplinary National Consortium on Violence Research. Its task is to uncover the multiple roots of violence and design policies to control violence based on research rather than improvising them in the light of ideological or electoral considerations. In short, the idea was to replace opportunistic political engineering with science-based social engineering. Whether this project will survive the current wave of contempt for intelligence and public welfare is uncertain.

Cross-disciplinarity, then is, the answer to the holist's charge that science alone cannot solve any social problems because it would be inherently sectoral (see, e.g., Vanderburg 2000). There is nothing scientific about the division of intellectual labour: its adoption is just a matter of expediency, not principle. When such division fails, scientists must cross disciplinary boundaries (see Hirschman 1981). A number of scientific enterprises, such as the study of materials, the weather, the mind, and social inequality, have become cross-disciplinary and are proving very fruitful.

In conclusion, the failure of any one science to address multidimensional problems only proves the failure of the sectoral approach. Whenever such failure occurs, the systemic approach should be tried before jumping to the conclusion that science has unsurpassable limits. It is an article of scientific faith that the only limits to science are those that individual researchers or their sponsors set themselves.

Concluding Remarks

It might be convenient, though perhaps no great fun, if at least the social sciences were reducible to one of them. But this won't happen, because every one of them accounts for only some features of society – which suggests that integration, rather than radical reduction, is the ticket. In fact, even the simplest of modern social systems, the family, must be accounted for at the same time in biological, psychological, economic, political, and cultural terms. Just think of twin birth, regarded as lucky in some tribes and catastrophic in others. Or of family planning without infanticide – a twentieth-century achievement made possible by a combination of biochemical research, the pharmaceutical industry, old-age pensions, and changes in morals, women's social status, and legislation.

Things do not get simpler in the upper social sublevels. For example, only a conjunction of technological, political, economic, and psychological circumstances may explain the sudden emergence and diffusion of e-commerce – as well as its subsequent catastrophic contraction. Likewise, only a combination of economic, political, ideological, and psychosocial considerations might explain the periodic resurgence of religious fundamentalism in all three Americas, most of whose founding fathers were free masons or agnostics; the persistence of huge military expenditures one decade after the end of the Cold

War; the slowness of Russia's recovery after the collapse of state socialism; or the adoption of conservative social policies even by nominally socialist governments the world over.

The complexity of social systems renders radical reduction illusory, and it invites the integration of the various social sciences. The same holds, *mutatis mutandis*, for the social technologies, from management science and normative macroeconomics to education and the law. The failure of wasteful unilateral social programs – such as literacy campaigns among the hungry, 'wars on poverty' among the illiterate, and 'wars on drugs' against indigent coca or poppy farmers – should have taught us that the various features of a social whole, though distinguishable, should not be detached from one another. What is surprising is that it is taking so long to realize the need for the merger of the social sciences and the social technologies if we are to understand social life and cure social ills. Is it possible that philosophical inertia has something to do with such blindness to system, emergence, and convergence?

In sum, reduction, which results in vertical or inter-level integration, should be supplemented with merger or horizontal (same-level) integration. This dual research strategy should work because the world happens to be a system of systems rather than either a pile of disjointed items (the individualist tenet) or an unanalysable solid bloc (the holist dogma).

12

Functional Convergence: The Case of Mental Functions

Everyone knows that a car is a complex system composed of many modules, such as spark plugs, wheels, and measuring gauges. Each of these modules carries out a specific function that no other component can perform. However, we also know that, when a car is running properly, all of its essential components work synergistically as a higher-level unit – a self-moving vehicle. Most other machines, such as computers, TV sets, and hydroelectric plants, are parallel. Their components are modules, and while they can be dismantled, rebuilt, and even upgraded up to a point, they function together as higher-level units. That is, they exhibit functional convergence, a.k.a. synergy. This is the root of the emergence of the car's global properties, such as self-propulsion, manoeuvrability, acceleration, elegance – and the ability to kill and pollute.

Other systems are even more tightly knit, as a consequence of which they cannot be dismantled and reassembled without loss, much less with gain. When whole, they function synergistically or even synchronically; but once reduced to their *disjecta membra*, they cannot be reconstituted. That is, the assembly process that produced them is irreversible. This is the case with cells, organs, organisms, families, businesses, and nations – as Humpty Dumpty and many a corporation and some multinational states found out too late. The structure or 'architecture' of systems of this kind is said to be integral rather than modular.

Still, some systems can function as higher-level units regardless of whether their structure is modular or integral. That is, they can exhibit synergy or functional convergence. We shall presently apply this idea to the case of the brain and its specific functions, that is, the mental functions. In particular, we shall discuss briefly the thorny problem of explaining the unity of the self, in particular self-consciousness, despite (or perhaps because of) the large number and variety of brain modules.

This problem is just an instance of the old philosophical problem of the one and the many, and it is being actively discussed by psychologists, neuroscientists, and philosophers of mind. The question is of particular interest to the central concerns of this book, for two major reasons. First, because mental functions are systemic: they only emerge when many neurons get together and act synchronically to form a functional unit with properties that its constituents lack – that is, emergent properties (see, e.g., Mountcastle 1998).

The second reason for the relevance of the emergence of mind to the emergence-convergence question is that the answers to that fascinating problem are coming from the convergence of several previously disconnected disciplines: neuroscience, neurology, endocrinology, psychology, ethology, anthropology, linguistics, and sociology. This, then, is one more case of the merger of convergence with emergence.

In his influential book on vision, David Marr (1982) suggested that vision is the function of certain computational modules in the brain. He also surmised that these modules 'are as nearly independent of each other as the overall task allows' – hardly a testable assumption. Jerry Fodor (1983) generalized the modularity thesis to all of the perceptual modalities, language, and action – though, unlike Marr, he did not craft any mathematical models. Besides, Fodor's modules are mental, not neural, although they may be 'embodied' or 'instantiated' in brain subsystems – as Plato might have said. For him, as for other followers of the early Chomsky, the mind is immaterial, whence psychology has nothing to learn from neuroscience (Fodor 1975, 1983). This thesis of the modularity of the mind has been adopted by the evolutionary psychologists (see Barkow, Cosmides, and Tooby 1992) and broadcast by Daniel Dennett (1991), Steven Pinker (1997), and other popular writers.

I submit that the division of the mind into mutually independent mental functions or modules has been a serious obstacle to the understanding of the intimate connections among perception and emotion, thinking and feeling, learning and motivation, perception and anticipation, attention and volition, planning and doing, and more.

In the present chapter I shall state and defend the thesis that, although the brain is composed of anatomical modules, its structure is integral rather than modular, as a consequence of which the mind, and particularly consciousness, is unitary. Even though this thesis seems self-evident, it has often been ignored or even challenged. Indeed, it is well known that traditional psychology divided the mind into mutually independent 'faculties,' compartments, or modules, such as motivation, emotion, perception, attention, memory, ideation, imagination, speech, volition, and consciousness.

1 Informationist Psychology

The cognitive psychology that emerged in the mid-1960s, in the wake of the exhaustion of behaviourism and the computer revolution, disregards the brain and conceives of the mind as a collection of fixed computer programs. Thus, it erects a barrier between the brain (hardware) and its mind (software) – which is like studying walking while disregarding the legs. Therefore information-processing psychology is a version of psychoneural dualism, even though many of its adherents believe themselves to be monists.

Furthermore, this school regards mental processes as (unspecified) 'operations over symbolic codes' (see, e.g., Lepore and Pylyshyn 1999). It overlooks the common knowledge that symbols are not natural items but conventional artefacts; that there is no mind without a brain; that the mind matures along with the brain; that there is no cognition without motivation; and that motivations, inhibitions, feelings, and emotions are anything but computable. Think of all the intellectual, affective, and motor functions that must converge to produce an article of clothing, a song, a scholarly paper, or just a shriek of anguish or joy.

The computationist and immaterialist approach has adopted at least one of the tenets of the old faculties psychology: the strict detachment of cognition from motivation and emotion. Such separation is an unavoidable consequence of the hyper-rationalist dogma that all mental processes are computations performed in accordance with precise algorithms. Ironically, this view – initially proposed as the alternative to behaviourism – may be regarded as a refinement, rather than a rejection, of behaviourist psyche-less psychology. Indeed, instead of the behaviourist's direct and mindless stimulus-response link, it is now proposed that the stimulus triggers an information-processing (or computational) chain that ends up in an (overt or silent) response. This response is allegedly predictable in being the end state of an algorithm-ruled process. Moreover, it is claimed that this process can be variously 'embodied,' 'instantiated,' or realized, not necessarily in living nervous tissue. Not only brains, but also computers and even ghosts, would qualify.

The advantage of this widespread view is obvious: since its practitioners do not poke at brains, or even peek into them, they do not have to manipulate laboratory paraphernalia or wear lab coats. They do not even have to read neuroscience journals. All they have to do is 'to study what is in the mind – their own.' And they profess to know beforehand the kind of thing they will find, namely symbols and operations on symbols – just as when typing or operating a computer.

The flaws in this view should be obvious. First, it confuses the mind (or the brain) with its conceptual models and with computer simulations. It is as if physicists claimed that light computes its trajectory – that is, integrates the pertinent equations – as it moves. Or as if explorers expected to find maps, rather than plains, hills, and rivers, when venturing into uncharted lands.

Second, the thesis that people are pure symbol processors ignores the well-known fact that, in trying to understand what we read, we place it in its proper context and resort to the non-grammatical or 'soft' constraints between symbols. For instance, we accept 'She parsed the first sentence,' but reject 'She parsed herself to death', because we know that 'parsing' goes with 'sentence' but not with 'death.' If the semantic, pragmatic, and emotive tendrils attached to a linguistic expression are cut, intelligibility vanishes. In other words, language-processing emerges from subsymbolic processes (Miikkulainen and Mayberry 1999).

Every mathematician knows that symbol-processing is not the same as thinking. For instance, the great mathematician Leonhard Euler is said to have noted in the mid-eighteenth century: 'My pencil knows more than I.' He meant of course that he could perform automatically a number of algorithmic operations, such as addition and differentiation. This should not be surprising, since algorithms are crafted precisely to allow us – or our artificial proxies – to carry out symbolic operations in a preconscious way and without regard for meaning. Closer to us, John Searle (1980) has shown persuasively, with his famous Chinese Room thought experiment, that symbol-processing according to rule is neither constitutive of understanding nor sufficient for it.

Third, computationism raises the paradox that we are supposed to compute computations in addition to movements, perceptions, and perhaps feelings as well. How is such iterated computation to be conceptualized? And where in the brain should we look to find out how computing calculations works and how it may be repaired in case of error?

Fourth, the view in question overlooks the interactions among the various modules or faculties. In particular, it overlooks the attention-motivation and motivation-cognition connections, so well known to any teacher. It also ignores the finding that the cerebellum and the basal ganglia, which are parts of the motor-control system, are involved in decision-making processes. The view in question also overlooks the fact that the brain, far from being a dry Lego toy, is awash in hormones that alter mood and effect connections among distant subsystems. Consequently, in addition to synaptic transmission, which is local, there is long-distance diffusion ('volume transmission') in the extra-cellular space (see Agnati et al. 2000). This process, which is of the Brownian movement type, has no computer analogue.

Fifth, computationism ignores the intense two-way traffic between the cerebral cortex and the seat of emotion, namely the limbic system (hypothalamus, amygdala, cingulate gyrus, hippocampus, etc.). Hence, it overlooks the strong reason–passion interaction, which explains both enthusiasm for some ideas and fear of others – something that no social psychologist, politologist, or culturologist can afford to ignore.

Last, but not least, information-processing psychology can at best describe some mental processes. It cannot explain them, because to explain is to exhibit mechanism, and only material systems, such as brains, can possess mechanisms (see, e.g., Bunge 1999; Machamer, Grush, and McLaughlin 2001). In fact, information- processing psychology is just classical psychology translated into Informationese. (Bunge and Ardila 1987). The occurrence in it of boxes and arrows should not fool us. The boxes are immaterial, and the arrows symbolize processes that are not being specified: functionalism is dualist.

In conclusion, informationism is scientifically untenable. Worse, it has blocked the advancement of psychology by discouraging the search for the neural processes called mental processes.

2 Mark II Model: Connectionism

In the 1980s, owing in part to Marr's work on vision (1982), informationism started to be displaced by connectionism. This is the attempt to model – not replace – mental processes with computational models that use selected neuroscientific findings rather than pure psychology, let alone folk psychology. The idea is to model parts of the brain as networks composed of interconnected mock neurons, that is, hypersimplified models of neurons (see, e.g., Churchland and Sejnowski 1993).

These artificial neural networks are systems and, as such, they possess some emergent global properties. Among these are the ability to learn to carry out some simple mental functions, such as word recognition, as well as dysfunctions, such as hallucination (see, e.g., Ruppin, Reggia, and Glanzman 1999). Such modelling is of course to be applauded, just as the mathematical modelling of physical, biological, or social processesis. However, one should be clear about the nature and limitation of such models. We shall examine this question in the light of our composition-environment-structure-mechanism model of systems (chapter 2, section 5).

The gist of connectionism is that the structure of a system (the mode of connection) is more important than either its composition (the nature of the connected elements) or its mechanisms (the peculiar processes that make it tick). This view, so natural to a repairman, is at variance with physics, chemistry, and biology. Indeed, all of these sciences teach that structure, composition, and mechanism go hand in hand. For instance, whereas the atoms of some species enter into ionic unions, the atoms of other species form covalent compounds. Mechanism too depends critically upon composition. Thus, neurons communicate through chemical messengers, whereas chips only communicate through electric signals.

Second, connectionist modelling is no substitute for laboratory research in cognitive and affective neuroscience: If good, 'dry' work only supplements 'wet' research. (For the dry/wet distinction, see Kosslyn and Koenig 1995.) In general, theoretical work is no substitute for experimental work, and each functions best when checked and enriched by the other. (Physicists learned this lesson four centuries ago; neuroscientists have yet to learn it.) And while thought experiments, in particular computer simulations, may have some heuristic power, they prove nothing: they just unpack hypotheses.

Third, the models in question are specific (rather than general) mathematical theories based on highly simplifying assumptions. In this regard they do not differ from elementary physical theories. Because they are mathematical, these models are precise, and their flaws can be pinpointed and corrected. And because they can be processed on computers, they lend themselves to simulated experiments, and thus to modelling neurological deficits and lesions. Remove a mock-neuron here, alter an input there, record the ensuing changes, and compare them with real brains.

For example, experimenting with an artificial network consisting of 148 mock neurons grouped into four different layers gave rise to two interesting findings (Hoffman and McGlashan 1999). One of them confirms a piece of conventional neuroscientific wisdom, namely that synaptic pruning (as it occurs in real brains at several stages of development, in particular adolescence) enhances energy efficiency and improves the signal-processing capacity of the system. (The reorganization of a system of any kind, whether neuronal, corporate, political, or other, is sometimes facilitated by removing some of its components.) Another finding is that, when excessive, pruning produces hallucinations, those typical symptoms of schizophrenia – a severe condition that affects about 1 per cent of the population during adolescence, a period of intense synaptic pruning.

Remember, however, that the whole thing is metaphorical: the 'neurons' in question, as well as their connections, are not the real things but simulates thereof. Moreover, the artificial networks are like isolated monads, whereas the brain modules are interconnected by nerve filaments and hormones, on top of interacting with the external world. Besides, the existing neural-network simulations are rather primitive, in being input-output systems – that is, black boxes rather than translucent ones. (See Bunge 1967a for this distinction.)

Furthermore, neural networks can only learn in the way behaviourists believe we learn, since they are designed on the strength of neo-behaviourist psychology. Hence, they make no room for spontaneous (stimulus-independent) activity at various levels, from neuron to neuronal system. They therefore fail to account for creativity, from coping with entirely new situations to inventing radically new ideas, and thus fail Claparède's criterion of intelligence as the ability to solve new problems.

Like all scientific models, computer models deliberately oversimplify some features of nature, and omit others altogether. Yet unlike traditional mathematical models, these computer models also contain artefacts, such as the activation of each neuron and the weights (strengths) of the inter-neuronal connections. Far from representing natural laws, these formulas specify artificial features of the neural network in question. (For example, one of them may specify the strength of the connection between two mock-neurons rather than that of two real ones.) Hence, it is not obvious how the merits should be apportioned between the realistic and the conventional components of such a model. In other words, how do we distinguish here between fact and fiction, or between truth and falsity? Regrettably, this unavoidable flaw of computer modelling is seldom if ever discussed in the literature. Such is Computer's prestige.

In any event, the fact that the models in question are tailored to computers is a matter of expediency, not principle; in fact, before the personal computer became popular, the mathematical models in psychology and in neuroscience were worked out with pencil and paper. Hence, whatever success the new computer models of cognitive processes may attain does not prove that mental experiences are computations – any more than the success of a computer simulation of a chemical reaction proves that chemicals compute.

By the same token, such success does not prove that computer science is part of cognitive psychology, any more than it is part of astronomy or chemistry. Computer science, like mathematical statistics, is a branch of applied mathematics as well as an auxiliary all-purpose discipline. Therefore, its inclusion in cognitive science on a par with psychology is a serious conceptual error, much like including paleontology in geology because fossils help date geological strata, or incorporating nuclear physics into archaeology because the latter uses radiocarbon dating. We shall return to this issue in section 5.

3 Localization of Mental Functions

The localizationist hypothesis about brain structure was proposed by Galen two millennia ago, and resurrected by the phrenologists nearly two centuries ago. This hypothesis has been amply confirmed by contemporary neuroscience, particularly with the help of brain imaging techniques such as fMRI (functional magnetic resonance imaging). For instance, it has been known since the mid-nineteenth century that there is a brain centre (Wernicke's 'area') for speech production and another (Broca's 'area') for speech delivery. Hence, if the two centres become anatomically disconnected, the patient can still understand and form linguistic expressions, but cannot utter them. This is only one of many 'disconnection syndromes' that result from brain lesions, such as those caused by strokes, infections, or hard blows.

An even better-known case of disconnection is this: If the corpus callosum, which bridges the two brain hemispheres, is either absent from birth or surgically cut, the subject cannot name the objects present in his left visual field, because these project onto his right visual cortex, which has become disconnected from the language centres, which in males are normally located in the left cerebral hemisphere.

The dual of such disconnection, namely the strengthening of the inter-hemispheric bridge, is every bit as interesting. For example, whereas amateurs listen to music with only their right cerebral hemispheres, professional musicians use both halves. And the anterior half of the corpus callosum is larger in musicians with early commencement in music than in non-musicians: it grows with practice. As a matter of fact, all the brain 'centres' involved in making music develop with practice (see Zatorre and Peretz 2001). In general, we sculpt our own brains as we learn and do: the function builds the organ just as the organ performs the function.

Brain specialization varies enormously with tasks. Some 'centres' or 'areas' in the human brain specialize in 'processing' (perceiving) faces, and others places. The removal or temporary inactivation of certain small cortical 'areas' results in linguistic impairments such as the inability to utter short words, or else words denoting living things, artefacts, and so on (see, e.g., Warrington and McCarthy 1987). And neurotransmitter imbalances in the prefrontal cortex may result in impaired cognitive control (e.g., distractibility and socially inappropriate behaviour).

Not only large neuron assemblies but also individual neurons can specialize. For instance, it has been known for nearly three decades that the so-called 'feature neurons' react only to the presentation of vertical lines, others to horizontal lines, and still others to diagonal lines. Again, a certain neuron increases its firing rate when the subject sees a given object presented earlier, and also when he recalls the same object with closed eyes, but is insensitive to different objects. The localization is so precise, that '[b]y observing the activity of this neuron, it is possible to predict with rather high accuracy what the subject was viewing' (Kreiman et al. 2000: 360).

Evolutionary neurobiology confirms the hypothesis of brain modularity.

Indeed, it has been found that the evolution of different parts of the mammalian brain, as of other parts of the body, has been mosaic-like rather than concerted (Armstrong and Falk 1982; Barton and Harvey 2000). Paradoxically, such mosaic evolution is a consequence of the functional interdependence of the various brain subsystems. These do not evolve in step because they do not have the same functions and are not subject to the same developmental constraints. Yet, each of them evolves in partial response to the evolution of other parts of the brain. The evolution of the human jaw and teeth is a clear example. Our wisdom teeth witness to the unequal evolution of our jaws and crania. Indeed, as the cranial capacity enlarged, the jaws shrank till not enough room was left for the back molars: the mother's birth canal imposed a constraint upon the newborn's total head bulk. (Incidentally, this is further evidence for the hypothesis that evolution, like human history, is opportunistic rather than programmed.)

In sum, at first sight the structure of the minding brain is modular. That is, there would be one brain 'centre' for every mental 'faculty.' This one-to-one correspondence fits the reductionist ideal of one module–one function, exemplified by the one gene–one trait, and the one molecule–one disorder hypotheses. However, the fate of these simple conjectures should be a warning. Indeed, we have learned that genes come in networks; that a deficiency or excess of molecules of a kind may be necessary but not sufficient to cause a disease; and that the genome-behaviour chasm must be bridged by the brain. Things, and therefore their functions, come in bunches and levels. Therefore, to skip systems and levels is to invite barrenness.

4 Functional Interdependence of Neural Modules

The brain is thus a supersystem constituted by systems, every one of which performs at least one specific function in addition to such generic functions (processes) as synthesizing proteins and hormones, and taking part in metabolism. However, those systems are not mutually independent. On the contrary, the activation of any one 'centre,' 'area,' or 'structure' is likely to induce the activation of several other centres, whether near or distant. For example, when we speak, the Wernicke and Broca 'areas' become active, jointly with the prefrontal cortex, the thalamus, the cerebellum, and other organs. All of them must act in a coordinated fashion for adequate speech delivery. Most of them, even the cerebellum, are also recruited during inner or silent speech. This is because, as Rafael Lorente de Nó discovered in 1938, no neuron is an island unto itself. The brain neurons are so densely interconnected that the neural activity triggered in any region by a stimulus ricochets around for a while, stimulating other brain areas. Functional interdependence results from anatomical connectedness.

The interdependence of mental functions was a central theme of the Gestalt psychologists – Wertheimer, Koffka, and Köhler – as well as of Piaget and Vygotsky. All of them held that mental processes are wholes and, more particularly, functional systems. However, they did not offer a clear definition of this important concept, which may be elucidated as follows (Bunge and Ardila 1987: 101).

A *functional system* is a collection of properties of a material system, such that, given any member of the collection, there is at least another member of it that depends upon the former. In self-explanatory symbols, $F = \{P \in \mathbb{P} \mid (\exists x)(\forall P)(\exists Q)[x \text{ is a concrete system } \& Px \& Q \in \mathbb{P} \& (Px \Rightarrow Qx)]\}$, where $\mathbb{P}$ designates the totality of properties of the system in question. Emergence, a peculiarity of any *Gestalt*, is referred to tacitly above in the mention of a concrete system.

A newly discovered and striking example of neural systemicity is this: The perception of an event as menacing activates the amygdala, the organ of fear and other emotions. In turn, the amygdala signals back to the orbitofrontal cortex, which evaluates the threat, and 'commands' the motor strip to activate the limb muscles to execute the appropriate behaviour, such as cringe, wince, flight, or fight. When at least one of the two organs is anatomically or physiologically impaired, the perception of emotionally salient events is impaired, and accordingly the response is likely to be inadequate (see, e.g., Anderson and Phelps 2001).

Thus, it has been conjectured that extreme aggressiveness, as in psychopaths, may result from amygdala dysfunction (see, e.g., Blair et al. 1999). But it would be foolish to blame only the amygdala for any crime: this organ is only one of the many brain subsystems – cognitive, affective, and motor – involved in that complex process. The prefrontal cortex too is involved in cold-blooded murder, particularly its planning. Besides, only a few crimes, all of them on the small scale, can be explained in neurological terms. Large-scale crimes, such as wars, genocides, colonizations, ecocides, and urbicides, are likely to be the work of groups of neurologically normal individuals. After all, successful large-scale crime takes talent.

5 Consciousness: From Mystery to Scientific Problem

Any mental process, such as the recognition of a face, the identification of a sound, the recall of an episode, or the completion of a figure, engages a number of brain modules. Imaging studies have shown repeatedly that this number increases with the complexity of the task and the level of consciousness of the operation. It is also well known that practice reduces the required level of consciousness: think of driving on a familiar highway, which is ordinarily done in a semi-automatic fashion. The level of consciousness rises only in the face of an emergency or a demanding task.

Unconscious mentation, once the preserve of psychoanalytic fantasy, is coming under scientific scrutiny. The scientific study of unconscious mental processes began a couple of decades ago with observations on split-brain patients. Since then, the various brain-imaging techniques, such as PET (positron emission tomography) scanning and functional MRI, have made it possible to ascertain whether someone feels or knows something even though he or she does not know that he feels or knows it. Moreover, these techniques make it possible to localize such mental processes in a non-invasive way. A recent example is the paper by Morris, Öhman, and Dolan (1998) – which, unsurprisingly, does not cite any psychoanalytic studies. Let us peek at it.

The amygdala is the tiny brain organ that feels such basic and ancient emotions as fear and anger. If it is damaged, a person's emotional and social life will be severely stunted. The activity of the amygdala can be monitored by a PET scanner: this device allows the experimenter to detect a subject's emotions, and even to locate them in either side of the amygdala. However, such neural activity may not reach the conscious level. In such a case, only a suitable brain scanner can help.

For example, if a normal human subject is briefly shown an angry face as a target stimulus, and immediately thereafter an expressionless mask, he will report seeing the latter but not the former. Yet, the scanner tells a different story. It tells us that, if the angry face has been associated with an aversive stimulus, such as a burst of high-intensity white noise, the amygdala is activated by the target even though the subject does not recall having seen it. In short, the amygdala 'knows' something that the organ of consciousness (whichever and wherever it is) does not.

Ordinarily, a heightened level of consciousness is only required for unexpected novelties, and by the deliberate effort to formulate the strategies called for by non-routine problems, such as the search for new information sources, new problems, or new conjectures. Such tasks, from the perception of an ambiguous figure such as a Necker's cube to moral self-criticism, require not just awareness but self-consciousness. All such self-conscious tasks call for sharply focused attention and strong motivation in addition to the activation of a number of cortical modules such as the anterior cingulate cortex. Incidentally, fMRI studies show this to be the organ of cognitive conflict-detection, as when one is asked to read the word *red* displayed in green (Botvinick et al. 1999). An even more remarkable phenomenon is the ability to extinguish an innate reflex, such as the vestibulo-ocular reflex (rotating the eyeballs as the head rotates in the opposite direction), through sheer imagination (Melvill Jones et al. 1984).

Perhaps the most amazing and moving feat of self-conscious control was the self-immolation of Buddhist monks in protest against the Vietnam war. The dualist will claim that these actions exemplify the power of mind over matter. However, since the immaterial mind cannot be detected in the laboratory, the cognitive neuroscientist will prefer to regard those portents of willpower as examples of the effect of the prefrontal cortex upon the rest of the brain, in particular the motor strip.

The foregoing considerations can be summed up in two hypotheses proposed by Dehaene, Kerszberg, and Changeux (1998), which Dehaene and Naccache (2001) state as follows:

H1 Connection and coordination of modular activities: '[B]esides specialized processors, the architecture of the human brain also comprises a distributed neural system or "workspace" with long-distance connectivity that can potentially interconnect multiple specialized brain areas in a coordinated though variable manner' (Dehaene and Naccache 2001: 13).

H2 Consciousness emerges under attention: '[T]op-down attentional amplification is the mechanism by which modular processes can be temporarily mobilized and made available to the global workspace, and therefore to consciousness' (ibid.: 14).

The subjective (or phenomenological) unity of consciousness is thus (tentatively) explained in terms of the interconnection of a number of specialized neuronal modules (or circuits) in the neuronal 'workspace.' The modules in this supersystem are not necessarily fixed and adjacent. They might be itinerant (Bindra 1976; Bunge 1980). Consequently, 'the contours of the workspace fluctuate as different brain circuits are temporarily mobilized, then demobilized' (Dehaene and Naccache 2001: 14).

This explains not only consciousness but also unconscious processes: those occurring in neuron assemblies that remain isolated from the 'workspace' – as is the case with subliminal perception (in particular blindsight). It also explains the active role of consciousness – by contrast with mere awareness – as the top-down control of low-level neuron assemblies by the high-level workspace. (Some dualists have called this 'downward causation,' claiming that it is a case of 'mind over matter.') In short, we have now the kernel of a plausible theory of consciousness that accounts for both its passive or monitor function, and its active or control role.

The above explanation is at variance with the view that consciousness is the activity of a single supervisor or 'central executive' (see Silvia Bunge et al.

2000). This view is inconsistent with an extensive body of neurological data about patients who have sustained lesions in any of the brain centres, yet have retained the ability to perform mental operations requiring not only a number of 'faculties,' but also mental effort and concentration – that is, consciousness.

At this early stage of the scientific investigation of consciousness, we need not commit ourselves to the hypotheses we have just discussed. There are a number of plausible alternative conjectures (see, e.g., Gazzaniga 2000, Damasio 2000, Llinás 2001). However, to qualify as scientific, any hypothesis about mental states must construe these as states of neuron assemblies: there is no more a brainless mind than there is heartless blood circulation, gutless digestion, or a faceless smile. There are several reasons for this.

One reason is that there is plenty of empirical evidence for the hypothesis that mental processes are brain processes. For example, 'hot' peppers and other spicy foods feel hot because they activate the same neurons that detect heat, and menthol feels cold for the same reason; every time a subject reports seeing a face, his fusiform face area is activated; and decisions are processes in the prefrontal cortex. Another reason is methodological: immaterial objects, such as disembodied souls and mathematical objects, are experimentally inaccessible. Only material objects are changeable and can act upon material tools such as measuring instruments. A third reason is philosophical: psychoneural dualism is at variance with the ontology of science, which is monistic and attaches function to organ. In other words, any psychology or philosophy of mind that postulates that mind is detachable from matter is as anomalous as vitalist biology.

6 Two Convergence Processes

Psychologists learned long ago that they are not the only ones interested in mental processes. Neuroscientists, linguists, social scientists, and AI (Artificial Intelligence) experts too are interested in the mind. The need for a synthesis became apparent around 1960. Eventually two quite different mergers came about, namely cognitive science, and cognitive and affective neuroscience (or psychobiology). The former is a synthesis of cognitive psychology, linguistics, and AI. By contrast, cognitive and affective neuroscience is the fusion of neurobiology, neurology, psychology, ethology, linguistics, psychiatry, endocrinology, and immunology. Shorter: Whereas cognitive science is brainless, its rival is brain-centred (see Bunge 2001a: 93–6).

A main objection to so-called cognitive science – as opposed to cognitive and affective neuroscience – is that its practitioners overlook the brain, which happens to be the organ of the mind. This exclusion is like planetary astronomy without consideration of the Sun. Such a discipline would conceive of the planetary orbits as straight lines, not as ellipses; it could not explain why the solar planets happen to be together; and all of its predictions beyond a few seconds would be false.

The philosophy underlying cognitive science is idealism, which – like most religions – holds that mind is detachable from brain. By contrast, the materialist hypothesis that mental processes are brain processes has inspired the rival synthesis. Incidentally, this hypothesis, usually called 'identity theory,' is misunderstood every time it is discussed outside a broad ontological framework. For example, Searle, who prefers to call himself a biological naturalist rather than a materialist, states (1997) that the brain causes mental states – which is like saying that the digestive system causes digestive states, or that wheels cause rotational states. Only events (changes in the state of concrete things) can cause other events – and this by definition of 'causation' (see, e.g., Bunge 1959a, 1977a).

The (emergentist) materialist hypothesis, which underlies cognitive and affective neuroscience, is that 'neurophysiological events in the brain do not cause mental events, but rather [that] mental events are a feature of neurophysiogical systems with certain properties' (Zeki 1993: 345).

The doubling of psychology into brainless and brain-centred forms is similar to the division that Wilhelm Dilthey proposed in the late nineteenth century (1883), between *Naturwissenschaften* (natural sciences) and *Geisteswissenschaften* (cultural sciences). His idea was that social life is dominated by the spirit (*Geist*), both individual and collective, and hence social scientists ought to focus on ideas rather than on the so-called material basis of social life (see chapter 13).

This idealist approach has proved to be rather barren, since the farmer has got to grow wheat, the miller has to turn wheat into flour, and the baker must bake some bread before the professor can eat a slice of bread. The solution lies not in ignoring either material things or ideas, but in relating them correctly. This holds for individuals as well as for groups. And that is what the psychobiological synthesis is expected to do. However, let us take a second look at it.

The critical reader will notice two holes in the psychobiological synthesis: AI and sociology. The first omission can easily be explained. First, we are dealing with basic science, not technology – and AI is a technology. Second, further advances in AI depend on the progress of psychology (both classical and neuroscientific), not the other way round, because the aim of AI is to simulate animal behaviour, and in order to imitate anything one must know something about it.

Third, AI focuses on intelligent processes – or rather their artificial imita-

Functional Convergence: The Case of Mental Functions 193

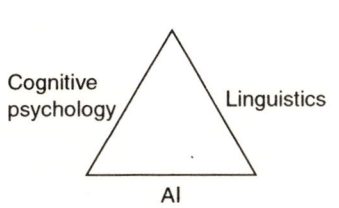
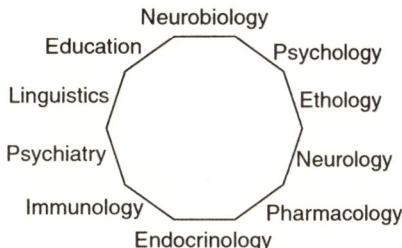

Figure 12.1a Cognitive science: Mind and machine without brain and in a social vacuum (dualism)

Figure 12.1b Psychosocial neuroscience: Mental processes as brain activities in a social context (materialism)

tions – and moreover on rule-directed (algorithmic) processes. Only these can be controlled and put in our service. Motivations, feelings, emotions, spontaneity, creativity, and their kin are neither algorithmic nor useful in a machine. We want machines to be what they are: smart perhaps, but unfeeling and obedient slaves that will perform in seconds, and without complaining, what may take us many months and some complaining.

The second omission, that of sociology, cannot be explained away. It has always been known that, to some extent, we all feel, think, and do what we learned at home, in school, in the workplace, on the street, and other places. Surely, we have some free will, and exercise it whenever we invent, criticize, rebel, or choose to go our own way in some respect or other. Yet social coexistence constrains our freedom in some regards just as it elicits originality in other ways. All these and many more facts prompted the foundation of scientific social psychology in the 1920s. Remember, for instance, the pioneer work of Asch and Sherif on the effects of group pressure on perception and judgment; that of Luria and Goody on the effects of education upon inference; or that of Melzack on the high pain threshold of dogs reared in isolation.

The next problem was to study the neurophysiology of the brain's social inputs and outputs, which has been tackled by social psychobiology (or cognitive and affective neuroscience) since the 1980s (see, e.g., Cacioppo and Petty 1983; Ochsner and Lieberman 2001). In sum, the synthesis prompted by psychoneural monism may now be pictured as a decagon, one of whose sides, psychology, is a biosocial science: see figure 12.1.

Which is preferable, the brainless triangle or the brain-centred decagon? We should choose not just on the strength of our philosophical views, for these might be obsolete or narrow, and therefore might hinder research rather than having a heuristic power. The choice should be based on performance and promise. Now, even a cursory look at the contemporary scientific literature will show lots of amazing findings in interdisciplinary fields, such as psycho-neuro-endocrino-immuno-pharmacology, that were not even suspected half a century ago.

When attention is narrowly focused on a single organ, with neglect of its partners and their common environment, there is no incentive to investigate the function-organ, organ-organ, and organ-environment connections. These come to the fore as soon as the function is regarded as what its organ does in different environments. That is, a shift from a sectoral to a systemic approach is bound to lead to new discoveries. Convergence *de re* calls for convergence *de dicto*, and the latter explains emergence.

Concluding Remarks

In sum, the brain is modular. But its structure is integral rather than modular, as a consequence of which the mind is unitary. We feel ourselves as units rather than as patchworks of memory, emotion, cognition, volition, and so forth because our selves are unitary. This is not to not say that there is a single mental function, let alone that brain and mind are parallel. The thesis defended above is that the brain modules do not work independently of one another: that none of them works in isolation from all the others.

For example, the normal visual perception of a thing involves the synchronic activity of at least four anatomically different cortical 'areas': one for shape, another for colour, a third for motion, a fourth for texture – and perhaps a fifth for the 'binding' of all four. As Zeki put it, 'the integrated visual image of the brain is the product of the simultaneous activity of several visual areas and pathways' (1993: 334). Likewise, playing music requires the coordination of many specialized systems scattered throughout the brain. It also seems clear that the state of self-consciousness is only attained when 'all (or rather nearly all) systems go at the same time.'

Another example: Some if not all cognitive, evaluative, and volitional processes are 'coloured' by emotion and motivation as well as by memory, expectation, and attention. Or, to put it in neuroscientific terms, some if not all cortical processes are influenced by limbic processes. And these are in turn influenced by the former. In particular, we can 'reappraise' our emotions. That is, we can alter the way we feel by changing the way we think. Work with functional magnetic resonance imaging has shown that this process takes place in the prefrontal cortex and nearby systems (Ochsner et al. 2002).

Here, as elsewhere, the functional specialization of the various modules or subsystems raises the problem of finding out how they can coordinate or integrate – that is, the ways in which they interact. The converse is true as well: The emergence of a whole of any kind poses the problem of finding out its constituents and the manners in which these interact. That is, analysis and synthesis are just two faces of the same coin. This holds for all the pieces of the furniture of the world, and consequently for all the branches of inquiry too. Synthesis or assembly, when it involves emergence, calls for convergence; and convergence requires analysis, though not necessarily reduction.

13

Stealthy Convergence: Rational-choice Theory and Hermeneutics

It is well known that the convergence, amalgamation, or syncretism of originally disparate ideas and practices has occurred more than once in history. The fusion of early Christianity with Oriental mysticism, and of the many ancient superstitions jumbled together in New Age mumbo-jumbo, are only two outstanding examples. Convergence has also occurred throughout the history of philosophy. A classic example is Thomas Aquinas's synthesis of Christian theology with Aristotle's basically naturalistic cosmology. Another is Marx's attempt to fuse Enlightenment principles with Hegel's idealism and Feuerbach's materialism. A third is the so-called Critical Theory of Adorno, Horkheimer, and Habermas – an amalgam of Hegel, Marx, Dilthey, Husserl, and Freud.

Occasionally, convergence is seen only dimly, until it emerges clearly upon analysis. A recent case of stealthy convergence is the rapprochement of two trends that for several decades were seen as antithetical: linguistic (or ordinary-language) philosophy and modern irrationalism, particularly of the hermeneutic variety. This convergence is seen in such different admirers of Wittgenstein's as Peter Winch (1958), who claimed that social science is reducible to linguistic philosophy; Georg Henrik von Wright (1971), who defended the role of *Verstehen* (understanding, empathy, or interpretation) in social studies; Richard Rorty (1979), who held that all investigation is a form of conversation; and Latour and Woolgar (1986), as well as other post-Mertonian sociologists of science, who assert that doing science is just making inscriptions and negotiating about them. Such convergence should not have taken anyone by surprise, since both schools, linguistic philosophy and hermeneutics, worship the word.

In the following I shall examine another case of crypto-convergence, at first sight even more surprising than the former ones. This is the closeness of two rather homogeneous and influential schools (or trends or paradigms) in the field of social studies, namely rational-choice theory and hermeneutics. I shall begin by recalling their well-known divergencies, shall go on to exhibit their usually overlooked common ground, and shall end up with a plea for a third way – the scientific approach to social facts.

1 Divergence and Convergence

The distinctive thesis of rational-choice theory is that social facts, whether demographic, economic, political, or cultural, can only be explained by focusing on individual actors, who are to be seen as attempting to maximize their expected utilities under the constraints inherent in their situations. (See, e.g., Becker 1976; Booth, James, and Meadwell 1993; Coleman 1990; Hechter and Opp 2001; Hogarth and Reder 1987; Homans 1974; Moser 1990; Olson 1971; Schelling 1978; and the journal *Rationality and Society*.)

By contrast, the hermeneuticists claim that all social facts are cultural, or even spiritual and symbolic, and must accordingly be interpreted like texts rather than accounted for as objective facts. (See, e.g., Baudrillard 1973; Bloom et al. 1990; Blumer 1969; Dilthey 1959; Fiske and Shweder 1986; Freud 1938; Gadamer 1976; Geertz 1973; Habermas 1967; Lacan 1966; Martin 2000; Outhwaite 1987; Ricoeur 1981; Schutz 1967; and Taylor 1971.) My inclusion of Freud and Lacan in the hermeneutic movement should raise no eyebrows, since both made much of symbols, arbitrary interpretation, and stories, but nothing of the scientific method.

In short, whereas rational-choice theorists regard humans as basically traders, hermeneuticists view them as basically symbolic creatures, more interested in words, rituals, and ceremonies than in work and social intercourse. For example, whereas the rational-choice theorists view education as trading homework for grades, the hermeneuticists see both teachers and students as only symbol-users. Learning, which is supposed to be the specific function of schools, is lost in both perspectives. Moreover, when confronted with a social situation, the rational-choice theorist will perform a cost-benefit analysis. By contrast, the hermeneuticist will wonder how best to 'interpret' the situation: that is, he will try and guess the intentions of the actors involved in it.

At first sight, then, the two schools in question are polar opposites. In fact, they do oppose one another in some important respects. To begin with, they belong to inimical traditions: rational-choice theory is an heir to the Enlightenment, whereas hermeneutics stems from the Counter-Enlightenment, in particular Romantic philosophy. (Adorno and Horkheimer [1972] made this point clear in their attack upon the Enlightenment as the source of all contemporary evils.) Indeed, rational-choice theorists extol rationality, both conceptual and practical, and attempt to explain social facts. By contrast, hermeneuticists swear by *Verstehen* – an ambiguous term that Dilthey understood as empathy, while Weber equated it with 'interpreting' (guessing, hypothesizing) the actor's 'meanings' – intentions, motives, or goals (see Albert 1994 and Bunge 1996 for this ambiguity).

Second, rational-choice theory is ahistorical: it assumes that human beings are essentially the same everywhere and at all times, namely smart and scheming utility maximizers, or even natural capitalists. By contrast, hermeneutics is historicist: it emphasizes the local and contingent, as well as the weight of tradition and the power of intuition; and it denies that there are any universal human features and patterns, or even such a thing as human nature.

Third, the rational choice theorists write clearly, in the tradition of Smith, Mill, Marshall, and Pareto. By contrast, the hermeneuticists prefer the Romantic *chiaroscuro* of their heroes – Hegel, Nietzsche, Dilthey, Husserl – or even of Heidegger, the Prince of Darkness. (Remember Nietzsche's complaint against Mill's clarity.) Hence, unlike the former, they invite endless reinterpretation and inconclusive controversy. On the other hand, the hermeneuticists overlook the legitimate operation of interpreting ambiguous social data such as social indicators. For example, a quick rise in income inequality may indicate either rapid modernization or, on the contrary, de-industrialization or repression of the labour unions. Social scientists resolve such ambiguity through further investigation of reality, not by attempting to read the actors' minds – particularly when we have no access to them.

In the fourth place, because of its recourse to numbers, rational-choice theory attracts pro-science individuals. By contrast, people steeped in the literary tradition prefer hermeneutics because of its conceptual imprecision and innumeracy, in addition to its appeal to symbolic factors – an empty appeal when the doctrine is put in the service of a political movement. (Most rational-choice theorists favour political democracy, if only because they defend rational debate. By contrast, fascists and some neo-Marxists tend to sympathize with hermeneutics.)

The unwary reader is thus left with the impression that, whereas the hermeneuticists live on a humanistic or even literary high ground, the rational-choice theorists dwell in the lowly scientific or even managerial camp. Yet, if one scratches underneath styles and position papers, some remarkable similarities between the two schools on important matters emerge.

Indeed, I will argue that the intersection between both schools is sizeable, for it includes

1 methodological individualism, or the bottom-up (individual → society) procedure;
2 emphasis on mental processes, such as evaluation, decision-making, and

'interpretation,' at the expense of environmental features, work, power relations, and, in general, social structure;
3 a strict separation between the social (or cultural) and the natural sciences, with regard to both objects and methods;
4 arbitrariness in the attribution of actors' motives;
5 lack of interest in macrosocial issues, such as economic crises, war, overpopulation, unemployment, poverty, social inequality, gender discrimination, and political oppression;
6 disregard for empirical testing, in particular for statistical data;
7 conceptual fuzziness: witness the ambiguity of the word *Verstehen* and the untranslatability of Weber's 'verständliche Sinnzusammenhang' in the case of hermeneutics, and the ill-defined concepts of subjective utility and subjective probability in that of rational-choice theory;
8 lack of methodological sophistication, particularly regarding the problem of social indicators, and the difficuty of the inverse problem Behaviour → Intention that both schools attempt to tackle;
9 frequent resort to arguments from alleged authorities – such as those of Hegel, Dilthey, Husserl, or Freud in the case of the hermeneuticists, and neo-classical microeconomists in that of rational-choice theorists; and
10 lack of social sensitivity, hence moral responsibility, when recommending social policies.

This rather vast common ground is the reason that a mix of the two schools – what Martin (2000) calls 'methodological pluralism' – is doomed to perform just as poorly as each school separately. For the same reason, many a thinker can be characterized as both a rational-choice theorist and a partisan of *Verstehen*. This applies to Pareto (1935), who noted that even the superstitious can be said to act rationally since they justify their actions in terms of received beliefs – as when the ancient sailor offered a sacrifice to Poseidon when the seas got rough. It also applies to Popper (1985), who claimed that an action is rational, even when not backed by a utilitarian calculus, if it is appropriate to the situation in which the agent finds himself – a condition met even by an electron subject to a magnetic field. In sum, diluted practical reason is hardly distinguishable from hermeneutic meaning.

Let us next examine the common ground covered by the two schools in question, and then sketch an alternative to them.

2 Methodological Individualism

Rational-choice theory and hermeneutics share methodological individualism. That is, they hold that, since social facts are in the last resort a result of individual actions, they are to be explained exclusively in terms of individual interests, preferences, intentions, and decisions. Both focus on actions that, though socially constrained, are not socially embedded: the individual would precede her social circles or networks. In other words, both the hermeneuticist and the rational-choice theorist proceed from the bottom up, and thus disregard the complementary top-down way.

In focusing on individual desires and goals, methodological individualism misses the larger picture even when it admits institutional constraints. It exaggerates selfishness at the expense of cooperativeness; it underrates irrational behaviour, although this is pervasive even among the educated (see, e.g., Sutherland 1995); it ignores the fact that the very goals as well as the means of an individual are to some extent shaped by her society (Merton 1968); it is puzzled by the existence of social wholes, such as firms; and, most importantly, it overlooks the three main pillars of society: natural resources, work, and social intercourse. The result is a shallow understanding of social facts, and an inadequate reaction to them.

Interestingly, Max Weber has been appropriated by both schools. This is because, in his methodological writings, he stressed that individual action is the point of departure of both social existence and social theory; and also because he extolled the *Verstehen* 'method' (which he got from Dilthey via Rickert). However, as several scholars have noted, Weber's methodological writings contradict his own scientific work, which attempts to explain such supra-individual institutions and movements as slavery, capitalism, the caste system, religion, rationalization (modernization), the state, and bureaucracy, without resorting to *Verstehen* (see, e.g., von Schelting 1934; Bunge 1996). What Weber did was not so much to try and reduce the macrosocial to the microsocial, as to relate these two levels. This point deserves further elaboration.

Inventions constitute an interesting class of processes that start at the individual level and have social repercussions. This holds not only for engineering inventions, such as the electric motor and TV, but also for social inventions. Think, for instance, of schools, hospitals, armies, organized religion, journalism, banking, foreign trade, insurance, mass vaccination, the welfare state, the modern corporation, the credit card, e-mail commerce, or the world wide web. In all these and similar cases the individual proposes and the market disposes. Then again, the market encourages or discourages, and the individual responds – or does not. In any innovation process there is an intense social give and take, but no social construction as conceived by the fashionable social constructivists. Only individuals can create or improve, because only individuals have brains: the market can only prompt or cull.

In sum, individualism, whether of the rational-choice or of the hermeneutic variety, only captures one component of social life, namely individual agency. But it does so incompletely, for it overlooks the social systems in which individual action is embedded. And it ignores the fact that the point of social interaction is to set up, use, or alter social systems through the initiation or control of social mechanisms of many kinds, such as mutual help, division of labour, taxation, voting, persuasion, and repression. Moreover, individualism smuggles in two unanalysed wholes: the institutional framework and the 'situation,' or state of society at a given moment. Hence, it is logically inconsistent on top of painting a shallow picture of the social world.

3 Subjective Process and Observable Behaviour

A methodological individualist is bound to focus on the subjective experiences of actors. If a hermeneuticist, he will focus on either empathy, like Dilthey, or on intention-guessing, like Weber. (In either case he will engage in psychological interpretation, or attribution of mental states, not in semantic interpretation, or pairing of symbols to concepts.) And if a rational-choice theorist, the individualist will centre his attention on the (alleged) probabilities of events and their attendant subjective utilities. That is, in either case the individualist will psychologize even if, like Weber and Popper, he may inconsistently (and dogmatically) deny that social science is reducible to psychology.

In either case the student of social life will tend to exaggerate the importance of individual motives and decisions, at the expense of natural resources, work, and capital – the three main factors of production according to the classical economists as well as Marx. Yet we all know how important these factors are in figuring out what to do in the face of a practical problem. Suppose, for instance, that a manager ponders over the possibility of initiating a new line of business, such as manufacturing computers, cars, or sophisticated pharmaceuticals, in a Third World country. Long before making any decision, he will inquire into the availability of the required natural resources, skilled workforce, and capital, as well as into the disposable income of his prospective customers, the likelihood of competing successfully with imports, and the possibility of obtaining tariff privileges and tax exemptions. However, such background is absent from the typical rational-choice model, which takes merchandises and their producers and consumers for granted. It is likewise absent from the hermeneutic speculations about symbols.

The methodological individualist claims to be able to find out what is usually supposed to be private, namely the hopes, fears, and intentions of the individuals, named or anonymous, that he claims to study. If the student is a rational-choice theorist, he will presume to know the utility functions of the decision-makers in question. And if he is a hermeneuticist, he will claim to read such subjective features off the actor's behaviour; he believes himself to be the privileged owner of a special faculty, *Verstehen* (understanding or interpretation), unknown to the natural scientist.

Yet how does the hermeneuticist know that he has 'interpreted' correctly the data at hand, that is, that he has hit on the true hypothesis? He does not, because he lacks precise and objective standards or criteria for choosing among alternative 'interpretations' (guesses). Nor is there any chance that a hermeneuticist will come up with such criteria, because the very concept of an objective and universal standard goes against the grain of any subject-centred philosophy.

A scientist would say that the hermeneuticist's choice is arbitrary, because he trusts his own intuitions and consequently does not put his guesses to the test. But this objection won't worry the hermeneuticist, because he holds that the cultural sciences, unlike the natural ones, cannot abide by the scientific method, which is centered on the requirement that hypotheses pass some empirical tests before being declared to be true to some degree. In sum, the hermeneuticist asks us to take him at his word – which is natural enough for an intuitionist but not for a rationalist. (See further criticisms of hermeneutics in Martin 2000.)

The rational-choice theorist, for his part, does not bother to guess the particular motives of his actors: he decrees them to be the same for everybody under all circumstances. He postulates that all motives are basically one, namely to maximize expected utilities. Whereas the hermeneuticist offers to cook à la carte, the rational-choice theorist offers a single tourist menu. (The moral philosopher will note the parallel between act-utilitarianism and rule-utilitarianism.)

Neither school explains any of the great social transformations that failed to involve either deliberate choices or symbols, such as the Neolithic revolution and the emergence of civilization. Thus, 'in each area of the globe the first people who adopted food production [instead of hunting-gathering] could obviously not have been making a conscious choice or consciously striving toward farming as a goal, because they had never seen farming and had no way of knowing what it would be like' (Diamond 1997: 105).

Whether rational-choice theorist or hermeneuticist, the methodological individualist faces the problem of figuring out ('inferring') the beliefs and intentions that may motivate his characters' actions. In psychological jargon (Premack and Woodruff 1978), he has to invent a 'theory of mind,' so as to be able to 'read' other people's minds. (Methodologically speaking, he transforms the inverse problem Behaviour → Intention into a family of direct problems Intention → Behaviour.) But of course his 'theory' of mind may be false.

(Worse yet, he may be unable to craft one because his medial prefrontal cortex is deficient, or damaged like that of autistic patients.)

Judging from the assurance of his pronouncements, the believer in accurate mind-reading does not seem to realize how formidable this task is. And yet this problem was familiar to the explorers and the early settlers in foreign parts. Thus, in 1798 the perceptive British Lieutenant Watkin Tench reported about the Australian Aborigines (whom he called 'Indians') he interacted with in Port Jackson (now Sydney): 'We had lived almost three years at Port Jackson ... before we knew that the word Bee-al signified no, and not good' (from an exhibit at the Museum of Sydney). And of course any married person knows how many misunderstandings arise daily from the naive belief that one can always read correctly a mate's intentions off their face, hands, or stray words.

Going from behaviour to intention is incomparably harder than going the other way round. In this process, *Missverstehen* (misunderstanding) is about as frequent as *Verstehen*. One reason is that social signs, such as gestures and linguistic expressions, are contextual, hence ambiguous. Another reason is a basic law of psychology: we are likely to react differently to the same stimulus when in different internal states. This holds, a fortiori, for different people: some will act from self-interest, others from commitment; some from habit, others under coercion; some rationally, others under the sway of passion; some out of fear, others out of greed; some for moral reasons, others from hypocrisy; some lucidly, others under self-deception – and so on and so forth.

In short, the behavior–motivation relation is one-to-many, hence the province of tentative guess rather than reliable intuition. To claim that it is one-to-one is to defy overwhelming empirical evidence and to oversimplify a huge pile of extremely complex problems.

4 Inverse Problems

Most problems in any walk of life can be classed into direct and inverse. See table 13.1. The investigation of a direct problem proceeds downstream, from premises to conclusions or from causes to effects. By contrast, working on inverse problems involves reversing the logical or the causal stream. A peculiarity of inverse problems is that, if soluble at all, they have multiple solutions. The social sciences tackle inverse problems of two main kinds: 'inferring' (actually guessing) individual behaviour from collective behaviour, and figuring out intention from individual behaviour. The two problems concatenate thus:

Collective behaviour → Individual behaviour → Intention.

Table 13.1 Sample of forward vs. backward problems. The arrow symbolizes the flow of research, from problem to solution.

Direct or forward	Inverse or backward
Cause → Effect(s)	Effect → Cause(s)
Code → Message	Message → Code
Disease → Symptoms	Symptoms → Disease
Equation of motion → Trajectories	Trajectories → Equation of motion
Legislation → Social behaviour	Social behaviour → Legislation
General statement → Particular case	Particular case(s) → Generalization
Individual behaviour → Mass behaviour	Mass behaviour → Individual behaviour
Input → Output	Desired output → Requisite input
Intention & circumstance → Behaviour	Behaviour → Intention & circumstance
Legislation → Social behaviour	Social behaviour → Legislation
Manufacture schedule → Production	Production → Manufacture schedule
Means → Goal	Goal → Means
Past → Present	Present → Past
Question → Answer	Answer → Question(s)
Stimulus → Response	Response → Stimuli
Theory & data → Predictions	Data → Theory
Transaction → Budget	Budget → Transaction

Whereas collective behaviour can often be gathered from social statistics, individual behavior calls for either direct observation or conjecture; and attribution of intention demands questionnaire or conjecture – as well as experiment whenever possible. Again, these are only special cases of problems of the forms

Observable → Unobservable, Output → Mechanism, and
Effect → Cause,

which occur in all of the sciences and technologies.

The reason that these problems are as difficult as they are important is of course that to explain a fact is to unveil its mechanism, and mechanisms are typically unobservable (recall chapter 1, section 5). Hence, the problem should not be written off in the name of the behaviourist dogma that all that matters is observable behaviour. At the same time, it would be foolish to ignore that the problem is hard: so much so, that in attempting to solve it we often mistake fiction for fact – as when taking an altruistic action for an instance of enlightened self-interest, or conversely.

Regrettably, philosophers have been remarkably silent on the subject of inverse problems and, indeed, on problems in general. Not even those who, like Popper, have emphasized that research is sparked off by problems rather than data have bothered with the logic, semantics, epistemology, and methodology of problems in general. (However, see Bunge 1967a.) Let us therefore spend some time exemplifying and analysing problems of the inverse type, ignored by most philosophers and social scientists, though well known to mathematicians, physicists, and engineers. (A special journal devoted to these problems, *Inverse Problems*, has been in existence since 1985, and the first international congress on them was held in Hong Kong in 2002.)

The Behaviour → Intention problem is similar to that of trying to figure out the shape of a body from the shadow it casts; to diagnose a sickness on the strength of its symptoms; to discover the authors of a crime knowing the crime scene; to come up with the design of an artefact given its desired functions; or guessing the premises of an argument from some of their conclusions. Any problem of this type has either multiple solutions or none. Let us peek at the problems of this kind.

To predict the behaviour of a consumer knowing his preferences and budget constraints is a direct problem. By contrast, to figure out the preferences and budget constraints of a consumer given his behaviour is an inverse problem. For instance, seeing an unknown person buying a pair of cheap shoes may be 'interpreted' as an indicator of either poverty, thriftiness, a temporary liquidity trouble, or the wish to give a present to a humble person. The only way to find out what the real motive of the buyer was is to investigate thoroughly his circumstances – a task that the psycho-economist and the marketing expert are seldom in a position to perform.

(Incidentally, although the social scientist is not equipped to investigate motivations, desires, intentions, and the like, the psychologist may be competent to tackle this task. He can tackle it provided he makes use of neuroscience – a natural science. In fact, neuropsychologists have been doing this for nearly a half century, using such techniques as implanting electrodes on the prefrontal cortex to record the neuronal discharges accompanying decision-making. They have also started to study the brain mechanisms that result in social behaviour as well as in its inhibition – such as, for example, the role of the amygdala and other brain 'centres' in causing both pro-social and antisocial behaviour. To accomplish this task, they resort not only to classical electrophysiological procedures, but also to brain-imaging techniques such as CT (computerized tomography) and fMRI. Thus, the neuropsychologist outreaches not only the social scientist but also the pre-biological social psychologist, because he treats mental processes as brain processes. The claims of the believer in the power of *Verstehen* sound like magical thinking by comparison.)

A second example: In principle, any public-policy goal, such as population control or the containment of the AIDS epidemic, can be achieved by different means. These range from education to economic disincentives to forced sterilization to massacre – or even to such cruel and unusual punishment as sexual abstention. This is a problem of reverse social engineering. And it is no rarity: most technological problems are of this kind. The reason is that the technologist's employer or customer typically requests the invention of a device performing a given function: he only specifies the output. Think of designing a firm capable of producing or selling a given good or service, as well as of yielding a given return on a given capital.

The inverse problems in the social sciences and technologies are tougher than in natural science and engineering, because of the scarcity of powerful mathematical theories in those fields. In other words, the dearth of solutions of direct problems in social studies makes it harder to guess solutions to the corresponding inverse problems.

5 Figuring Out Mediating Mechanisms

The most important and toughest of all inverse problems are those of figuring out the internal mechanism that mediates between the observable inputs and outputs of a system. This task cannot be performed without any extra hypotheses, if only because mechanisms are occult in the visible box. Mechanisms must therefore be guessed at even if one gets access to the interior of the box, because the world is full of entities and processes, from force fields to intentions, that are not directly observable. (Contrary to the popular saying, that 'What you see is what you get,' what you really get is mostly invisible.) Even solving an induction problem at the low level of an empirical generalization (as in curve fitting) does not get us any nearer to finding the true deep laws, for these describe things or processes that lie beyond observation. An inductive generalization only poses the problem of inventing the high-level law(s) that entail the former.

For example, the observation of a pendulum or an oscillator poses the problem of guessing the equations of motion. Human migrations result from similarly unobservable motivations shaped by economic and political circumstances. Stock-market fluctuations are attributable to a combination of real performance and accounting, which are indirectly observable, with the misinformation, fear, and greed of millions of people, none of which is readily observable. As for the mechanism of the business cycles, nobody seems to have found it. In particular, the neo-classical microeconomists do not have a clue about it, because they focus their attention on individuals operating in markets in equilibrium and in an ideal political vacuum. (See further criticisms in Bunge

1998.) An additional reason that they cannot find such a mechanism is that their theories contain only observable variables (quantities and prices), while they seldom if ever involve what physicists call 'the independent variable,' namely time. Incidentally, this limitation to observable variables is the reason Marx called them 'vulgar economists.'

Marx's own failure to find what he called 'the laws of motion' of the capitalist economy is beside the point. His point, that such laws lie beneath the phenomena (quantities and prices), was methodologically sound. Logic and semantics explain why we cannot proceed from observable trajectories to laws (inverse problem), but must proceed the other way round (direct problem). The logical reason is that valid (deductive) reasoning goes from the general to the particular, not the other way round. The semantic reason is that the high-level concepts occurring in the high-level laws – such as those of mass, energy, anomie, price elasticity, and popular participation – do not occur in the data relevant to them, or even in the low-level generalizations related to them.

The hermeneuticist does not seem to realize this elementary methodological point. Indeed, he believes that he can jump safely from observable behaviour to the wishes or intentions behind it. By contrast, the rational-choice theorist seems to realize that point, since he claims to be able to deduce behaviour from the alleged universal law about maximizing expected utilities. In other words, whereas the hermeneuticist claims to be able to discover causes from their effects, the rational-choice theorist claims that he knows a priori the putative mother of all effects, namely self-interest.

The problem with the rational-choice theorist is that he takes that alleged law for granted. However, the 'law' is not universally true. Indeed, the experimental economists have shown that we tend to minimize losses rather than maximize gains (Kahneman, Slovic, and Tversky 1982). Besides, most of us cannot afford the best: ordinarily we must settle for a far more modest goal, namely satisfaction, or even breaking even (March and Simon 1958). Consequently our revealed preferences seldom coincide with our secret preferences. Worse, sometimes we delude ourselves; at other times we do not know what we want; and we seldom know exactly why we do what we do. Knowledge of self is so imperfect that the ancient Greek adage 'Know thyself' is still pertinent. Thus, the claim of both the rational-choice theorist and the hermeneuticist, that they can find out other people's innermost motivations, parallels the psychoanalyst's conceit.

To conclude this section, neither the rational-choice theorist nor the hermeneuticist succeeds in explaining the behaviour of social wholes (systems) on the basis of imagined individual proclivities (or preferences) and intentions (or goals). They fail in this task because they refuse to admit the very existence of social systems, and are therefore bound to overlook the social mechanisms (processes) that make a system tick. (See Bunge 1999, Hedström and Swedberg 1998, Pickel 2001, and Tilly 2001 for the centrality of the concept of a social mechanism in social science.) And it so happens that an explanation proper – by contrast to a mere subsumption under a regularity – consists in laying bare the mechanisms (causal, stochastic, teleological, or mixed) underlying observable behaviour. No system, no mechanism; and no mechanism, no explanation. To clarify this point, and vindicate the scientific method, let us study a particular case.

6 Example: The Delinquency–Unemployment Connection

We start from the well-known sociological finding that, to a first approximation, crime is an increasing linear function of unemployment. This is an empirical generalization that cries out for explanation, all the more so since there are great differences in crime rates among countries. For instance, the crime rate is much higher in the United States than in countries with endemic high unemployment and underemployement, but far more cohesive at the lower social strata, such as India and Turkey. In general, crime is higher in divided societies, where everyone is expected to fend for himself, than in homogeneous societies, where everyone is expected to help and control others.

Suppose then that the clues to such differences are social inequality, anomie (the mismatch between aspirations and accomplishments), and solidarity. More precisely, consider the following two-tiered causal (or Boudon-Coleman) diagram:

Macro-level   Unemployment → Inequality → Crime   *Ostensive variables*
                                    ↓   ↑
Micro-level   Anomie ↔ Solidarity   *Hypothetical variables*

The upper-level variables are observable or nearly so; the lower-level ones are hypothetical constructs. The arrows symbolize causation. In all cases but the last, an increase in one of the features causes an increase in the dependent one; by contrast, stronger solidarity (or community support) decreases criminality.

All five variables are quantitative. (In particular, social inequality can be taken to be estimated by either the Gini or the Theil coefficient; solidarity by the rate of participation in volunteer work; and anomie as the average ratio of unfulfilled individual desiderata to the total number of desiderata.) Consequently a system of algebraic equations can be set up and solved. For example, it may be assumed that the crime rate is proportional to the total anomie, that is, $C = aA$; that there is a trade-off between inequality and solidarity, that is, $A = bI$

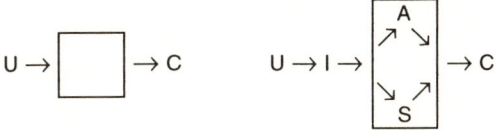

Figure 13.1 (a) Black-box description; (b) Translucent-box explanation

– cS; and that in turn I and S are linear functions of the unemployment rate. Merton would call this a 'middle range theory.' It can be checked using tables of social statistics to estimate the parameters.

The point of this exercise is to suggest that there is a way out of the rational choice–hermeneutics dilemma: that it is possible to tackle problems in social theory, in particular problems of the inverse type, with the ordinary tools of science. The clue is to regard society as a system with a number of interdependent features on at least two levels, among which there is a two-way traffic. This amounts to replacing the U input with the two-step U → I input, and filling the U → C black box with a mechanism composed of two coupled springs: A and S. See figure 13.1.

Even this model is likely too simple to be realistic (true). Besides, it only covers 'honest' crime, the type prompted by (real or perceived) inequity; it does not cover corporate crime. Yet one of the well-known advantages of mathematical (or at least mathematizable) models over the verbal ones is that their limitations and flaws are obvious and can be located precisely; consequently they can be tested and repaired. By contrast, the vagueness and ambiguity of verbal models may give rise to barren, endless, and acrimonious controversies.

Our model of the criminality-unemployment connection, as well as any refinement of it, is at variance with the rational-choice view of crime as a freely chosen career in the supposedly free labour market (e.g., Tsebelis 1990). The model is equally at odds with whatever ad hoc stories the *Verstehen* theorists and psychoanalysts may invent to account for the behaviour of particular delinquents. Neither group explains the fact that the vast majority of delinquents are petty thieves or occasional assailants, lacking in skills and acting mostly on impulse, rather than cunning and hardened criminals, expert at estimating expected utilities, and worthy of Sherlock Holmes.

Concluding Remarks

Raymond Boudon (2002) has held that current sociology is divided into four different ideal types: informative, critical, expressive, and cognitive. The first is limited to data; the second exhibits flaws in the social order; the third aims at arousing emotions; only the fourth attempts to explain social facts, and is therefore the only one that 'really matters.' I believe that Boudon's typology is illuminating, and agree with his indictment of the 'expressive' school, which is a branch of literature – even though the novelists Honoré de Balzac, Charles Dickens, Leo Tolstoy, Benito Pérez Galdós, Eça de Queirós, G. Bernard Shaw, and Mario Vargas Llosa were arguably more objective and insightful than Erving Goffman, Jean Baudrillard, or Clifford Geertz.

However, there is another partition of the social literature, which is at least as important as Boudon's, namely that into scientific (or rigorous or hard) and non-scientific (or imprecise or bland). The authors in the former class, such as Thucydides, Ibn Khaldûn, Tocqueville, Durkheim, Weber, Keynes, Merton, Dahl, and Coleman, engaged in scientific research and imagined testable hypotheses about the mechanisms underlying social facts. Regrettably, the vast majority of philosophers of science ignore this kind of social studies, and sympathize with either rational-choice theory or hermeneutics, neither of which is scientific.

At first sight, these two schools are mutually exclusive. For one thing, rational-choice theory looks scientific whereas hermeneutics is bookish or even literary. For another, whereas hermeneuticists stress locality, particularity, contingency, and historicity, rational-choice theorists emphasize economic 'rationality' (self-interest) as the universal and ahistorical prime mover. Accordingly, they propose a one-size-fits-all theory for ready use in every single field of social study, from anthropology and demography to political science and history. The economists among them also favour a single-size straitjacket, called 'the Washington consensus,' designed to control the economies of all countries regardless of their history, level of development, and aspirations.

In sum, there are clear differences between the two schools in question. The most outstanding of them are that the rational-choice theorists, unlike the hermeneuticists, are willing to tackle any problems and to engage in rational discussion. However, the common ground of these competing schools is far larger than their followers realize. For one thing, both schools are actor-based: both overlook social systems, and therefore social structures and mechanisms, as well as systemic crises and systemic reforms. Second, both schools are aprioristic, and both rely on authority rather than on observation. In other words, neither of them uses the scientific method. Third, and as a consequence of both features, neither of the schools in question has succeeded in explaining any macrosocial features, such as the increasing income inequality that has accompanied the rise in productivity, corporate profits, and globalization; the renewed intimacy between politics and religion; or the distortions of the democratic process by big business.

Nor has either school accounted for any megasocial processes and movements, such as the business cycles, the increase in income inequality that often accompanies the rise in productivity, the collapse of the Soviet empire, Americanization (a.k.a. globalization), the revival of ethnic nationalisms, the changing fortunes of democracy in the Third World, the one-sided continuation of the arms race, the commercialization of politics, the political exploitation of fear and racial prejudice, or the rise of religious fundamentalism in what is supposed to be the age of science and technology. It would seem obvious that no social theory incapable of grappling with such global issues is worth the candle. Yet, in the field of social studies tradition seems to weigh like a tombstone, because reality checks – tests for truth – are seldom carried out.

If, as I claim, both schools have failed to predict or even explain anything interesting, then we need to look elsewhere, beyond individuals. I have argued in earlier chapters that the proper approach is centred in systems and their mechanisms, that is, the processes that make them tick. The consistent adoption of this alternative, together with the scientific method, should bring social science in line with the strategy that has succeeded in mathematics, the natural sciences, and the technologies, all of which, except for elementary particle physics, study systems. Besides, this strategy should improve the efficiency of the social technologies, in particular public policy-making, since the point of all of them is to design or redesign social systems, and maintain or reorganize them.

However, in view of the periodic recurrence of unsuccessful attempts to account for everything social in terms of all-purpose (or 'grand') theories, it must be stressed that genuine systemism is not a theory but an approach to problems, and thus a foil for a whole family of theories. If preferred, it is a hypergeneral theory that – like automata theory, information theory, and general control theory – can only be tested through the testable theories that it spawns (see Bunge 1973b). Hence, it is no substitute for painstaking fact-finding and solid social theorizing. Indeed, the systemic approach is consistent with any number of theories that aim at explaining a whole by its parts, as well as the parts by the whole – or, if one prefers, alternative theories about micro-macro links.

Any science-oriented student knows that the choice among alternative theories, systemic or otherwise, about any given set of social facts, is a task for empirical investigation rather than philosophical or ideological controversy. The theorist proposes and the empirical researcher disposes. However, the latter won't engage in any interesting and promising research project unless he is enough of a realist to focus on individuals-in-their-systems rather than on either individuals or unanalysable totalities. After all, an unrealistic viewpoint prevents one from facing reality. The real sequence is, then, neither

Facts → Theory (classical empiricism)

nor

Theory → Facts (classical rationalism).

Rather, the real sequence is the one proposed by scientific realism (Bunge 1967; Mahner 2001):

Background Knowledge & Approach → Problem → Empirical or Theoretical Research → Revised Theory or New Facts.

14

Convergence as Confusion: The Case of 'Maybe'

So far we have been concerned with the rapprochement, correct or wrong, of clear and distinct ideas – approaches, theories, or entire research fields. This chapter will deal with a convergence of a different kind, namely conflation or confusion. More particularly, we shall examine the confusion among different members of the vast family of concepts behind the innocent-looking word 'maybe,' such as those of possibility, likelihood, probability, frequency, plausibility, partial truth, and credibility. These confusions, inherent in ordinary language, have given rise to some philosophical games and even entire academic industries.

Confusion is likely to occur at the beginning of a research line – as when the concept of energy was initially mistaken for that of force, and that of genetic structure for that of information. Yet confusion may also occur midway, after the precursor ideas have attained some degree of maturity. A clear case of this kind of convergence is the popular conflation between the seven concepts mentioned above. This conflation is not just a piece of carelessness that could be corrected with the help of a good philosophical lexicon. It is, instead, the source of interesting philosophical problems and even four academic industries: inductive logic and probabilistic logic, epistemology, and metaphysics.

In ordinary language one often conflates the seven nouns in question. For example, one sometimes says that a given event is probable, while actually meaning that it is likely, frequent, or even just possible; or that a given proposition is probable, while intending to say that it is plausible or even approximately true. It is also said of events that they come in three degrees: possible, probable, and certain. But of course these are different categories altogether: the first is ontological, the second mathematical and scientific, and the third psychological.

Some of these confusions are inherent in modal logic, where 'possibly p'

may mean that p is plausible or just conceivable. (In the same context, 'necessarily p' may mean that p is a valid consequence of some premises, or else that the fact reported by p is bound to occur in some fantasy world.)

The Bayesian (or subjectivist or personalist) school in statistics consecrates some of the same confusions. Indeed, according to it, every event and every proposition can be assigned a probability – albeit in a subjective or intuitive fashion, and therefore arbitrarily. Rational-choice theories, such as neo-classical microeconomics and game theory, compound the problem by multiplying subjective probabilities by subjective values (utilities) to obtain expected utilities. This mixture is particularly explosive owing to the human tendency to overrate the likelihood of both very valuable and very disvaluable events (Tversky and Kahneman 1982). This irrational tendency makes some people buy lottery tickets and others avoid plane trips.

My aims in this chapter are to draw attention to these confusions and to help avoid them by exactifying some of the concepts in question. The related concept of partial (or approximate) truth will be tackled in the next chapter.

1 Logical Possibility

Everyone accepts the centuries-old distinction between possibility *de dicto*, or logical possibility, and possibility *de re*, or factual possibility. I submit that the first concept is not an interesting object of research, for it is reducible to those of relevance and consistency. Indeed, one usually says of a construct that it is logically possible in a given context if it does not introduce any irrelevancies or contradictions in the context. Obviously, no special theory, such as any of the 256 logically possible systems of modal logic, is required to elucidate this rather trivial concept of possibility. Nor is any of these calculi necessary to elucidate the concept of possibility as consistency, since the latter belongs in standard metalogic.

I submit that the interesting concepts of possibility are ontological and epistemological, not logical. That is, only facts and our knowledge thereof are possible or impossible in an interesting way. To make this point, let me start by distinguishing facts from the propositions, in particular data, that describe them. (Regrettably, in ordinary language 'fact' and 'datum' are often equated.) Further, let me propose the convention that a fact is the existence of a concrete (material) thing either in a given state or in a process (change of state). In other words, I submit that only states and events are (really) possible (or impossible). For example, it is possible that this book will be read by someone, but impossible for an illiterate person to read it. Likewise, it is really possible for a gene to undergo a mutation, and for a neuron to discharge spontaneously. This ontological concept of possibility is not elucidated by modal logic. Indeed, both in science and in ordinary knowledge, 'Fact f may happen' is equivalent to 'Fact f may not happen.' By contrast, in modal logic 'Possibly p' is inequivalent to 'Possibly not-p.'

By contrast, concepts, propositions, classifications, and theories, when considered in themselves, are neither possible nor impossible: they are either available or not. And of course they are conceivable or inconceivable within certain contexts – but conceiving, like believing and doubting, is a mental process, hence one beyond the competence of logic and mathematics. In other words, the crafting of any construct is a process and, as such, it may occur in certain brains under certain circumstances. But logic and mathematics are not concerned with processes: they tacitly feign that all the relevant constructs are ready-made. (See Bunge 1997b for moderate mathematical fictionism.)

Much the same holds for talk of possible worlds, as when one says, following Leibniz, that tautologies (logical truths) hold in all possible worlds. Actually, this only means that the interpretation of the predicate variables occurring in the formulas in question is immaterial – or that the formulas hold under all interpretations of the predicate symbols. As Hughes and Cresswell admit, 'perhaps some such phrase as "conceivable or envisageable state of affairs" would convey the idea [of a possible world] better' (1968: 75).

Consider, for example, the difference between the conceivable and the real links among $N > 1$ items of any kinds, individuals, or sets, concrete or abstract. The number of conceivable dyads among them is $P = (1/2)N(N-1)$. By contrast, the actual number of links is some number A lying between 0 and P, which depends on the nature of the items. The connectivity of the given assemblage is defined as the ratio $\kappa = A/P$ of the actual number A of links divided by their possible number P. (Obviously, $0 \leq \kappa \leq 1$.) This ratio, well known to ecologists, sociologists, and other students of networks of various kinds, underlines the radical differences between actuality and possibility, as well as between fact and fiction. (Besides, it is a handy if simplistic measure or indicator of the systemicity, cohesion, or integration of a collection of items.)

In physics, the conceptually possible is often called 'virtual,' as in 'virtual displacement' and 'virtual image.' This distinction is at the heart of the most general law statements, namely the variational principles, such as Hamilton's principle of minimal action in mechanics, and Fermat's principle of minimal time in optics. Indeed, these principles single out the actual or real trajectory from the infinitely many conceivable (geometrically possible) paths between any two points (see, e.g., Lanczos 1949).

Nevertheless, some philosophers, starting with Saul Kripke four decades ago, and culminating at present with David Lewis, refuse to distinguish conceptual from real existence. That is, they distinguish neither constructs from concrete (material) things nor, consequently, conceptual possibility from real possibility. These conflations have led them to uphold 'the thesis that the world we are part of is but one of a plurality of worlds, and that we who inhabit this world are only a few out of all the inhabitants of all the worlds' (Lewis 1986: vii). How do they know? They could not possibly know because, by hypothesis, the 'parallel worlds' they fantasize about are causally isolated from one another, and therefore empirically inaccessible from the only world we know.

This is not the way of science. Indeed, science takes the uniqueness of the world for granted, and attempts to understand it; and technology designs possible worldly things. In conflating fact with fiction, and postulating a plurality of worlds, possible-worlds metaphysics qualifies as idle fiction. It would be more interesting and useful to imagine really possible worlds, such as worlds with amazing gadgets and social inventions, though without misery and war. Such mundane possible worlds are what the amazing geneticist, physiologist, and statistician J.B.S. Haldane thought up in his book *Possible Worlds* (1927).

Still, though useless to philosophy, possible-worlds metaphysics offers tantalizing theological possibilities, such as any number of many-God theologies. In particular, every one of the infinitely many parallel universes might (or would have to?) have its own Christ, complete with crucifiction, resurrection, apostles, church, popes, schisms, crusades, Inquisition, favourite torture method, etc.

The many-worlds interpretation of quantum mechanics (Everett 1957) does not fare any better. Indeed, it assumes that every time a measurement is performed, the universe branches out into a number of mutually independent worlds – as many as the possible results of the measurement. I regard this as a piece of science fiction. First, it conflates possibility with actuality. Second, it violates the principle of energy conservation. Third, it fails the testability test because the parallel universes cannot communicate with one another. Fourth, it is useless, because it does not predict anything over and above what the standard theory predicts. So, a quantum theorist can toy with this deviant theory during weekends, as long as he retains his sanity during workdays. And experimentalists need not concern themselves with this fantasy at all, if only because their research grants hardly suffice for them to tinker with one world.

As for the concept of logical necessity, it is certainly needed, but it is usually defined in non-modal terms, namely thus:

Proposition p is *L-necessary* in theory T if L is the logical theory underlying T, and p is deducible (provable) in T with the help of L. Briefly: $T \vdash_L p$.

What about the concepts of deducibility (or provability) and definability, which prima facie are irreducibly modal? They are so, indeed, in an epistemological context, but not in standard logic. Here they are elucidated in a non-modal fashion. For example, one says that proposition p is *L-deducible* in the set B of propositions if B contains premises that entail p in accordance with the rules of inference in L. Likewise, one says that the concept (predicate or set) C is (explicitly) *definable* in the body B if C is not a primitive (undefinable) in B, and B contains the definiens of C. Of course, the epistemic process of actually deriving or defining a construct occurs in someone's brain, and is therefore a fact. Hence, it is legitimate to attribute it real (factual) possibility – which hinges on the brain in question being suitably equipped, motivated, and socially situated.

The epistemic and methodological modalities, such as knowability and testability, can be demodalized in a similar fashion. For instance, one may say that a given item is knowable to animals of a given kind if these have access to information and means of certain types. Real possibility is tacitly involved here, because the animals in question may or may not make actual use of the information and the means in question. More on this anon.

2 Real Possibility

What I am proposing then is that the operand of the modal operator 'possible' be restricted to facts, whether in the external world or in the mind. That is, the f in the statement 'f is possible,' or 'Pf' for short, must name a fact, not a proposition. (In other words, P maps facts into modal propositions.)

In ordinary knowledge and the metaphysics based upon it, no specialized empirical knowledge is deemed necessary to determine which facts are really possible and which are not. Does not everyone know that sparrows can fly, whereas pigs cannot? Thus Lowe, who regards metaphysics as an a priori discipline, informs us that its concern is 'to chart the *possibilities* of real existence' (Lowe 2002: 11 Accordingly, metaphysicians would be competent to decide whether time travel, making gold from lead, centaurs, telepathy, and the like are really possible. We are being invited to go back to the Middle Ages.

By contrast, in the sciences one accepts tacitly the definition of factual possibility as lawfulness, that is, compatibility with objective laws, such as the law of conservation of energy, or the law that organizations decline unless revamped once in a while. Factual possibility is thus identical with lawfulness. Hence, only scientists can determine, say, whether a certain quantum jump, chemical compound, organism, or social process is really possible.

In obvious symbols:

f_b = State or change of state of thing b of a given kind,
L = Law of things of the given kind, as in $Lb = (Ab \Rightarrow Bb)$, an instance of '$\forall x\, Lx = \forall x\, (Ax \Rightarrow Bx)$.'
$Pf_b =_{df} \exists L\, (L$ is a law & $Lb)$.

Example: f_b = state or change of state of a closed material system b, and Lx = {Closed $x \Rightarrow$ [Energy of x = const]}.

An *impossible fact* is of course one that is not really possible. For example, the creation of energy is impossible because it would contradict the law of conservation of energy. Much the same holds for the quantum jumps ruled out by the so-called selection rules, as well as for the spontaneous unscrambling of eggs, human immortality, levitation, and telepathy. By contrast, a golden mountain, though unlikely, is not impossible: there is no law of nature against a gold lode as big as a mountain. Incidentally, this is one of the stock examples of the philosophers who, from Alexius Meinong to Richard Routley, have been fascinated by *impossibilia*.

Another shop-worn example is 'the round square.' But this case belongs in a totally different category, namely that of linguistic expressions that can be uttered or written, but which are either self-contradictory or meaningless. All impossibilia are certainly objects of discourse, but they belong in the fiction genre, along with Italo Calvino's halved viscount and the non-existing knight.

What about factual necessity? The concept of necessity often used in science and technology is this: A fact is *really necessary* (or will necessarily occur) if (a) it is really (nomologically) possible to begin with; and (b) the attendant circumstances of its occurrence (such as the initial and boundary conditions) obtain. In self-explanatory symbols:

$Nf_b =_{df} Pf_b\, \&\, Cf_b$.

For example, a coin will necessarily fall down if dropped in a gravitational field (circumstance) – not so if dropped in a space station. Again, we are all bound to die due to relentless wear and tear, limited self-repair ability, programmed cell death, and accidents both internal and external. (Incidentally, the existence of such mechanisms shows the inadequacy of inductivism, which feeds irrational hopes of immortality. 'All humans are mortal' is a law, not a mere empirical generalization with possible exceptions.)

Our definition of real necessity only holds for causal (non-probabilistic) laws. In the case of probabilistic laws, such as those of quantum physics and population genetics, the occurrence of favourable circumstances will only result in contingent facts. These happen only in a given percentage of cases when the right conditions obtain. We have thus the following partition of the totality of real facts (Bunge 1976: 20):

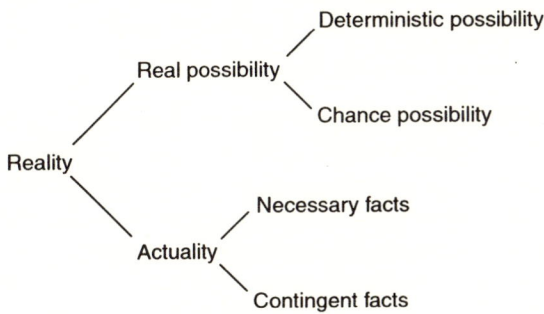

Interestingly, chance possibility and real (as opposed to logical) necessity are not mutually exclusive. Indeed, according to statistical physics, whatever is probable is bound to occur and recur – provided it belongs in the class of repeatable events to begin with. This is in stark contrast with the incompatibility between possibility and necessity in modal logic. But it is understandable, because the latter theory contains the concepts of neither recurrent event nor time. Being eventless and timeless, it cannot possibly be relevant to the understanding of the real world.

Obviously, factual necessity implies factual possibility: $Nf_b \Rightarrow Pf_b$. In other words, what is or will be the case was possible to begin with. This is an ontological version of Aristotle's law, shared by all the systems of modal logic, to wit, $p \Rightarrow \Diamond p$. The remaining formulas of the various modal logics are useless, because the ontological and scientific concepts of law and circumstance do not occur in them.

In particular, we do not need such Byzantine and hardly understandable formulas as 'Whatever is the case is necessarily possible,' 'Whatever is necessary is necessarily necessary,' and 'Whatever is possible is necessarily possible,' all of which are included in the S5 system of modal logic.

Unsurprisingly, modal logic has found no more use in factual science than in mainstream mathematics. It is just an academic game with fake diamonds and paper boxes. Nor is modal logic of any use to either technology or practical philosophy, both of which involve the concept of feasibility. This is the ability to bring to fruition an intention or a plan. Feasibility can be technical, economic, or moral. Technical feasibility equals compatibility with the present state of technology. Economic feasibility is affordability or cost-effectiveness.

And moral feasibility equals consistency with the morality that is taken for granted. Neither of these three concepts of feasibility is indebted to the concept of logical possibility.

In conclusion, whereas the scientific metaphysics project is promissory (Peirce 1965; Bunge 1971), non-scientific metaphysics is not.

3 Likelihood

Usually some events are more likely to occur than others. That is, of any two given events (of the same or different kinds), one of them may occur in preference to the other. For example, it is more likely for an old (and therefore internally stressed) glass container to shatter when hitting the floor than for a brand-new one to end up in the same state. To be sure, a long run of trials may show how much more likely the events of one type are than the other. That is, in principle we can observe, or rather count, their relative frequencies. And we say that, if facts of a given kind have occurred frequently in the past, then they are likely to recur – provided the things in question are still in existence.

However, in practice we usually employ a non-quantitative and comparative concept of likelihood, propensity, or inclination. And we admit more or less tacitly that this comparative concept has the same formal properties as the concept $\geq$ of partial order. That is, we assume that any set of possible events in a given thing (and relative to a given reference frame) is partially ordered. In particular, transitivity is assumed to hold for them: For any three possible events f, g, and h in a given thing and relative to the same reference frame, if $f \geq g$ and $g \geq h$, then $f \geq h$.

This comparative concept of likelihood is the one involved in the so-called Likert scales, widely used in psychology and social science. For example, the subjects in a survey may be asked about the likelihood – not just the possibility, let alone the probability – that the people in their neighbourhood will be willing to help others in an emergency. In fact, they are likely to be presented with what is called a five-point Likert-type scale: very likely, likely, neither likely nor unlikely, unlikely, and very unlikely.

How much more likely is an event than another? An answer to this question requires the technical concepts of either frequency, probability, or statistical association (e.g., linear correlation). All three exactify the vague pre-analytic notion of events of one kind being more likely than events of another. This is why we can be certain that the increase in the disorder of a closed complex system is far more likely than its decrease, or that being killed in a car accident is far more likely than winning the lottery.

In sum, likelihood is predicable of some facts, and it is rendered precise by the technical concepts of relative frequency, probability, and statistical association – none of which applies to propositions.

4 Relation between Frequency and Probability

Although the concept of relative frequency is ubiquitous in ordinary knowledge, science, and technology, it is often confused with that of probability. A reason for this conflation is that in a few cases frequencies are probability indicators or symptoms: we can use the former to estimate the latter. For example, the frequency of the clicks of a Geiger-Müller counter placed near a radioactive source is an indicator of the probability of radioactive decay.

However, we also count the frequencies of such non-chance events as plane departures, traffic accidents, robberies, births, suicides, the occurrence of certain words in texts, response accuracy in tests, and so on. Such frequencies are not probability indicators, because the items concerned are non-random. The same holds, though for a different reason, for the distribution of digits in the decimal expansion of the number π. It makes sense to ask, say, what is the frequency of a quintuplet of 9s, but not what is the probability of such an 'event.' The reason is that those numbers are generated by non-probabilistic timeless formulas defining π, not by stochastic devices.

Likewise, weather stations keep meteorological records that allow them to associate precipitations and the like with such variables as temperature, pressure, humidity, and wind velocity. This is a basis for weather forecasts. The word 'probability' that occurs in the broadcasts of such predictions is misleading, because they are not made on the strength of a probabilistic theory of weather changes – if only because no such theory exists. (In fact, mainstream meteorology is an application of classical fluid mechanics and thermodynamics.)

Actually, the numbers occurring in weather forecasts are likelihoods based on past observed frequencies under similar conditions, as well as on current satellite images. A forecast such as 'The probability of precipitation is 80%' should then be reread as 'The likelihood of precipitation is 80%' or, perhaps more modestly, as 'A precipitation is highly likely.'

It is well known that there have been several failed attempts, from John Venn (1888) to Richard von Mises (1931), to define probabilities in terms of relative frequencies. The philosophical rationale behind this project was empiricism: Whereas frequencies are observable, the concept of probability sounds metaphysical, for it evokes that of potentiality. The simplest such attempt is that of equating probability with the limit of the relative frequency as the number of trials (or possible cases) approaches infinity:

$$p =_{df} \lim_{N \to \infty} n/N.$$

But this formula is not mathematically well defined. This is because, if the process in question is genuinely random, then there is no *law* relating n to N, such as $N = an$, or $N = an - n(1 - b)n$, with a an arbitrary number and $b < 1$. In general, there can be no mathematical methods, such as functions or algorithms, for producing genuinely random numbers, such as those occurring in Tippet's famous table – which are actually pseudo-random. Only nature provides genuine randomness – as in radioactive disintegration and in random gene-shuffling during the mating of the eggs and sperm of parent organisms.

Besides, the above formula is semantically wrong, because every probability is a probability of a single (though not unique or unrepeatable) event, whereas a frequency is a property of a collection of events of some kind. (Incidentally, some of the philosophical literature on probability is marred by the confusion between single or individual, and unique or unrepeatable.) Finally, the formula in question is also ontologically wrong, because it purports to reduce potentiality to actuality, which amounts to collapsing the future onto the present or, rather, the past.

Von Mises's frequency theory of probability refined the above-mentioned pseudo-definition of probability as long-run relative frequency. However, his theory involved a mixture of conceptual and empirical concepts, which is bad form in pure mathematics, as we have known since Hilbert. Worse, Jean Ville (1938) has shown that this theory is mathematically flawed. Regrettably, Ville's book went unnoticed, perhaps because it was written in French and appeared on the eve of the Second World War. In any event, it did not catch the attention of philosophers, most of whom have continued to follow either the frequency or the subjectivist school, as if no third stand were available.

The only regularity found in relative frequencies is a marked *tendency* to become more or less constant for large values of N. However, this is a trend, not a law. Indeed, a relative frequency of chance events fluctuates in an irregular and therefore unpredictable fashion as N increases, even though in the long run it does approach the corresponding probability – albeit, only irregularly and approximately (see, e.g., Cramér 1946: 142).

The previous statement connects the concepts of probability and frequency: it does not reduce the first, a theoretical construct, to the second, an empirical one. Hence it is a bridge between probability theory, a chapter of pure mathematics, and mathematical statistics, a branch of applied mathematics. As such, that bridge provides a method for the empirical test of some probability distributions. Indeed, a distribution of a certain property is true provided it matches (to within the experimental error) the corresponding statistical histogram.

(Moreover, whereas most theoretical probability distributions are continuous functions, histograms are necessarily discontinuous, since every set of data is finite.)

However, the above is not the sole bridge between probabilistic formulas and reality. Statistical physics contains further such bridges, none of which involves the frequency concept. For instance, the probability density of the velocities of the molecules in an ideal gas (the Maxwell-Boltzmann distribution) is a function of the gas temperature and the molecular velocities. This formula can be checked by measuring the gas temperature as well as the velocities of the individual molecules that escape through a hole in the container. Likewise, the quantum-theoretical formula for the probability of the radiative decay of an atom from one energy level to a lower level is checked by measuring the wavelength and the intensity of the light emitted as a result of such decay. No relative frequencies occur in either case.

The pride of place still assigned to frequency in teaching and in applied probability is a relic of the times when probability theory focused on games of chance. But in nature and society, coins, dice, roulettes, and lotteries are exceptional. And of course they do not occur in the mathematical theory of probability, except in its application to examples, such as the urn model. Besides, from a logical point of view, probability theory precedes mathematical statistics, not the other way round. Indeed, given a probability distribution, the basic statistical parameters, such as averages and mean standard deviations, are uniquely determined (direct problem). By contrast, any given set of statistical parameters poses the problem of guessing the underlying probability distribution(s). This is an inverse problem, with an indeterminate number of solutions. (For the concept of an inverse problem see chapter 13, section 4.)

In short, the concepts of probability and frequency are interrelated but not inter-definable.

5 Probability, Chance, and Causation

The standard elementary theory of probability, which deals with the particular case of countable sets of possible point events, is due to Kolmogorov (1956 [1933]). In keeping with the mathematical format so eloquently advocated by Hilbert (1918), Kolmogorov proposed an axiomatic definition of probability. That is, he specified the conditions (axioms) that a function must satisfy to qualify as a probability function defined on a certain family of sets of 'events' (which in this theory are just nondescript sets). Alternative axiomatic definitions are possible and have, indeed, been proposed, among them A. Renyi's, whose basic notion is that of conditional probability.

Any scientific application of the mathematical theory of probability is assumed to refer to random facts, such as the relative orientations of the spins of the atoms in a chunk of a non-magnetic substance or of the fibres in a clump of paper pulp; the disintegration of the atoms of a body of radioactive material or the scattering of a beam of elementary particles by a target; the thermal noise in a conductor or the accidental errors of measured values; the shuffling of genes occurring during egg fertilization or the indiscriminate mating of most lower invertebrates. No stochastic mechanism, such as random shuffling and blind choice, no probability.

Non-random events, such as meteoritic impact, cell division, the onset of sicknesses, wedding, bankruptcy, as well as changes of sex and government, do not qualify as arguments of probability functions: they are essentially causal, though subject to minor random perturbations. Such events may be assigned non-quantitative likelihoods or propensities, and sometimes even frequencies, but not probabilities proper. For example, poor people are more prone to getting sick than well-off people, but no probability can be assigned to such propensity. Likewise, *coups d'état* are more frequent in Third World countries than in others, but there is nothing random about them: they are all plotted.

Propensities or dispositions, then, are of two very different kinds: causal like fragility and suggestibility, and random like radioactivity and genetic mutability (Bunge 1976). Only chance propensities can be identified with probabilities. The confusion between the two concepts in favour of causal propensity, a notion familiar from ordinary knowledge, has led Humphreys (1985) to the quaint conclusion that the theory of probability is not the correct theory of chance. What is true is that, to be applied, the mathematical theory of probability must be enriched with non-mathematical hypotheses, such as the symmetry of the proverbial dice and their being either shaken or chosen in a random fashion.

The confusion of likelihood with probability occurs in some of the creationist arguments against the evolutionary hypothesis of the spontaneous emergence of the first living beings from abiotic materials. One of those arguments reads as follows. For a cell to emerge spontaneously, at least a dozen conditions must have been met: the formation of a planetary system, the existence on one planet of the requisite elements (carbon, hydrogen, nitrogen, etc.), the prevalence on that planet of favourable pressure, temperature, and humidity conditions, and so on. Now, every one of these conditions is highly improbable; hence the probability of the meeting of all such conditions, which is the product of a dozen tiny probabilities, is derisory. (Some bold writers have offered numerical estimates of such probabilities.) Therefore, the emergence of life must have been the product of divine design – whence the sanctity of life, or at least that of the believer.

The snag in this argument is that none of the said events is a chance event: every one is the result of a deterministic process, such as the accretion of cosmic dust under the action of gravitation and the combination of complex molecules out of atoms. Hence it makes no sense to assign them probabilities. What is true is that some of those events are unlikely, and their conjunction was a lucky accident. Still, organisms may have emerged spontaneously more than once both on Earth and elsewhere. This hypothesis has become highly plausible because we know now that (a) there are many extrasolar planetary systems; and (b) complex biomolecules, such as DNA, emerge spontaneously when their constituents are mixed under suitable physical conditions. In sum, the spontaneous emergence of life may have been unlikely, but it was neither probable nor improbable in the technical sense of the term, because it was not a chance event. It was bound to happen given the right circumstances.

A related example of pseudo-probabilistic reasoning is one used by some creationists: the Drake formula for the expected number of advanced civilizations in our galaxy. It purports to give that number as a product of a dozen or so frequencies or probabilities, among them that of planetary systems including an inhabitable planet, the fraction of inhabitable planets on which life could emerge, and the fraction of those inhabited planets where civilizations may have emerged. There are at least two problems with the Drake formula: (a) it assumes that the facts in question are both random and mutually independent – neither of which hypotheses is plausible; and (b) it postulates the given numbers instead of deriving them from either data or theories enriched with data. Besides, whichever number may result, it proves nothing. In fact, even if the number in question is extremely small, it cannot be used by the creationists, because a small probability does not rule out actual likelihood. Rather than continuing speculation along the Drake line, it would be more interesting to design instruments capable of detecting intelligent signals coming from the fifty or so extrasolar planets certified at the time of writing.

In sum, only random events can correctly be assigned probabilities; so much so that randomness can be equated with having a probability distribution. For example, unlike systematic errors (or biases), the accidental errors made in performing a sequence of high-precision measurements fit a bell-shaped curve. Likewise, contrary to the notes produced by a pianist playing a classical music piece, white noise is random.

As indicated above, two kinds of propensity must be distinguished: causal and stochastic. Causal propensities may be measured by the frequency of actual occurrence, as in 'The relative frequency of bankruptcies of small firms in this country over that period was such and such.' By contrast, chance propensities are probabilities interpreted in factual terms, as in 'The probability of the transition of an atom of species K from energy level E_1 to energy level E_2 within the time interval T is p.' These probabilities can sometimes be measured by the corresponding frequencies: in these cases, the latter are probability indicators.

Incidentally, the difference between causal and stochastic propensity should provide a caveat against the rather popular idea that causation is a particular case of probability (e.g., Suppes 1970). Although the two concepts in question are occasionally related, they are not inter-definable. Indeed, the concepts of cause and effect do not occur in the definition of 'probability,' and the latter is not involved in the definition of 'causation' as either energy transfer or event trigger. What is true is that, as indicated above, the two concepts in question are often related. For instance, in the quantum theory of scattering one calculates the probability that movement through a force field (a cause) will deviate an incoming particle beam within a certain solid angle.

In sum, just as only the exact concept of energy should be employed, so only the exact (mathematical) concept of probability, which is coextensive with that of chance (or randomness), should be used. Expanding its use to have it coincide with the intuitive notions of likelihood and plausibility, as is sometimes done, poses a number of problems. One of them is that the imprecise notions of likelihood (of events) and plausibility (of hypotheses) are treated as if they were exact. Arbitrary numbers are thus used to confer scientific respectability upon mere hunches. Consequently, the pitfalls of intuition are ignored, and thus conceptual sloppiness and disastrous practical decisions are risked.

6 Credibility

Verisimilitude and credibility are often equated with probability. (Actually in German the same word, *wahrscheinlich*, designates both the ordinary-knowledge and the technical concepts. By contrast, French and Spanish have different words for the two concepts.) This conflation of an epistemological category (verisimilitude), a psychological one (credibility), and an ontological one (probability) is a root of the subjectivist or Bayesian theory worked out by Ramsey (1931), de Finetti (1937), Jeffreys (1937), Good (1950), and Savage (1954), among others. This theory is somewhat popular among statisticians, and even more so among rational-choice theorists and philosophers. However, it lies outside both mainstream probability theory and mathematical statistics.

Like von Mises's frequency theory, the Bayesian one involves a non-mathematical notion, this time a psychological one. Indeed, the theory conceives of the probability of an item (whether event or proposition) in terms of the credibility, credence, or degree of reasonable belief assigned to it by a subject. The original motivation of Bayesianism seems to have been ontological: If the world is strictly deterministic, as it looked in the eighteenth century, then chance, like ugliness, must be in the eye of the beholder.

The subjectivist view of probability is open to the following objections (Bunge 1981b).

1. Probabilities cannot be assigned to propositions. These can be more or less true, but not more or less probable, because there is nothing dicey about them. The expression 'the probability that a proposition be true' is 'a meaningless collection of words' (Du Pasquier 1926: 197).

2. The most firm scientific finding concerning the credibility–likelihood relation is that 'people do not follow the principles of probability theory in judging the likelihood of uncertain events' (Kahneman and Tversky 1982: 32). Moreover, subjects with different backgrounds, personalities, or even just different expectations or moods are likely to assign different subjective probabilities to the same event. For example, pessimists tend to exaggerate the 'probabilities' of infrequent disasters. Hence, the concept of credibility is psychological, not epistemological.

3. Since the 'probabilities' in question are subjective, there can be no objective criteria for assigning them. Hence stating, for instance, that the probability of a given verdict by a jury is such and such is every bit as arbitrary as assigning numbers to the beauty of the participants in a beauty contest. (Incidentally, this is why the famous Condorcet jury theorem is pseudo-scientific.) Such numbers are phony. In general, subjective probabilities are pseudo-exact. Therefore, their use can have disastrous practical consequences.

4. The standard mathematical formalism of the probability calculus makes no room for a variable representing a probability appraiser. Indeed, in physics, genetics, and other sciences, one writes formulas of the form '$Pr(e) = p$,' where e names an event and p the numerical value of the probability, not '$Pr(e, s) = p$,' where s stands for a subject. If one were to write formulas of the second type, one would have to add axioms for the behaviour of subjects of different personality types. And such axioms, absent from the mathematical calculus, would have to be proposed and tested by psychologists.

5. Subjective estimates of any quantities, though often necessary in practice, are neither mathematically well defined nor reliable. To use them in preference to postulation or calculation is like replacing geometry with the carpenter's eyeballing of distances and angles. This ability comes in handy in the workshop, but not in mathematics. As for the intuitive estimates of proba-

bilities, psychological research has shown that they are unreliable for several reasons, such as misconceptions of chance and failure to appreciate the role of sample size.

6. There is no way anyone could guess correctly, without performing any calculations, however approximate, any of the tiny probabilities that occur in atomic and nuclear physics, many of which are smaller than 10^{-24}. Such extremely small numbers defy intuition.

7. The subjectivist interpretation of Bayes's theorem involves regarding the two unconditional probabilities occurring in it as a priori, that is, assumed before collecting any relevant data. The formula in question is $Pr(h|e) = Pr(e|h) \cdot Pr(h)/Pr(e)$, where e is a datum relevant to hypothesis h. The priors are $Pr(h)$ and $Pr(e)$. How, if not arbitrarily, does one evaluate such prior probabilities? In practice one may sometimes estimate conditional probabilities by counting the corresponding frequencies, without hazarding arbitrary prior probabilities. However, this only holds for random events.

It makes no sense to ask, for example, what is the probability of the phylogenetic tree (evolutionary phylogeny) of a group of organisms, since the descent relation, far from being random, is determined by genome and environment. Nonetheless, Huelsenbeck et al. (2001) have used Bayes's formula to estimate the probability of a phylogeny given a set of data. But since they do not know the prior probabilities of such trees, they take them to be a priori equally probable – a good example of both the confusion of plausibility with probability and the fallacy of trying to deduce knowledge from ignorance.

By contrast, the likelihood of an item of one kind given the occurrence of an item of another kind (and conversely) makes sense whenever the two are related. Moreover, sometimes such likelihood can be estimated by counting the corresponding frequencies. For example, it is important to know the likelihood of having breast cancer when a mammogram is positive, as well as the likelihood that mammography reveals the presence of cancer. But it would be wrong to insert either in Bayes's formula, because neither cancer nor mammography tests are chance events. (More in chapter 16.)

In sum, probability, likelihood, and plausibility must not be confused with credibility, if only because the latter, unlike the other two, is relative to some subject or other; so much so that one may occasionally have to lie for the sake of credibility, though never for the sake of plausibility.

7 Probabilistic Epistemology

The scholars who write about Bayesian or probabilistic inference are legion (see, e.g., Kyburg 1961 and Franklin 2001). Could they all be wrong? Why not? Let us examine a typical example, that of the so-called proportional syllogism. But let us first recall its counterpart in deductive logic. The following is an instance of the most important rule of inference in elementary logic:

All swans are white.
This is a swan.
∴ This is white.

The inductive counterpart of this valid inference would seem to be this:

99% of all swans are white.
This is a swan.
∴ The probability that this is white is 0.99.

However, the whiteness of a swan is not a matter of chance, but a biologically necessary consequence of its having certain genes characteristic of its species.

A correct reasoning (though still one beyond deductive logic) would be this:

99% of all swans are white.
This is a swan picked at random out of the swan population.
∴ The probability that this is white is 0.99.

The key words here are 'picked at random out of the (entire) swan population.' They show that in this case randomness, hence probability, inheres in the sampling process, not in either the swan population or any propositions referring to swans. And the reasoning itself, though not countenanced by any logical theory, is fertile. Rather than a rule of inference, it is a heuristic for estimating probabilities from frequencies.

Lastly, the subjective probability theory is rife with paradoxes. One of them is the following – which, according to legend, nearly wrecked a theoretical biology conference in 1966 (Bartlett 1975: 101–2). Of three prisoners, Matthew, Mark, and Luke, two are to be executed, but Matthew does not know which one. He believes that his own chance of being executed is 2/3. He asks the warden to tell him in confidence the name of one man, either Mark or Luke, who is going to be executed. The jailer informs him that Mark will be executed. Subjectivist that he is, Matthew feels somewhat relieved: he believes that this piece of information lowers his chance of being executed from 2/3 to 1/2. Is he right? No, because one of the givens of the problem is that the victim has already been chosen: it will be either Matthew or Luke. New information about a fact does not alter the latter.

This problem is alien to chance, hence talk of probability is unjustified. In fact, only if all three prisoners were to be selected at random, would the prior probability of Matthew's being chosen have been 2/3. And his chance would indeed drop to 1/2 after Mark was condemned (and whether or not Matthew learned about this fact), only if his jailers decided to draw lots between Matthew and Luke. But this is not a datum of the problem. See Block 14.1.

Datum: Prisoners A (Matthew), B (Luke), C (Mark)
Probabilties before warden's revelation
 $P(AB) = P(AC) = P(BC) = 1/3$
 $\therefore P(A) = P(AB) + P(AC) = 2/3$

Probabilities after warden's revelation (C is out of the game)
 $P(AB) = 1$
 $\therefore P(A) = P(B) = 1/2$

Block 14.1 Matthew's calculation is wrong because he includes Mark from the start, while actually his fate was not left to chance – as revealed by the warden.

The popular Monty Hall Problem of the three doors is essentially the same: the Grand Prize and the Booby Prize were placed behind two of three doors before the game started. Hence no chance is involved, and consequently it makes no sense to ask what is the probability of winning when opening a given door. The moral is that attaching numbers to guesses does not render these respectable, because it does not enrich knowledge. It only exemplifies pseudo-quantitation. (For the latter see Bunge 1999.)

However, since so many people claim to think in terms of subjective probabilities, and to make important decisions with their help, it would be legitimate to try and exactify them. In fact, the project of building an objective calculus of subjective (eyeballed) probabilities is as intriguing and legitimate as any other study of subjective experience – the central task of human psychology. Still, such project in cognitive psychology has not even been written down, and this presumably for two reasons.

The first is the popular confusion between a probability and its subjective estimate. In fact, this confusion has generated the false belief that the mathematical theory of probability doubles as the psychological theory of credence, and hence the project in question is unnecessary. The second reason is the array of formidable difficulties such a project encounters. Let me mention only two such obstacles.

In the first place, when examining a fan of possibilities, normally we know or include only the known available options. Hence, we have no right to demand that the probabilities of the mutually incompatible possible outcomes add up to unity. Second, suppose that probability estimates satisfy the psychophysical power law: $\pi = ap^b$, where p is the probability and π its estimate by a subject, whereas a is a positive constant characteristic of the subject, and $b \neq 1$ is a positive number peculiar to probability estimates. (Both a and b are to be determined experimentally.) Then the estimated probability of the occurrence of either of two mutually exclusive events would not be the sum of the partial subjective probabilities, since $\pi = a(p_1 + p_2)^b \neq \pi_1 + \pi_2$. This confirms the experimental finding that probability estimates are unreliable. Probabilities must be calculated or measured, not eyeballed – particularly when they are as tiny as those involved in nuclear physics.

Until recently, controversies over non-deductive inference and subjective probability were only of academic interest. But since the explosive expansion of the rational-choice approach around 1970, the matter has acquired practical importance. Indeed, such theories are sometimes being used uncritically in policy-making, management, and even medicine.

I submit that these practices are unwarranted because, as Poincaré emphasized a century ago, serious probability calculations presuppose some knowledge instead of allowing us to dispense with it. For example, in quantum physics the probability densities result from solving such problems as finding the possible states that an atom can be in or the possible outputs of a scattering experiment. The only way to cope with uncertainty is to engage in research.

Decision theory is often advertised as the guide to action under uncertainty. The basic idea is that life is full of uncertainties, and that these uncertainties can be tamed upon identifying them with probabilities, since these obey a mathematical theory. However, the alleged probabilities are in fact mere qualitative likelihoods, and these are estimated intuitively, not rigorously. Further, such estimates are unreliable: most of us are very bad at eyeballing likelihoods and probabilities. For example, it is well known that most people believe that the irregular sequence of coin throws HTHTHT has a greater chance than the sequence HHHHHH, while actually their probability is exactly the same, namely $(½)^6$.

Further, experiment has shown that subjects have a tendency to overrate the 'probabilities' of unlikely events, and hence tend to exaggerate the corresponding low risks (Kahneman, Slovic, and Tversky 1982). Kahneman and Tversky also found that, when asked to rank the 'probability' (likelihood) of certain possible events, most management students at Stanford University assigned a greater probability to compound events than to their constituents. (Incidentally, Kahneman and Tversky too conflated likelihood with probability.)

Medicine has not escaped the probabilistic fad. Indeed, some standard medical textbooks (e.g., Wulff 1981) adopt the Bayesian approach to medical diagnosis and treatment. They write about the probability that a symptom indicates a cause and conversely, and they claim to estimate the expected utilities of alternative treatments (usually in terms of additional life expectancy, with no concern for the quality of life). They thus ignore the fact that the cause–symptom relation is causal, not stochastic; they assign prior probabilities intuitively; and they remain satisfied with them instead of attempting to reduce uncertainties through further tests and biomedical research into the mechanisms behind the diagnostic signs. All of which is as alarming as Russian roulette. (See Eddy and Clanton 1982; Murphy 1997; Bunge 2000d.)

In sum, possibility and chance are ontological categories, whereas uncertainty and credence are psychological and epistemological ones. Hence, to put them all in the same bag is to perpetuate a category mistake. Since there are four different concepts in question, we need altogether at least four different theories. So far, only one is available, namely the mathematical theory of probability, or calculus of chances, as it used to be called.

8 Plausibility or Verisimilitude

Facts are more or less likely to happen, and facts of a certain kind, namely chance events, can be assigned probabilities. What about the corresponding propositions? If pertinent and clear, such propositions will be more or less plausible or verisimilar. Reasonings are parallel: if they lead to plausible 'conclusions' (derived surmises), they will be more or less valid or plausible. But such plausibility is purely conceptual and has nothing to do with probability.

For example, the probabilistic formulas in quantum physics are highly plausible even when they concern extremely small probability values. And none of those formulas can be assigned a probability. At least, no physicist does it. Likewise, the plausible arguments that a mathematician constructs while tackling a new problem or exploring a novel domain are just tentative and more or less promising gropings towards demonstrative reasonings (Polya 1954).

One reason for refusing to regard probability as quantitating plausibility is this. All plausibilities are contextual. If a priori, they are relative to the background knowledge; and if a posteriori, they are relative to both background knowledge and fresh findings. For example, in the context of today's cognitive neuroscience, self-consciousness and free will are plausible, whereas telepathy and psychokinesis are not. Now, if we were to identify plausibility with probability, we would need objective rules for the assignment of probability values to propositions – and no such rules are known. Worse: We would have to admit not only expressions such as the conditional probability $P(h|B)$ of hypothesis h

given background knowledge B, but also $P(B|h)$, the probability of B, which might be bulky and solid, in the light of a single conjecture h. And this would be lunacy, because B is weighty whereas h may be wacky. We need then a non-probabilistic theory of plausibility. An embryo of one such a theory follows.

Plausibility is a qualitative and relational property of propositions (in particular hypotheses), beliefs, and inferences. A hypothesis that has not yet been checked, or the evidence for which is inconclusive, may sound plausible in the light of some body of knowledge. The same holds for theorems that have been guessed but not yet proved: they can be more or less plausible. How plausible? There is no way of knowing, as long as no tests are conducted. But once the tests have been carried out, if they are conclusive we say of the hypothesis that it has been confirmed (or falsified), so that it may be pronounced true (or false) – at least for the time being. Thus, plausibility (or verisimilitude) is potential truth, not potential fact.

Likewise, a proved theorem is no longer just plausible. That is, after a conclusive test we no longer need the concept of plausibility. And before testing we cannot (or ought not to) try and measure the degree of plausibility. In this case, the most we can say is that the conjecture in question is or looks plausible or implausible in the light of some body of knowledge, or that one hypothesis is or looks more plausible than another in the same context.

More precisely, let p and q designate co-referential propositions, and B a body of knowledge relevant to both p and q. Suppose that B can be split into an essential part E and an inessential part I, that is, $B = E \cup I$, with $E \cap I = \emptyset$. (Typically, E will contain generalizations with a good track-record, whereas I will contain only narrow hypotheses and empirical data.) I stipulate that

p is *plausible* w.r.t. $B =_{df} p$ is compatible with every member of B.

p is *more plausible than q* w.r.t. $B =_{df} p$ is compatible with more members of B than q.

p is *essentially plausible* w.r.t. $B =_{df} p$ is compatible with every member of E.

The definitions of the dual concepts of implausibility and essential implausibility are obvious.

9 Towards a Plausibility Calculus

The following axioms for a future calculus of plausibility seem to capture some of our intuitions about the matter. Assuming a fixed body B of background knowledge, and interpreting '$p \geq q$' as 'p is more plausible than q,' we postulate

A1 $\neg (p \geq q) \Leftrightarrow (q \geq p)$
A2 $p \vee q \geq p$
A3 $p \geq p \wedge q$
A4 $(\exists x)Fx \geq (\forall x)Fx$

Some logical consequences follow. A2 entails

Theorem 14.1 $(p \Rightarrow q) \geq \neg p$.

A3 entails that, of two theories differing only by a finite number of axioms, the simpler is the more plausible. However, because the simpler theory is the less comprehensive and deep one, it benefits from a lesser number of possible confirming instances. Hence, the more plausible theory is not necessarily the more promising one: it is only the one that should be checked first.

A hypothesis may be said to be *empirically plausible* with regard to a set of data relevant to it if the overwhelming majority of the data confirm it. And a hypothesis may be said to be *theoretically plausible* if it is consistent with the bulk of the background knowledge relevant to it. Normally, only theoretically plausible hypotheses are put to the empirical test, and only empirically plausible hypotheses are judged to be worthy candidates for some theory. This is how research proposals are written, evaluated, and funded. Nobody would be rash enough to claim a definite probability – rather than a strong plausibility – for the hypothesis he intends to put to the test.

Finally, beware the equating of plausibility with either probability or improbability. Both identifications are wrong if only because, as remarked earlier, there is no objective criterion for assigning probabilities to propositions. Besides, even if propositions could be assigned probabilities, these would measure neither plausibilities nor degrees of truth. One reason is that the probability of two mutually independent items equals the product of their partial probabilities. But if one of them is highly plausible, and the other is very implausible, their conjunction should be half-plausible; in fact, the probability of such conjunction (supposing it made sense) is nil.

However, though different, the two concepts are related in a straightforward way, namely thus: If a and b are random events, and a is objectively more probable than b, then obviously

Event a will happen $\geq$ Event b will happen.

In particular, if two events are equiprobable, then the corresponding assertions are equiplausible.

In short, let us not confuse the plausibility (or verisimilitude) of an idea with the probability of the corresponding fact. The former is in the head, whereas the latter is in the external world. Poisson knew this as far back as 1837; so did Cournot, who in 1851 described probability as 'the measure of physical possibility.' Maxwell, Boltzmann, Einstein, and other workers in statistical physics took this realist interpretation of probability for granted. Smoluchowski (1918), Du Pasquier (1926), Bunge (1951), Fréchet (1955), Popper (1957), and a few others have argued explicitly that objective probabilities are properties of real things just as much as their energies. Were it not so, probabilities would neither occur in scientific theories nor be measurable.

Concluding Remarks

The following table summarizes the preceding.

Concept	Referents	Methodological status
Possibility *simpliciter*	Anything	Imprecise
Possibility *de dicto*	Propositions	Imprecise
Possibility *de re*	States or events	Qualitative, precise
Feasibility	Human actions	Qualitative, precise
Likelihood	States or events	Qualitative, precise
Relative frequency	States or events	Quantitative, precise
Mathematical probability	Sets	Quantitative, precise
Objective probability	States or events	Quantitative, precise
Subjective probability	Anything	Pseudo-quantitative
Plausibility	Propositions	Qualitative, precise
Partial truth	Propositions	Quantitative, precise

In closing, let me offer a few heuristic suggestions.

1. Distinguish possibility *de dicto* from possibility *de re*. Whereas the first concerns propositions, the second is attributable to facts, in particular actions. That is, while possibility *de dicto* equals plausibility, possibility *de re* equals either likelihood or technical feasibility.

2. Be sparing in the use of the word 'probability,' which designates a technical concept applicable solely to random events.

3. Do not gamble with either health, truth, or justice.

4. Do not assign probabilities to propositions, because there is nothing random about them.

5. Do not confuse chance, an objective state of affairs, with uncertainty, a

state of mind. Chance may generate uncertainty, but the converse is false. So is, consequently, the statement that the two categories are mutually equivalent.

6. Take into account that, whereas the concept of probability is theoretical and predicable of the individual members of a random collection, that of frequency is empirical and represents a property of collections of factual items of any kind, whether random or not.

7. Do not attempt to assign either a probability or a frequency to one-off events such as the emergence of the solar system, the first cell, or the first university, for they were not chance events; and also because, such events being not just single but also unique, it would be impossible to check empirically the hypothesized probability.

8. Recall that a plausible construct may turn out to be incorrect, and that most scientific and technological truths are approximate – whence the need for a true (or at least plausible) theory of partial truth. This is an important unfinished business of exact semantics and epistemology.

9. Do not equate credibilities with probabilities, for experimental psychology has shown the former to be subjective and not to satisfy the laws of the probability calculus.

10. Remember Francis Bacon's remark, that confusion is worse than error because the latter, if detected, can be corrected.

15

Emergence of Truth and Convergence to Truth

What is factual truth, as exemplified by 'It is true that rain wets'? What type of fit, matching, or correspondence is involved in a statement of the form 'That hypothesis fits the facts'? Further, what can be truth bearers: propositions, sentences, pictures, or all of them? Are truth and falsity born or acquired? That is, are propositions true or false whether or not we know it, or do their truth values emerge sometimes from their test? Can a sequence of partial truths converge to a total truth? And what is the point of seeking the convergence of data, hypotheses, methods, approaches, and research fields? Just coherence, or truth as well? These are some of the questions to be investigated in this chapter. The questions are very old, but some of the answers may be novel.

1 The Nature of Truth

The first truth about truth is that it is multiple. Indeed, there are logical, mathematical, factual, moral, and artistic truths. For example, 'Here we are' is a logical truth because, by definition, 'here' is wherever we are. A multiplication table is a collection of mathematical truths beyond logic, though consistent with it. 'Free traders practise protectionism' is a factual truth these days. 'It is wrong to take advantage of the weak' is a moral truth. And 'Don Quixote is generous' is an artistic truth.

The concepts of logical and mathematical truth have been elucidated by logicians and metamathematicians. Consequently, philosophers have little if anything to say about them. The logical concept of truth coincides with that of logical truth or tautologousness. In mathematics, the coherence (or consistency) concept of truth reigns. However, mathematical truth comes in two varieties: the truth of abstract formulas, such as those of set theory, and that of interpreted formulas, such as differential equations. The former, or model-theoretic concept, may be summarized as follows: A formula is said to be true in a model (example) if it is satisfied in it. For example, the formula for the commutativity property, 'x ⊗ y = y ⊗ x,' is true for number and set product, but false for vector and matrix multiplication. This concept of formal truth is elucidated in model theory, a branch of logic in the broad sense. The second mathematical concept of truth coincides basically with that of theoremhood: A formula F is true in a theory T if F is deducible in T with the help of the logic underlying T. This concept of truth is elucidated in proof theory. Any of the three concepts of truth examined so far – as tautologousness, satisfiability, and theoremhood – deserves to be called 'formal,' by contrast with 'factual.' (Leibniz called them *vérité de raison* and *vérité de fait* respectively.) Whereas formal truths are invented – though not as one pleases – factual truths are discovered, though not without the help of formal truths.

The concept of factual truth, or correspondence *theory* of truth as it is called optimistically, is another matter. It may be compressed into the following definition: 'A proposition p referring to a fact f is true $=_{df} f$ is really (actually, in fact) the case as described by p.' Note the occurrence of three heterogeneous items: a fact f, its propositional representative p, and the metaproposition 'p is true.' Note also that truth is predicated of propositions, not sentences. The latter may be said to be vicariously true or false, that is, having a truth value by virtue of designating propositions. This is because languages, unlike bodies of knowledge, are and should remain alethically neutral (recall chapter 4).

The correspondence 'theory,' central to scientific realism, is thus far just a definition crying for a research project. In fact, no detailed and generally accepted correspondence theory proper (hypothetico-deductive system) seems to exists. Let me offer a few preliminary ideas on this subject in the present section and the next. However, the reader uninterested in technicalities is invited to skip them, jumping to section 5.

A convenient point of departure is naive realism, which includes what Marxists call the reflection theory of truth, and Wittgenstein called the 'picture theory of language.' According to it, all ideas 'reflect' facts out there. Regrettably for common sense, this view is false because most factual truths refer to selected aspects of facts rather than to whole facts. Besides, some propositions are irrelevant or false, because they refer to non-existent facts. Furthermore, negative propositions do not refer to negative facts, for there are none such. Likewise, disjunctions do not refer to disjunctive facts, for there are none such either. (On the other hand, there are conjunctive facts, such as the simultaneous or successive occurrence of two events.) Last, but not least, general propositions, particularly if somewhat abstract, like the scientific laws, cannot picture anything, since only particulars are picturable. Nor is it true that true theories (in particular theoretical models) are isomorphic to their referents. A system of propositions does not resemble a set of facts, any more than the spoken or written word 'table' resembles a table.

All this is and more is realized the moment one stops thinking of the correspondence in question as a bijection or one-to-one mapping. Bijection is a very special and therefore uncommon case of correspondence, namely, the one wherein every fact under consideration is described correctly by a single proposition and, conversely, every proposition under consideration describes adequately only one fact. In this case, no false propositions are considered. But of course a true theory of truth will cover falsity along with truth, partial truth as well as full truth, and as-yet untested predictions along with tested ones. (Recall Aristotle's problem of the outcome of tomorrow's sea battle.)

The proposition-fact pairing is as varied as sexual pairing: in addition to monogamy there are polygamy, engagement, and singlehood. A false factual proposition is one without a factual counterpart. And a proposition about the future is seen as a proposition with only possible partners: the actual pairing occurs only if and when the prediction is borne out. In other words, all factual truths, however plausible a priori, arc a posteriori.

In fact, there are five jointly exhaustive cases: one-to-one (one proposition for every fact and conversely); many-to-one (different descriptions of the same fact); one-to-many (same description of different facts); factual gap (propositions without corresponding facts); and propositional gap (facts without corresponding propositions):

A One-to-one	B many-to-one	C one-to-many	D factual gap	E propos. gap
$p \to f$	$\left.\begin{array}{l}p \\ q\end{array}\right\} f$	$p \left\{\begin{array}{l}f \\ g\end{array}\right.$	$p \to f$	$p \to f$
$q \to g$			q	g

Example of A: Correct enumeration of a temporal sequence of events.
Example of B: Multiple descriptions of a single fact.
Example of C: Confusion or stylization of different facts.
Example of D: Irrelevant, false, or as yet untested proposition.
Example of E: Incomplete description of a domain of facts.

2 Towards an Exact Concept of Correspondence

Let $\mathbb{F}$ denote a set of possible facts, and $\otimes$ their conjunction (or concatenation, or conjunction). We assume that the concatenation $f \otimes g$ of any two facts f and g in $\mathbb{F}$ is a third fact – rather than, say, a fiction. We also assume that factual concatenation is associative: $f \otimes (g \otimes h) = (f \otimes g) \dot{\otimes} h$, for any f, g, and h in $\mathbb{F}$. And we define the null fact O as that which, when concatenated with an arbitrary fact f, leaves it unchanged: $O \otimes f = f \otimes O$. That is, O plays the role of the unit element. Clearly, $\langle \mathbb{F}, \otimes, O \rangle$ is a monoid (or semigroup with identity). Note that we do not assume that facts can disjoin: real things and changes in them (events or processes) conjoin but do not disjoin. They do not practise negation either: negation, like disjunction, is a conceptual process.

Next, let $\mathbb{P}$ stand for the set of propositions about the facts in $\mathbb{F}$, and $\wedge$, $\vee$, $\neg$ for the standard propositional connectives. It is well known that $\langle \mathbb{P}, \wedge, \vee, \neg \rangle$ is a complemented distributive lattice. (The rumor that quantum mechanics requires it to be non-distributive stems from the confusion of propositions with operators. The logic underlying quantum mechanics is as classical as the mathematical formalism of this theory.)

We stipulate that a map from $\langle \mathbb{F}, \otimes, O \rangle$ into $\langle \mathbb{P}, \wedge, \vee, \neg \rangle$ formalizes the intuitive concept of representation of facts by propositions. (This map is the inverse of the map $\mathcal{R}$ that pairs propositions to their referents, defined in Bunge 1974a.) In turn, a partial map from $\langle \mathbb{P}, \wedge, \vee, \neg \rangle$ into the unit interval [0,1] of the real line is a truth-valuation function $\mathcal{V}$. (This map is partial because not every proposition in $\mathbb{P}$ can be assigned a truth value: think of the untested and the undecidable propositions.) The two maps should combine as follows:

Representation $\mathcal{R}^{-1}$ *Truth valuation* $\mathcal{V}$
$\langle \mathbb{F}, \otimes, O \rangle$ $\rightarrow$ $\langle \mathbb{P}, \wedge, \vee, \neg \rangle$ $\rightarrow$ [0,1]

The theory should specify the maps $\mathcal{R}^{-1}$ and $\mathcal{V}$ in such a way that
1. all the propositions representing the null fact O are false: If $\mathcal{R}^{-1}(p) = O$, then $\mathcal{V}(p) = 0$ for all p in $\mathbb{P}$;
2. for some $f \in \mathbb{F}$, $\mathcal{R}^{-1}(f) = p \in \mathbb{P}$, and $\mathcal{V}(p) = u \in [0,1]$;
3. for some $f, g \in \mathbb{F}$, $\mathcal{R}^{-1}(f \otimes g) = (p \wedge q) \in \mathbb{P}$, and $\mathcal{V}(p \wedge q) = v \in [0,1]$.

Note the occurrence of 'some' instead of 'all' in the last two clauses. This is due to the gaps in both $\mathbb{F}$ and $\mathbb{P}$. These gaps make room for the hypotheses (e.g., predictions) that have not yet been tested, whence their truth values are unknown – as will be argued in section 4.

The facts-propositions map $\mathcal{R}^{-1}$ can be analysed into two maps: facts-thoughts, and thoughts-propositions. (A particular thought is construed here as a particular brain process, whereas a proposition is conceived of as an equivalence class of thoughts: see Bunge 1980a. No two thoughts are strictly identical even if they consist in thinking of the same proposition.) The analysis in question is the composition of two maps: imaging, or $\mathcal{I}$, from facts $\mathbb{F}$ into thoughts Θ, and conceptualization, or C, from thoughts to propositions $\mathbb{P}$:

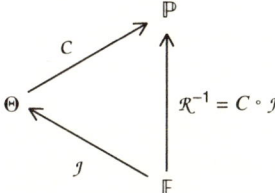

As long as the representation map $\mathcal{R}^{-1}$ remains undefined, we have no right to talk about a correspondence *theory* of truth. We can only talk about *projects* of crafting such a theory.

3 Partial Truth

Most philosophers still believe that there are only two truth values: truth and falsity. However, craftsmen, businessmen, scientists, and technologists have known for about three millennia that, in addition to fully true propositions, there are approximately true ones. Examples: the statements that π equals 3, and that our planet is spherical. We should therefore admit many, perhaps infinitely many, truth values between the extremes of complete truth and complete falsity.

This is what approximation theory (pioneered by Archimedes) and the calculus of errors (fathered by Gauss) are all about. A methodological consequence of the thesis that truth is graded is that so is falsification (or refutation). For instance, the popular opinion that Newtonian mechanics has been falsified is false. In fact, that theory is an excellent approximation for medium-size bodies in slow motion. This is why physicists, astronomers, and mechanical engineers continue to use classical mechanics when appropriate. For the same reason, not all negative empirical findings are as conclusive as falsificationists like Popper suppose.

It might be thought that the notion of graded truth requires replacing ordinary logic, which is two-valued, with many-valued logic. This is not so, because logic deals with deducibility, not truth; so much so that in constructing an argument, one may assert or deny just for the sake of the argument. And it may be assumed, for instance, that $\pi = 3$, provided the area of a plane circle is taken as $3r^2$ rather than πr^2. Graded truth is thus compatible with ordinary logic as long as truth values, whatever they are, be preserved under deduction.

In other words, one assumes, usually tacitly, that there is a truth-valuation function $\mathcal{V}$ from some set $\mathbb{P}$ of propositions into a numerical interval, which for convenience may be taken to be the unit real interval $[0,1]$. For example, if one takes π to equal 3, one commits the relative error 0.14/3.14, so that one may set $\mathcal{V}(\pi = 3) \approx 1.00 - 0.14/3.14 \approx 0.95$. In general, one may set $\mathcal{V}\colon \mathbb{P} \to [0,1]$, with the understanding that $\mathcal{V}$ is a partial function, since it is not defined for the untested or undecidable propositions in $\mathbb{P}$. Our problem is to come up with a plausible system of conditions (postulates) defining $\mathcal{V}$.

We want such postulates to make room for such half-truths as 'Aristotle was a Mexican philosopher,' and '$\pi = 3$.' The following is a plausible, yet tentative, set of desiderata for $\mathcal{V}$.

D1 If p is a quantitative proposition that has been found to be true to within the relative error ε, then $\mathcal{V}(p) = 1 - \varepsilon$.

Example: $p =$ 'There are 9 people in this room,' while the actual count shows that there are 10 people. Relative error $= \varepsilon = 1/10$, whence $\mathcal{V}(p) = 1 - 1/10 = 9/10$, which is a pretty good approximation.

D2 If p is not the negate of another proposition, then

$$\mathcal{V}(\neg p) = \begin{cases} 0 \text{ iff } \mathcal{V}(p) = 1 \\ 1 \text{ iff } \mathcal{V}(p) < 1. \end{cases}$$

Otherwise, that is, if q is the negate of a proposition p, which in turn is not the negate of another proposition, then

$$\mathcal{V}(\neg p) = \mathcal{V}(q).$$

Example: if p is the example of D1 above, then $\mathcal{V}(\neg p) = 1$. That is, the statement that there are not 9 people in the room is completely true – a cheap truth, though.

D3 For any two propositions p and q : if $p \Leftrightarrow q$, then $\mathcal{V}(p) = \mathcal{V}(q)$. This is not offered as a deep insight but as an obvious control.

D4 If p is not the negate of q, then

$$\mathcal{V}(p \wedge q) = 2^{-1}[\mathcal{V}(p) + \mathcal{V}(q)].$$

Otherwise, $\mathcal{V}(p \wedge \neg p) = 0$.

Example: $p =$ 'Aristotle was a Mexican philosopher.' This is the conjunction of two propositions, one true and the other false. Hence, $\mathcal{V}(p) = 1/2$.

Note that, if $q = p$, then $\mathcal{V}(p \wedge q) = \mathcal{V}(p)$, as it should be. Note also that the valuation function jumps at $q = \neg p$. This discontinuity reflects the abruptness of negation. If the case $q = \neg p$ were treated as any other, we would obtain $\mathcal{V}(p \wedge \neg p) = (1/2)\mathcal{V}(p)$, so that inconsistencies would be just half-truths. A similar proviso must be made for disjunction:

D5 If p is not the negate of q, $\mathcal{V}(p \vee q) = \max\{\mathcal{V}(p), \mathcal{V}(q)\}$. Otherwise, $\mathcal{V}(p \vee \neg p) = 1$.

Example: $p \vee q$ = 'Heidegger was a philosopher or a scribbler.' $\mathcal{V}(p \vee q) = 1$. A corollary of D5 is $\mathcal{V}(p \Rightarrow q) = \max\{\mathcal{V}(\neg p), \mathcal{V}(q)\}$. In particular, if $\mathcal{V}(p) = 1$, then $\mathcal{V}(\neg p) = 0$, and $\mathcal{V}(p \Rightarrow q) = \mathcal{V}(q)$; and if $\mathcal{V}(p) < 1$, then $\mathcal{V}(\neg p) = 1$, and $\mathcal{V}(p \Rightarrow q) = 1$.

In this system negation does not come in degrees: like death, it is abrupt, equalizing, and cheap. This is why critics can be right more often than their targets. However, not all contradictions are utterly worthless: some of them function as alarm bells, and others trigger research projects. For example, whereas 'Angels are winged and angels are wingless' is barren, 'The universe is infinite and the universe is finite' is interesting and still a challenge to cosmology. That is, if one of two mutually exclusive options may (though need not) be true, then their conjunction has a heuristic value. Besides, without contradictions we would not be able to use the *reductio ad absurdum* principle strategy.

Despite these redeeming virtues, contradiction is a bane if it stops reasoning in its track or paralyses action. Yet, it is not as bad as confusion, let alone nonsense. Indeed, a contradiction can be 'solved' (eliminated) by just dropping one of its constituents; and confusions can be clarified by analysis, whereas nonsense is intractable. The correct semantic ranking is this:

Nonsense < Confusion < Contradiction < Partial truth < Total truth.

And the correct methodological ranking is

Meaningful statement < Plausibility judgment < Test < Truth-value assignment.

(Meaning precedes test, and untested propositions have no known truth values: see Bunge 1974b. This thesis turns the Vienna Circle's verification thesis of meaning upside down.)

The next task is to craft a consistent postulate system incorporating some or all of the above desiderata. Three caveats are in order. The first is that, if truth and falsity are regarded as mutually complementary, one may be tempted to postulate that $\mathcal{V}(\neg p) = 1 - \mathcal{V}(p)$. However, this assumption entails that the negation of a half-truth (= half-falsity), which should be completely true, is worth the same as its assertion. Besides, jointly with D4, it leads to the unacceptable result that conjunctions and disjunctions have the same truth value. Allow me to repeat: Denying is far cheaper than asserting. This is why falsificationism is only good for weeding out error.

The second caveat is that one should resist the temptation to define partial truth in terms of probability – a temptation that neither Reichenbach nor Popper resisted. One reason is that truth and probability are not inter-definable, if only because truth is predicated of propositions, whereas probabilities can only be predicated of facts of a certain type (recall chapter 14). A second reason is that the concept of truth is logically prior to that of probability because, when checking probabilistic statements, whether theoretically or experimentally, we take it for granted that they can be true to some extent. Finally, truth values do not combine like probabilities. For instance, the truth value of the conjunction of two independent propositions with the same truth value equals the latter, whereas the probability of the conjunction of two independent equiprobable events equals the product of their probability, which is less than their individual probability.

The third and final caveat is that the theory should include the concept of reliability of the truth source, such as the testing technique. Indeed, it often happens that we assign high truth values when employing a coarse method, only to discover that the truth values are actually lower when a more exacting procedure is employed. This suggests adding the following desideratum:

D6 If a proposition *p* can be assigned different truth values on the strength of tests with different reliabilities $r(p)$, choose the assignment that maximizes the product of the two values:

$$r(p) \cdot \mathcal{V}(p) = \max.,$$

where the reliability $r(p)$ ranges between 0 and 1.

So much for the desiderata for the valuation function. The enumeration of these conditions is proposed as a research project that consists in finding at least one set of axioms satisfying most of the above desiderata. This project will not be attempted here.

Another research project, to be sketched presently, is that of setting up an alethic metric, that is, of defining the distance between any two propositions with regard to their truth values. A possible alethic metric is the function δ: $\mathbb{P} \times \mathbb{P} \to \mathbb{R}$ such that, for any *p*, *q*, and *r* in $\mathbb{P}$,

(a) if $p \Leftrightarrow q$ then $\delta(p, q) = 0$;
(b) $\delta(p, q) = \delta(q, p)$
(c) $\delta(p, q) + \delta(q, r) = \delta(p, r)$;
(d) if $p =$ '$M(a) = r$,' and $q =$ '$M(b) = s$,' then $\delta(p, q) = |r - s|$.

If the distances between two propositions *p*, *q* and a baseline *b* are given, postulates (b) and (c) above allow to one compute the alethic distance between the given propositions:

$$\delta(p, q) = \delta(p, b) + \delta(q, b).$$

These formulas should be useful for comparing measurement results.

Finally, one may define the concept of relative truth, or truth relative to a baseline b taken to be completely true, as $\mathcal{V}(p, b) = \delta(p, b)$. For example, if p is an approximate solution, whereas b is the exact known solution, then the distance between them equals the error of the former relative to the latter: $\delta(p, b) = \varepsilon$. The knowledge of ε helps evaluate the approximation method.

4 The Emergence of the Knowledge of Truth

According to Platonism, ideas are self-existent, and so are their truth values. That is, propositions would be true or false whether or not we know it. Aristotelians and materialists deny the independent existence of ideas. But they need not take sides in the debate over whether truth and falsity are inherent in propositions or must be found out. Only a methodological analysis of the test (or verification) process can contribute to this debate. I submit that such an analysis suggests that (a) propositions, if regarded as brain processes, are true or false to some degree the moment they are thought up; but (b) truth values must be discovered by analysis, in the case of formal propositions, or by empirical procedures, such as counting and weighing, in the case of propositions about the real world. To indulge in a biological metaphor, truth is innate, but its explicit knowledge is acquired.

In other words, weighing the worth of a factual proposition p is the end-state of an eight-step process:

1 Formulating p as precisely as possible.
2 Checking the background knowledge to find out whether p is compatible with it and likely to be new or so far untested.
3 Checking whether p is testable in principle.
4 Making sure that p is worth being tested or retested.
5 Devising a procedure for testing p.
6 Setting up a procedure for testing p.
7 Carrying out truth-tests of p.
8 Evaluating the truth value of p in the light of both the background knowledge and the test findings; i.e., stating the metaproposition $\mathcal{V}(p) = v$.

In the sciences and technologies, some truth-tests are inconclusive. In such cases one must withhold judgment for a while on truth values, hoping to be able to repeat the test, perhaps modifying the techniques involved. For example, until a few decades ago the evidence for the hypothesis that smoking causes cancer was inconclusive; likewise, a number of medical procedures have been discontinued in the light of unfavourable new tests.

However, factual truth, as a correspondence between a fact in the inquirer's brain and some other fact, is an objective relation that holds or fails to hold regardless of any test procedures. To realize this consider the following simple example. I suddenly hear a booming noise, but I do not know what caused it. So, I make up the following conjectures:

A That was thunder.
B That was a supersonic shock wave.
C That was a cannon shot.
D That was a dynamite explosion.
E That was a firework.

I look, listen, and ask around, in an effort to find evidence for or against these five conjectures. In undertaking this task I assume tacitly that only one of them is true. Being a realist, not a constructivist, I assume that factual truth – the matching of ideas to the facts they refer to – is not a construction, let alone a social one. So, I look up at the sky and, as it happens, I see dark clouds moving rapidly, and a flash of lightning. So, I 'conclude' (hypothesize) that A was true, although I did not know it when I thought it up. Of course, further inquiries may show that A is false: actually a supersonic jet caused the boom. But the point is that such inquiries are conducted to discover whether or not the 'correspondence' held.

In sum, the knowledge of truth emerges from tests, it comes in degrees, and in principle there are ways of improving the accuracy of calculations and measurements. In particular, approximation theory, a branch of mathematics, contains methods of successive approximations that in some cases allow for the uniform convergence to the truth. For better or for worse, no such methods are available in factual science. Consequently, it is pointless to think of infinite sequences of factual propositions with increasing truth values. The reasons for this are the following (Bunge 1963: 124):

1 Standards of rigour and error estimates are not fixed once and for all, but change as aims and techniques alter.
2 Although the history of knowledge does exhibit improvement in approximations, it fails to show a uniform convergence to a limit, if only because progress is often achieved by changing course, and by replacing hypotheses with others containing entirely new concepts.
3 The concept of infinity, involved in the definition of a uniform convergence to the truth (or to anything else), does not apply to finite sets of empirically testable propositions.

The most we can hope for is that almost all the interesting quantitative factual propositions can be improved in accuracy. This is an article of faith of all factual scientists.

Our next problem is to find out whether ethics and ideology too could and should be turned from dogma into scientific disciplines.

5 Truth-centred Ethics and Ideology

According to popular wisdom, all values and all moral norms are subjective – a matter of feeling, taste, or vested interests. I propose to challenge this view. I submit that morals can be scientific, in the sense that the rules of a viable and fair morality can be made compatible with what is known about human nature and social life. This thesis has been formulated and discussed at length elsewhere (Bunge 1989). A few examples will have to suffice here.

The first example is reciprocal altruism, or quid pro quo. Contrary to the selfishness preached by utilitarians and neo-liberals, that norm has a firm basis in social science. Indeed, it is conducive to social justice and social cohesion, and thus to both social peace and social progress. (Ironically, Brazil's device, *Ordem e progresso*, is still topical although it was inspired by the long-dead philosophy of the later Auguste Comte.)

Second example: Unlike traditional pedagogy, its modern successor holds that learning is its own reward. Consequently, instead of using punishment as an incentive, it uses reward and the withholding of reward. This reorientation has two roots. One is the moral thesis that we are not condemned to suffer (as Luther taught), but may enjoy life. The other is the finding of modern psychology that children respond better to reward or its withholding than to punishment. Both ideas are alien to the myths of original sin and eternal damnation.

A third example is the case for responsible procreation, starting with planned parenthood. I submit that it is cruel to procreate children who, being unwanted, will neither be loved nor, consequently, be properly raised and educated. And, since cruelty is abominable, opposition to family planning is starkly immoral. The immorality of condemning the use of condoms is compounded by the current AIDS epidemic, since unprotected sex transmits the HIV virus that causes that appalling sickness.

Another example is the ban on the research and use of embryonic stem cells to replace diseased or dead tissue. This ban is likewise cruel, hence immoral, because it condemns to incapacity and premature death the victims of such neurodegenerative diseases as Parkinson's, Alzheimer's, and Huntington's. The ban in question is based on a falsity: that a five-day-old human embryo, the blastocyst – which contains only about 150 cells – is a person. The truth is that we are not persons from conception: we become persons gradually, as we develop from fertilized egg to embryo to fetus to new-born baby to toddler and beyond.

On the other hand, human cloning should be suspended until shown to be safe – and until a legal instrument is invented to avoid litigation stemming from the inheritance rights of individuals who are at once children and brothers of the cell donor. The reasons in support of a moratorium cannot be moral, because identical twins are natural clones of each other – and only a few backward tribes prohibit raising identical twins. In any case, there are overriding scientific and legal reasons against human cloning.

The main scientific reasons at present are these. First, an overpopulated world like ours is in no need of artificial reproduction. (Far from being an endangered species, ours is the supremely endangering one.) Second, far too many artificial clones have severe genetic defects, among them a shortening of telomeres (tails of chromosomes). This suggests that artificial clones, unlike their natural counterparts, may be born old, hence prone to suffering old-age ailments when young or even from birth. Either of these reasons should suffice to ban human cloning for the moment. In both cases the moral norm is based on scientific considerations: truth shows the way to the good and the right.

Let us now jump from cases to generalities. A moral code can be either traditional or updated with the help of science and technology. If traditional, a code will ignore or even reject important truths found in recent centuries – such as that social inequality is not a genetic trait, and that normal child development requires love in addition to good nutrition, hygiene, and instruction. Consequently, such a code will condone serious mismatches between morals and modern life, thus contributing to the unhappiness of many people. This is why organized religions, attached to the past as they are, are unhappiness factories: just think of gender discrimination and the bans on trial marriage, divorce, contraception, abortion (even in cases of rape), assisted suicide, and eventually gene therapy.

A scientific morality, by contrast, would place more value on enjoying life than on observing dogmas that may have been justified several millennia ago to protect the viability of a few small and backward tribes in hostile environments. In sum, whereas traditional morals are bound to be obsolete, oppressive, and divisive, a scientific moral code would match modern life and would be liberating and inclusive in being based on objective truth. Indeed, the (scientific) truth (or rather its recognition) shalt make thee free!

Arguably, it is also possible to craft a scientific political ideology, that is, one based on social science. Take, for instance, egalitarianism, which in some form or other has been preached (though never consistently practised) by classical liberalism and democratic socialism, as well as by Spinoza, Locke, Rousseau, and Kant. Why should social (not just legal) equality be desirable? Because (a) social psychology has shown that people are dissatisfied not just when they suffer deprivation, but also when they are ostensibly worse off than their neighbours (Merton's well-confirmed reference-group theory); (b) sociology has shown that social cohesion increases with social participation, and decreases with social exclusion; and (c) political science has shown that deeply divided societies in the process of modernization are politically turbulent. Thus, there is some solid social science to back up political programs that pursue egalitarianism, particularly in one of its meritocratic versions, such a Rawls's.

Concluding Remarks

The concept of truth is central to all walks of life, not just to science and technology. However, truth never comes alone: it should always be flanked by work and love, or at least by concern for others. The search for truth requires the love of it as well as hard work. Work without truth is either inefficient or exploitative; and without love it can be drudgery. Love without truth can be deceitful; and without work it is unsustainable. However, there are conflicts, and therefore trade-offs, among the three values in question. When truth is the overriding goal, as it is in scientific and humanistic research, we may have to sacrifice work or love. But whenever work or fairness is the goal, truth will be a means to attain either. So truth, just like work and fairness, though sometimes goal and at other times means, is always central. And yet, paradoxically, no one seems to have produced a detailed, true, and generally accepted theory of objective and partial truth. We only have a few insights into the nature of factual truth, the way its knowledge emerges from tests, and the confluence of truths attained in different fields.

16

Emergence of Disease and Convergence of the Biomedical Sciences

When in good health we may enjoy joking about egregious medical mistakes. But when resurfacing after a successful medical treatment, we may be willing to admit that physicians often solve amazingly tough problems. Further reflection shows that medical failures are often due to the intrinsic difficulty of the problems they tackle rather than to professional incompetence. I submit that there are three main reasons for such inherent difficulty.

First, physicians are expected to inspect, understand, and repair the most complex and one of the most vulnerable of all systems in the universe: the live human body. This system crosses all the levels of organization, from the atomic to the social. For example, deficiency in the neuropeptide oxytocin causes autism, hence withdrawal; by contrast, epinephrine causes euphoria, and thus sociality. This is why physicians have to resort to a rich panoply of means, from drug to lab test, from lancet to speech, and more. Second, most of the problems physicians face are inverse and therefore tough. Indeed, they must conjecture disease from signs that are often ambiguous. Third, because of this difficulty and the complexity of humans, physicians have to use findings of research in many disciplines – from biophysics and biochemistry to psychology and sociology – some of which are not taught in medical school.

Let us glance at a few problems central to iatrophilosophy, or the philosophy of medicine: the multilevel nature of the human body, the peculiarities of the disease process, and the perplexities of medical diagnosis and treatment.

1 Multilevel Systems and Multidisciplinarity

Unlike traditional medicine, modern medicine is characterized by its systemicity, and employs the scientific method. This is the procedure whereby guesses are checked. It can be sketched as follows in the case of medical diagnosis and treatment:

Background biomedical knowledge → Diagnostic or therapeutic problem → Diagnostic hypothesis or therapy → Observational or experimental test → Evaluation of hypothesis, method, or therapy → Enlargement or revision of background knowledge → Revision of medical practice.

Ancient and medieval anatomy and medicine regarded the human body as an assemblage of mutually independent organs and humours. Modern medicine started in the seventeenth century with William Harvey's discovery of the cardiovascular system. The discoveries of the musculo-skeletal, respiratory, cardiovascular, digestive, nervous, endocrine, and immune systems followed. It was later discovered that the various body systems interact with one another, constituting supersystems such as the neuro-endocrine one. This interaction is one of the sources of the difficulty of medical diagnosis and treatment. Indeed, if subsystems A and B interact, then a symptom exhibited by A may have originated in B, and a treatment of A is bound to affect B as well. Eventually pharmacologists discovered an additional interaction, that among drugs. For example, it has recently been discovered that St John's wort, a popular herbal remedy for depression, reduces dramatically the effectiveness of dozens of prescription drugs. So much for sectoral therapies.

Before the time of Rudolf Virchow and Claude Bernard, medicine was isolated from all the other branches of reliable knowledge. By contrast, a large number of disciplines converge into modern medicine: physics, chemistry, biology, psychology, sociology, and the corresponding hybrid sciences, such as biochemistry, psychoneuro-endocrinology, and medical sociology – not to mention pharmacology, an applied science. Think, for instance, of the long and messy process that led in 1922 to the discovery that insulin cures diabetes. It was the culmination of a convergence process involving discoveries and inventions in physiology, endocrinology, biochemistry, clinical medicine, surgery, and more, made by uncounted researchers over more than two decades – not counting the support of a number of universities in several countries (see Bliss 2000).

Why such dependence of medicine upon so many other disciplines and interdisciplines? First, because the human being is nowadays conceived of as one more animal – though one with a rich mental life and embedded in complex and changing social networks. Second, because the life processes can be analysed into physico-chemical ones – which does not mean that they lack emergent biological properties. Third, because humans occupy a number of levels of organization, from the physical to the social, and hence they cannot be understood, let alone controlled, if attention is focused on a single level. Fourth, because the human body is increasingly seen as a system of systems, none of which can be understood on its own. Fifth, because modern medicine is conceived of, and practised as, a science-based technology, not as a craft using only empirical rules of thumb. Sixth, and consequently, because biomedical research is so many-faceted and so labour-intensive, requires such sophisticated equipment and so many clinical trials, and can be so profitable, it must be funded by governments and private firms.

Let us examine briefly some of the conceptual problems posed by medical diagnosis and treatment. However, before doing so we should be clear about the nature of disease: is it a thing, state, process, or social construction to be 'deconstructed' rather than cured?

2 What Kind of Entity Is Disease?

If we were to trust ordinary language, we would think that disease is either a thing or a state. Indeed, when we say, for instance, that someone *has* a cold, we seem to treat colds as things that one carries and in some cases can pass on to others the way one carries or can pass on a bad coin. Likewise, when we say, 'He is sick,' we seem to treat sickness as a state, as when one says, 'He is feverish.'

When intent on analysing the concept of a disease, the exact philosopher views the matter in a different light. Instead of saying that so-and-so *has* a sickness, or *is* sick, he introduces the verb 'to sick,' and says that so-and-so *sicks*. Indeed, one may write Sx. (To be sure, this locution violates what Noam Chomsky calls 'the native speaker's linguistic intuition.' But this is beside the point, since ordinary language is anything but exact, and intuition is at best a starter.)

Modern medicine treats disease as a process in the organism, that is, a sequence of bodily states, with a start and a finish. Consequently, to treat a patient is not to try and wrestle something out of her – such as an evil spirit – but to interfere with the disease process until stopping or at least abating it. Moreover, it is not an instantaneous state that can be dealt with at one stroke: disease, like the succession of seasons, is a sequence of states.

(The concepts of state of an organism and process in it are best analysed in terms of the state function, or list of biological properties of organisms of the species under consideration. Calling $P_1, P_2, ..., P_n$ the n known properties of such organisms, their state function is the list or n-tuple $F = \langle P_1, P_2, ..., P_n \rangle$. This list may be construed as a vector in an n-dimensional Cartesian space, every axis of which represents one of the n properties in question. This space is called the *state space* of organisms of the given kind: call it S_K, where K names the biospecies in question. Organisms of different species have different state spaces. However, since all of them are phylogenetically related, all the organismic state spaces have a non-empty intersection. As time goes by, some or all of the properties of the organism change, and accordingly the tip of the vector moves in the state space of the species. But as long as the organism is alive, its state vector remains confined within a particular box: call it L_K. Moreover, the healthy states are inside a smaller box H_K included in L_K. In short, $H_K \subset L_K \subset S_K$. Along with the disease process, a large number of new axes are bound to sprout, and others are likely to be pruned. The net effect will be a distortion of the state space and its two subspaces.)

The view that a disease is a real process is only a skeleton to be fleshed out by biomedical research. And the findings of such research, though objective, are bound to be somewhat inaccurate, since we shall never know accurately all the properties of an organism and their mutual relations. In other words, such findings are at best partially true. It is up to biomedical researchers to improve on the accuracy of our knowledge of health and sickness.

This objective view of disease is at odds with the opinion, fashionable among postmoderns, that disease is a social construction, in particular an invention of the medical profession (see Fleck 1979). For instance, according to this view, without Koch there would be no TB. This quaint opinion, social constructivism, is part of the anti-science movement that is currently sweeping the faculties of arts in the industrialized countries. This view is patently false and dangerous. It is false because people usually get sick before they are diagnosed and treated by physicians. And the view is dangerous because it leads to treating diseases by controlling the medical profession, or perhaps suppressing it altogether. Anti-realism, then, is hazardous to your health. It should therefore be a subject of concern to biomedical ethics.

3 Diagnosis as an Inverse Problem

All medical people reason, and reasoning can be logically valid or invalid, regardless of the content of the premises. When a reasoning that precedes a medical decision is logically invalid, the corresponding course of action may be ineffective or worse. For example, from the fact that a treatment has proved ineffective it does not follow that the patient's disease is not organic. It only follows that the diagnosis has been incomplete or incorrect, that the patient has disobeyed his physician's instructions, or that no effective therapy has yet been devised.

Let us peek at the logic of medical diagnosis. This logic is complex because it raises the tricky problem of indicators – a key question that nearly all philosophers of science and technology have ignored. The basic premise is that every disease eventually manifests as a syndrome or cluster of signs, which in turn are observable features. The adverb 'eventually' is intended to include the case of the so-called asymptomatic diseases, which are discovered only when they are advanced.

In obvious symbols, the premise reads: *If D, then S*, where *D* describes a disease and *S* is the conjunction of a number of corresponding symptoms. More explicitly:

For every x, if x has disease D, then x exhibits syndrome S, or $\forall x(Dx \Rightarrow Sx)$. This is the correct logical form of the disease-syndrome hypothesis, for the latter fails only if there is disease without symptoms. Hence, the way to check for the truth of the hypothesis is to observe or provoke the disease, and check whether the expected symptom occurs. (Still, as will be seen below, it is possible to strengthen the *D–S* relation.) On the other hand, the converse conditional, $\forall x(Sx \Rightarrow Dx)$, is incorrect, because it is false only when there is syndrome without disease.

(Caution: if only some of the symptoms are available to the diagnostician, he may rightly hesitate between two or more possible causes. In this case, *D* will equal the disjunction D_1 or D_2 or ... or D_n of *n* propositions, where $n \geq 2$. Only further tests, or even further research, may discard all but one of the disjuncts. Moreover, obviously every one of the hypotheses in question should be plausible in the light of the available body of knowledge. This requirement disqualifies trial-and-error procedures as well as magical and religious hypotheses.)

Let us now apply the preceding. Assuming a hypothesis of the form *If D, then S*, the basic rule of inference called *modus ponens* allows us to reason as follows. If a particular patient *b* suffers from disease *D*, then she will exhibit syndrome *S*. That is, the logically valid argument is

Law	$\forall x(Dx \Rightarrow Sx)$
Datum	*Db*
∴ Conclusion	*Sb*

However, this valid piece of elementary logic is utterly useless for diagnosis. (It is only useful to forecast symptoms that have not yet appeared.) In fact, the typical diagnostic problem is not the *direct* problem of inferring syndrome from disease but the *inverse* problem of guessing disease from symptoms: recall chapter 13, section 4, and see figure 16.1.

In fact, the physician starts by observing a few signs. If these are insufficient to form a diagnostic hypothesis, he performs or orders a battery of tests allowing him to enlarge the list of symptoms, reasoning as follows:

Law	$\forall x(Dx \Rightarrow Sx)$
Datum	*Sb*
∴ 'Conclusion'	*Db*.

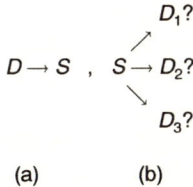

Figure 16.1 (a) *Direct problem*: Given (finding or assuming) disease D, search for or forecast syndrome S. The arrow symbolizes both the causal relation and the direction of the inference. (b) *Inverse problem*: Given syndrome S, guess the disease D – a problem that may have more than one solution. In this case the arrows symbolizes only the diagnostic process.

However, this argument is logically invalid if heuristically fertile (or misleading). Indeed, the 'conclusion' Db is not entailed by the premises, but is only suggested by them; so much so that further tests, or even a different initial hypothesis, might result in a different 'conclusion.' That is, the physician can only *conjecture* that his patient suffers from the disease in question. In short, Db is not a conclusion proper but another *hypothesis*. And of course a hypothesis remains at best plausible as long as it has not passed rigorous tests. (Recall chapter 14.)

The occurrence of guesses in the diagnosis process has to be emphasized to counter the widespread opinion that good doctors do not guess, but only go by hard data. As a matter of fact, good doctors gather data over and above what their patients supply, and they usually do so in the light of one or more hypotheses. They must do it because the patients cannot observe or feel the pathological processes that occur at the cellular or molecular levels. Regardless of their competence, doctors guess all the time (see Murphy 1997). But of course good doctors tend to make more or less plausible hypotheses on the strength of their periodically updated biomedical knowledge. And they check their guesses, revising them if need be, in the course of the treatment.

Moreover, the treatment provides new data and may suggest correcting earlier hypotheses. For instance, if treatment T for disease D does not work for a given patient, then the physician will conjecture either that the patient does not have D, or that, although he is afflicted by D, he also suffers from an additional condition that renders T ineffective, such as deficiency in certain enzymes or hormones. In either case he will have to come up with one or more plausible new treatments. If either of these works, nothing is being logically proved, particularly since the new treatment may act as a placebo. Still, this purist objection has little bite, because the medical practitioner, unlike the biomedical researcher, is more interested in results than in truth. To establish that the patient does suffer from a certain disease may require further tests – or even a post-mortem.

In short, good doctors cannot afford not to make guesses. However, unlike shamans, practitioners of alternative medicine, and psychoanalysts, they make educated and testable guesses, not wild ones. Yet, of course, even the most plausible educated guess may turn out to be false. When this happens, the responsible physician tries an alternative guess. Doctors are unwilling to switch guesses only if they belong to some sect or if they are sued for malpractice every time they make a mistake.

The preceding considerations help explain why different physicians may come up with different diagnoses of the same patient. Indeed, different people are likely to make different guesses even if they start from roughly the same data, simply because they have somewhat different background knowledge and somewhat different brains. In turn, this helps explain why consultation is often successful. The members of a medical team debate the merits of different hypotheses, until – hopefully – consensus is reached concerning the most plausible hypothesis. However, in the long run only further biomedical research can reduce doubt, by exhibiting the disease mechanisms – on which more anon.

4 Knowledge of Mechanism Strengthens Inference

It is possible to strengthen diagnostic inference. This is achieved not by multiplying the data or by changing the logic, but by deepening the underlying biomedical knowledge: by unveiling the *mechanisms* underlying the disease–symptom relation. For example, hepatitis, which in the 1950s was believed to be a single disease, has turned out to be a cluster of six viral mechanisms, from A to F, with different effects and prognoses.

A few mechanisms, such as cuts, bruises, and discolorations, are visible. But most, such as infection, blood clotting, organ dysfunction, or cell degeneration, are invisible. If hypothesized, they can often be checked in the lab. For instance, if an X-ray shows a shadow in a lung, the physician will suspect cancer, TB, a fungus, or scar tissue, and may order a suitable test, such as a biopsy. Once the test result is in, one of the competing hypotheses may be eliminated and the other confirmed – provided the test is reliable and its result inequivocal. (Remember the ever-present possibility of false positives and true negatives.)

Now, the physician has some plausible knowledge of the mechanism that may have caused the observed symptoms. He is reasonably certain that the said mechanism is both necessary and sufficient for the given symptoms to show: that it is their cause, and that no other process might have caused them. However, only biomedical research can establish that this is indeed the case. Likewise, the physicist trusts his instrument readings because he knows how the instrument works; and he knows this because he knows that the instrument has been designed on the basis of one or more well-confirmed physical theories. For example, a thermometer reading is reliable because it is based on a well-corroborated linear relation between the length and temperature of the thermometric column: that length is a visible and reliable indicator of the invisible temperature.

In general, an indicator hypothesis cannot be trusted unless justified by some well-corroborated theory (hypothetico-deductive system). Regrettably, there is a dearth of true theories in medicine. This theoretical scarcity renders medical diagnosis almost as dicey as economic prophecy. An additional source of uncertainty is the huge diversity ('variability') of humans, as a consequence of which different individuals, though suffering from the same disease, may show somewhat different symptoms, as well as different reactions to a treatment. Still, this only shows that there are several different mechanisms for any given disease-syndrome pair.

In any event, if a disease mechanism is plausible, the diagnostician can use two laws instead of one: one linking disease to mechanism and the other linking the latter to syndrome. Moreover, each of these new laws is logically stronger than the initial conditional *If D, then S*. Indeed, the new laws are

D if and only if M,

and

S if and only if M,

where M describes the sickness mechanism (infection, stroke, runaway cell proliferation, or what have you). Both laws are double-arrowed conditionals, that is, biconditionals, not just single-arrowed conditionals. In other words, every side of the formula is now both *necessary* and *sufficient* for the truth or the falsity of the other.

If the physician hits on the sickness mechanism, he reasons as follows. First he derives the disease-syndrome law from two deeper laws: the disease-mechanism and the mechanism-syndrome laws:

$\forall x(Dx \Leftrightarrow Mx)$ & $\forall x(Mx \Leftrightarrow Sx)$ $\therefore$ $\forall x(Dx \Leftrightarrow Sx)$.

In turn,

$\forall x(Dx \Leftrightarrow Sx)$
Sb
$\therefore Db$.

And this is a logically valid conclusion, not an uncertain guess. It is firm because research has found, not only that D has caused S, but also that no other mechanism could have caused S.

Unveiling the mechanism has paid off: no uncertainty is left – at least until the next discovery in biomedical research. In other words, the inverse problem has been solved by transforming it into a direct problem. Correspondingly, a shaky plausible inference, or an uncertain conjecture, has been converted into a solid deductive argument. There is still guessing, but only about the premises, not about the inference. (The premises are hypothetical, hence only plausible at best, even if they have been well confirmed under experimentally controlled conditions; and they are doubly so in medical practice because the physician may have wrongly chosen the relevant hypotheses.) This gain in inferential power has been achieved at a price: that of biomedical research aiming at unveiling previously hidden disease mechanisms.

What holds for diagnosis also holds, *mutatis mutandis*, for therapy and prognosis. Empirical therapy, prescribed on the strength of mere disease–symptom correlations, may succeed in simple and well-known cases. By contrast, complex cases are likely to require scientific therapy. And scientific therapy tampers deliberately with a known or conjectured mechanism. For example, the surgeon will eliminate a blood clot, replace a heart valve, suture a muscle, or cut off a slice of thyroid gland. Suitable medication will kill germs, dilate an artery, relax a muscle, have a diuretic effect, facilitate bowel movement, inhibit or stimulate the synthesis of a protein, or relieve anxiety. Moreover, social medicine will recommend providing sewers or drinking water, eliminating pollution sources, immunizing a community, or improving housing conditions or the work environment.

An experimental treatment tries out a hypothesis concerning the mechanism that accounts for the observed symptoms. A treatment continues to be experimental even after having been tried successfully on a statistically reasonable sample of patients. The treatment is experimental in that it subjects a mechanismic hypothesis to tests; and in that, if no positive results are obtained, an alternative hypothesis is invented and tried out. It is hoped that, if the unsuccessful hypothesis candidates are successively eliminated in the course of the treatment, the truth will finally emerge – provided the patient has not been lost in the process. In other words, the successive experimental treatments have been means to test and discard mechanismic hypotheses: they have been used to deepen the knowledge of the case. Only the last treatment has a purely practical import. (For the methodology of experiment and inference see Bunge 1967a, 1983b.)

5 Bayesian Number Juggling

There is no question but that, though seldom explicitly used by physicians, epidemiological statistics are valuable in medical diagnosis. For instance, since tuberculosis is endemic in Calcutta, it is reasonable to suspect TB as the cause of an hemoptysis in a Calcuttan. However, only further inquiry, such as lung X-rays and search for Koch bacilli in blood, can establish the truth value of the hypothesis. In any case, epidemiology yields frequencies, not probabilities, since the onset of disease is a matter of causation, not chance. This is why physicians speak of etiology, not tychelogy. In other words, morbidity and mortality statistics give estimates of *likelihoods*, such as that of lung cancer among fifty-year-old male smokers. This likelihood is not a matter of opinion or uncertainty, since the mechanisms whereby smoking causes lung cancer are known at least in outline.

However, the medical theorists and epidemiologists of the Bayesian school favour the use of the calculus of probability in medical diagnosis and treatment. The underlying ideas are these. First, when uncertain we resort to (subjective) probability (an opinion criticized in chapter 14, section 2.4). Second, the relation between a syndrome S and its cause D, as well as that between a treatment T and its outcome O, are probabilistic because of the attendant uncertainties. That is, the physician would be called upon to evaluate opinions with definite probabilities. Let us examine this procedure.

To avoid unnecessary duplication of formulas, we shall unify diagnosis and treatment, regarding both disease and treatment as causes C, and syndrome and outcome as effects E. We will then have four alleged probabilities: the prior probabilities $P(C)$ and $P(E)$, and the conditional probabilities $P(E|C)$ and $P(C|E)$ – the last two of which Bayesians call 'likelihood' and 'posterior probability' respectively. The first two will then be read as 'the probability of cause C' and 'the probability of effect E' respectively. The remaining symbols will be interpreted as 'the probability of effect E given cause C,' and 'the [inverse] probability of cause C given effect E' respectively. Sometimes the same arguments of the probability function P are interpreted in terms of hypotheses and data. For instance, it is said that $P(E|C)$ is the probability of the evidence or datum E given the hypothesis C, whereas $P(C|E)$ would be the inverse probability of hypothesis C given the data or evidence E.

The said probabilities are related by Bayes's theorem,

$$P(C|E) = P(C) \cdot P(E|C)/P(E).$$

This is the centrepiece of Bayesian medical diagnosis and statistical inference. While it is a correct formula of the calculus of probability, there are serious objections to its interpretation in terms of causes (or hypotheses) and effects (or data). The first is that the probability calculus refers to *chance* mechanisms, whereas the disease process is essentially causal, whence the disease–symptom and treatment–outcome relations are causal. For instance, excess cholesterol in blood causes the clogging of arteries, and bypass surgery restores normal blood flow: no randomness here. A second objection is that we seldom know the prior 'probabilities' (actually frequencies) of disease (or treatment) and syndrome (or outcome), that is, $P(C)$ and $P(E)$. Consequently, the Bayesian makes them up; but if they are made up, no science is being done. Quantitated opinion is not more rigorous than qualitative opinion.

It may be rejoined that the said probabilities can be estimated by frequencies. True, sometimes epidemiologists can tell us the frequency of a disease in a given population, which would give an estimate of $P(C)$. However, the three remaining frequencies are usually unknown. The most that medical reports and textbooks tell us is that a given E is 'usually,' 'typically,' or 'very frequently' associated with C (Eddy and Clanton 1982). Consequently, Bayesian diagnosis is practically impossible on top of being conceptually wrong because it assumes that the disease–syndrome relation is random.

At first sight, these objections can be overcome if C and E are interpreted as propositions: C as the hypothesis that the given disease (or treatment) is present, and E as the datum that it is accompanied by the syndrome (or outcome) in question. However, this reinterpretation does not work, because propositions cannot be assigned probabilities, any more than they can be attributed velocities or viscosities: recall chapter 14. Only facts (states and events) can be assigned probabilities, and even so provided they are random.

However, none of the above invalidates the epidemiological search for the disease-syndrome associations. There are two such associations: the *likelihood* of syndrome S when disease D is present, and the *plausibility* of the hypothesis that syndrome S indicates disease D. However, such likelihoods and plausibilities are produced by medical statistics, not by Bayes's formula. For example, the plausibility of breast-cancer diagnosis based on a positive mammograph may be 0.40, whereas the likelihood of testing positive given cancer may be close to 1. (Remember the possibility of false negatives.) These numbers, and the figures for the incidence of cancer and positive mammography test, do not combine in accordance with Bayes's formula, as is readily seen from Eddy's (1982) tables. Even more disturbing is Eddy's finding that 95 per cent of American physicians confuse the likelihood and the plausibility in question, which are actually very different both conceptually and numerically. Clearly, Bayesianism is hazardous to your health.

In conclusion, probability theory should not be used for medical diagnosis or treatment, because both procedures rely on causation rather than chance. Chance intervenes in medicine only when checking null hypotheses or when assigning patients to either the experimental or the control group in clinical trials. Besides, gambling with health is immoral as well as silly. It will be argued next that this maxim applies to therapy as well as to diagnosis.

6 Decision-theoretic Management of Therapy

Having failed in military strategy, political analysis, management, and other fields, decision theory is now being tried out in medicine. The idea is to choose the course of action most likely to maximize the expected utility of the outcome. This is the product of the probability and the utility of the option in question. For example, if one is faced with two options, such as two rival medical treatments of a given disease, one starts by estimating their respective probabilities and utilities. In the medical case, the probabilities may be set equal to the frequencies of success, whereas the utilities may be equated with the increases in lifespans. This is how one is supposed to proceed:

Treatment T_1: Frequency $= f_1$, utility $= u_1$, expected utility $= f_1 \cdot u_1$
Treatment T_2: Frequency $= f_2$, utility $= u_2$, expected utility $= f_2 \cdot u_2$
Decision rule: Prefer T_1 to T_2 if and only if $f_1 \cdot u_1 > f_2 \cdot u_2$.

Let us put this method to the test with two limit cases: those of a high-success treatment with a low gain, and a low-success rate with a high gain. Occasionally the numbers combine in such a manner that the expected gains from both treatments are the same, as in this example:

T_1 $f_1 = 9/10$, $u_1 = 1/10$, $f_1 \cdot u_1 = 9/100$ Low risk, low gain
T_2 $f_2 = 1/10$, $u_2 = 9/10$, $f_2 \cdot u_2 = 9/100$ High risk, high gain

In this case, the decision rule accommodates both the gambler and the risk-averse person, so that the choice between alternative treatments becomes a matter of temperament. Since this is anything but a rational choice, I suggest that the above decision rule is inadequate.

Hence, a more discriminating decision rule is needed. Besides, the number of extra years one gains is not a suitable measure of medical utility: the quality of life too is important. One year of active life is surely preferable to five years on a respirator.

It may be objected that the preceding case is too simple to be realistic, since any given treatment is likely to have more than one outcome, such as a central benefit and some bad side-effects. In the case of multiple possible effects, we have to calculate the expected utility using the formula $\sum_i P(O_i|T_j) \cdot V(O_i)$ for $j = 1, 2, \ldots$, where O_i is the outcome of treatment T_j, and $V(O_i)$ the corresponding utility (subjective value or gain). Let us apply this recipe to the case where the physician can choose between two treatments with two possible outcomes each:

T_1 with probabilities $P(\text{partial cure}|T_1)$ and $P(\text{no change}|T_1)$
T_2 with probabilities $P(\text{cure}|T_2)$ and $P(\text{death}|T_2)$

Any prudent (risk-averse) physician is likely to reject T_2 out of hand, even if the chance of death in T_2 is small. On the other hand, decision theory recommends weighing both alternatives, and choosing the greatest of the following two expected utilities:

$P(\text{partial cure}|T_1) \cdot V(\text{partial cure}) + P(\text{no change}|T_1) \cdot V(\text{no change})$
$P(\text{cure}|T_2) \cdot V(\text{cure}) + P(\text{death}|T_2) \cdot V(\text{death})$

Even assuming that both physician and patient can estimate such probabilities, how are they to assign utilities? We do not even know the range of values of the alleged utility function V. In particular, what are the values of no change and of death? Suppose, for instance, that T_1 has a 50% chance of effecting a partial cure, whereas the chance of T_2 effecting a total cure is 90 per cent. And assume that cure, partial cure, no change, and death are assigned the values 1, 1/2, 0, and −1 respectively. Then we obtain,

for T_1: 1/2.1/2 + 1/2.0 = 1/4, and
for T_2: 9/10.1 + 1/10.(−1) = 8/10 = 4/5,

which is 3.2 times that of T_1. Thus, decision theory recommends definitely the aggressive treatment T_2. I submit that no responsible physician would assume such a high (10 per cent) death risk, except perhaps in a terminal case.

The overwhelming majority of physicians (and most patients) are risk-averse rather than gamblers, because they are committed to Hippocrates' injunction not to harm. I also submit that responsible business managers and military strategists attempt to minimize losses rather than maximize gains. However, neither of them is likely to pin definite albeit phony numbers to their options. They are satisfied with ranking them.

In sum, whether in designing treatments or in making diagnoses, the competent physician uses maximally reliable data from many sources, and imagines and tries out several more or less plausible hypotheses. He never calculates probabilities, because he knows that the disease mechanisms are causal, not stochastic. This is not to exclude a priori the possibility of quantitative methods of diagnosis and therapy. Presumably, these will emerge along with adequate quantitative theories of disease processes.

7 Medicine between Basic Science and Technology

Three different aspects of medicine must be distinguished:

Biomedical research – an applied science like materials science;
Therapy design – a technology like mechanical engineering; and
Medical practice – a craft or service like plumbing, only far more sophisticated.

However, these distinctions do not involve separations. Indeed, competent medical practice is based on both true biomedical findings and successful therapies. All three depend on pharmacology (drug design and test) and the pharmaceutical industry. And they combine in a cycle: *Lab bench → Clinic → Factory → Clinic → Bench.*

Now, the design, test, improvement, and mass production of even the simplest and best-known drug calls for technical expertise based on some body of basic (chemical and biological) knowledge. In particular, the drug designer needs to understand the mechanism that the drug triggers or blocks in the organism. Otherwise he will be unable to recommend adequate dosages or tamper successfully with it, and so hope to come up with a more efficient drug – for instance, one with less severe side-effects. (Caution: So far, most of what is currently called 'drug design' is actually blind trial and error tests, at a rate of about half a million new molecules per year.)

The invention of antibiotic therapies – a fascinating story that has been told many times – will help clarify the science-technology connection. It all started in 1928, when the bacteriologist Alexander Fleming discovered accidentally that a culture of staphylococcus, the pus-producing bacterium, disappeared as a certain green mould grew in a Petri dish. Fleming examined this mould and found that it was constituted by spores of a fungus of the *Penicillium* genus. And, far from remaining satisfied with recording this surprising fact, he conjectured that the spores produced a substance that killed the staphylococcus bacteria, and which he called *Penicillium rubrum*.

A possible logical reconstruction of Fleming's reasoning is this:

Confimed hypothesis (law): Penicillin kills staphylococcus, or $P \Rightarrow K$. [1]
Guess: What kills staphylococcus cures pus-infections, or $K \Rightarrow C$. [2]

These two premises jointly entail the practical consequence

$P \Rightarrow C$, that is, Penicillin cures pus-infections. [3]

It is likely that Fleming and his co-workers, Howard Florey and Ernst Chain, jumped to an even more general and therefore more ambitious hypothesis, namely

Penicillin cures all bacterial infections, or $P \Rightarrow B$. [4]

Premise [1] belongs in bacteriology, a basic science. Premise [2] belongs in biomedical science, an applied science. Only the third and fourth propositions belong in therapeutics, a branch of medicine, which is essentially a technology since it aims at altering things with a view to improving them.

Note that, contrary to appearances, [3] and [4] do not follow directly from [1], if only because the concept of an infection occurs in [2] but not in [1]. In other words, [1] and [2] are logically mutually independent. Hence, they are methodologically independent as well. That is, the experimental confirmation (or else weakening) of either does not support (or undermine) the other. Hence, they must be tested separately.

However, we are still far from the prescription stage. Physicians would not know what to prescribe, since none of the above guarantees the commercial availability of the first antibiotic. Two more steps had to be taken: from the lab bench to the pharmacological bench, and from the latter to the pharmaceutical plant. Indeed, the substance originally isolated by Fleming and his co-workers proved to be at once unstable and not potent enough for therapeutic purposes. Moreover, it is well known that what works *in vitro* may not work *in vivo*. One reason for this discrepancy is that the drug in question, while effective against the target germ, may interfere with some vital functions, such as the synthesis of certain key proteins.

Further applied research over a period of thirteen years overcame these drawbacks. In fact, basic and applied researchers at Oxford University succeeded in preparing samples of stable penicillin. These they injected into patients with severe infectious diseases – with the well-known sensational results. These clinical trials completed the pre-industrial phase. The latter was undertaken in 1941 in the United States, and culminated with the purification, stabilization, and mass production of the first industrial antibiotic – also the most effective drug since quinine and aspirin.

The technology underlying the mass production of penicillin is based on the following therapeutic

Rule: Use penicillin to treat bacterial infections. [5]

Since this is a rule or prescription, it is more or less efficient rather than true or false. Moreover, it is not completely efficient. Indeed, some bacterial infections call for different antibiotics, of which there is now a large family. This is because the hypothesis [4] is unduly optimistic. Still, the later antibacterial drugs would not have been sought and developed if [4] had not first been invented, put to the test, and shown to fail in many cases.

The pharmaceutical industry was powerfully motivated by [5]. But it still had no clue as to how to manufacture purified and stable penicillin on a large scale, and moreover in the form of user-friendly capsules. This tough task was reserved to the industrial chemists and chemical engineers. Furthermore, even having penicillin in hand was not the final solution until the correct dosage was established. And this was a task for further pharmacological and clinical tests, a task complicated by the facts that different people react differently to any given drug and that some bacteria become drug-resistant owing to an overuse of antibiotics.

Yet, this is not the end of the story. As usually happens in science and technology, the solution to a problem allows researchers to ask further questions. In this case, the next scientific problem of interest was this: How do antibiotics work? That is, what mechanisms do antibiotics trigger in the sick cell? This question is both scientifically and medically important because, once a mechanism is unveiled, it can be tampered with – for better or for worse.

This question motivated a number of research projects over several decades. It was eventually found that some antibiotics act on the bacterial cell wall by inhibiting the synthesis of a molecule that is a necessary component of such wall, which consequently breaks down; other antibiotics interfere with protein synthesis, either by inhibiting it or by producing abnormal proteins. Still other antibiotics alter the metabolism of nucleic acids, thus changing the rate at which proteins are synthesized.

The R&D processes we have just described may be summarized thus:

1 *Trial phase*: Bacteriology → Biochemistry → Pharmacology → Therapeutics.
2 *Practical phase*: Therapeutics → Industrial chemistry → Chemical engineering → Industry.
3 *New research phase*: Cell biology → Molecular biology → Pharmacology → Better drugs → Clinical trials → Eventual drug re-evaluation.

The first phase started with a scientific discovery and culminated in a technological invention. The latter triggered further technological innovations, which ended up in a whole new branch of industry. In turn, the huge practical (medical and industrial) success of antibiotics encouraged further basic work, namely into the mechanisms of antibiotic action. And some of the results of this research suggested looking for more effective and efficient antibiotics.

In short, what began as a harmless disinterested question, namely 'Why did that green mould attack staphylococci?,' triggered an enormous R&D process that has yielded immense medical and pecuniary benefits, which are still being reaped after more than half a century. Yet the antibiotic technology and business did not drop ready-made from the basic science: both had to be constructed with ingenuity and great effort.

This success story is not unique. Look around and you will see plenty of products, most beneficial and some noxious, of the marriage of science, technology, and business – from sanitation and purified water to vaccines and surgical procedures, to electric power and refrigeration, to TV and computers, to automatic tellers and e-commerce, and so on and so forth.

A moral of this story is, Distinguish science from technology if you want to know more, but connect them if you wish to do better.

Concluding Remarks

Anyone with even a superficial acquaintance with medical practice knows that physicians, like anyone else, make mistakes. Some medical errors are unavoidable owing to holes in the body of medical knowledge, notably the dearth of medical theories. Others are avoidable because they are of the logical kind: they depend on training in logical thinking, not on biomedical research. If the medical school deans knew this, they would include elementary deductive logic, statistics, and scientific methodology in the pre-medical curriculum. Early training in these disciplines is bound to improve the quality of medical diagnosis and treatment. In turn, this would lower both the morbidity rate of the population and the number of malpractice suits. There are riches of all kinds – in knowledge, health, and money – to be had from the convergence of disciplines on the study and management of the emergence and submergence of disease.

17

The Emergence of Convergence and Divergence

The history of human knowledge is that of the search for truth (science and the humanities) or efficiency (technology). This search is punctuated by events of two types: branching out or specialization (or divergence), and branching in or integration (or convergence); see figure 17.1. Specialization is required by the diversity of the world and the increasing richness of our mental tools, whereas integration is called for by the contrast between the fragmentation of knowledge and the unity of the world.

In this chapter we will examine a few telling examples of both divergence and convergence, with a view to understanding why both are needed, and how each of them emerges to correct limitations or excesses of the other. We will also analyse the necessary conditions for the fusion or synthesis of theories or entire disciplines, since not all conceivable hybrids make sense or are viable. For example, so far molecular ecology is not a synthesis of two disciplines, but just the use of molecular techniques, such as DNA fingerprinting, to ascertain sex and kinship – much as art historians use X-rays to unveil paintings underneath paintings. By contrast, social cognitive neuroscience is a genuine synthesis, because it investigates the brain functions and dysfunctions that regulate social behaviour, a task that necessitates building bridges between the three founder sciences. The historiography practised by the innovative *Annales* school is another such synthesis, for all of the social and biosocial sciences converge in it.

1 Divergence

One of the characteristics of the modern period is the proliferation of the sciences. Four centuries ago, at the beginning of the Scientific Revolution, there was a single factual science proper, namely physics. Moreover, only one

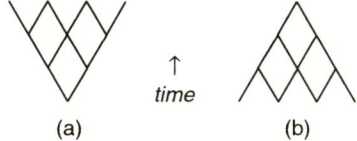

Figure 17.1 Two parallel processes: (a) specialization or divergence of disciplines; and (b) integration or convergence.

branch of it, mechanics – inclusive of astronomy – had high standards. Other studies, such as chemistry and biology, were at a protoscientific stage; and entire fields, from psychology to social studies, were at best in their philosophical womb. Two centuries later, around 1800, there were at least four factual sciences in the modern sense of the word, namely physics, chemistry, biology, and economics.

Since then, each of these fields has split into hundreds of subfields, and entire new disciplines and cross-disciplines, such as psychology, sociology, biochemistry, and socio-economics, have emerged. This vast diversity of interests is accompanied by poor communication among the corresponding scientific communities. Thus, the BEC community (of the specialists in Bose-Einstein condensates) would not know what to tell the members of the apoptosis community – those who study programmed cell death. Today there are several thousand disciplines and subdisciplines, along with the corresponding scientific communities. Worse, the barriers between them are so high that few attempt to cross them. Try jumping from high-energy physics to psychology, or from molecular biology to sociology!

As if this division of knowledge according to subject matter were not enough, some of the prospective students are confronted with a deep methodological cleavage, particularly in the field of social studies. Indeed, these are split into three main schools: scientific, pseudo-scientific, and literary. For example, most of the statistical studies of social inequality are scientific; by contrast, the sociobiological and rational-choice accounts of such human attitudes and activities as altruism and war, love and religion, are ingenious fantasy; and the wordy and opaque musings of the hermeneuticists and other postmoderns are at best literary.

Worse, the scientific camp itself is divided into rival approaches. Long after Newton there are still debates between mindless data hunters and gatherers and those who engage in hypothesis-driven research. And much of what passes for theory is either a search for short-lived statistical correlations or general-purpose fantasizing about rational choice.

No wonder then that the old ideal of the unity of the sciences is nowadays being doubted and even ridiculed, and that the disunity of science is being proclaimed (see, e.g., Dupré 1993; Galison and Stump 1996). The underlying rationale is that the universe is much too varied to be accounted for by a handful of disciplines. However, this insistence on disunity seems to be only a backlash against the reductionist project. If this project is abandoned, then unification through merger, rather than through reduction, remains a viable and worthy goal, as some philosophers have argued (e.g., Bunge 1983b; Bechtel 1986; Kincaid 1997; and Morrison 2000).

2 Convergence

By definition, all of the factual sciences study facts, whether actual or really possible. And all of them, even the social sciences, are expected to study them in a scientific manner, that is, in accordance with the scientific method rather than by navel contemplation, crystal-ball gazing, trial and error, or discourse analysis. That is, beneath appearances, the sciences are ontologically and methodologically one: all of them study putatively real things and their changes, in a distinctive manner that is quite different from the way theologians, literary critics, shamans, or even craftsmen proceed. This is why philosophers, from Whewell (1847) until recently, have praised all efforts to unify theories and even whole sciences. They used to praise Newton, Maxwell, and Darwin for having unified entire realms of facts.

Regardless of the fears and hopes of philosophers, the borders between the sciences are blurring. In other words, the 'sciences are becoming more unified, not less' (Medawar 1984: 72). In fact, unification has been proceeding in the sciences and technologies along with diversification. Witness the flourishing of cross-disciplinary research in all major fields. One of the best-known cases is the synthesis of Darwinism and classical genetics effected in the 1930s and 1940s. However, there have been many other cases of integration, if less glamorous. One of them is contemporary geology, a synthesis of petrology, mineralogy, stratigraphy, orography, and seismology. Another is toxicology, the confluence of biochemistry, pathology, epidemiology, and pharmacology – a field, moreover, that lies at the intersection of basic science and technology.

More recent and equally outstanding cases of convergence were the emergence of cybernetics and operations research in the 1940s, and of molecular biology in the 1950s. Cybernetics arose from the need to understand, design, and manage communication and control systems of all kinds, whether physical, biological, or social (see, e.g., Wiener 1948). Operations research arose from the need to manage large sociotechnical systems, such as naval convoys

The Emergence of Convergence and Divergence   271

and corporations (see, e.g., Churchman, Ackoff, and Arnoff 1957). And molecular biology emerged as a synthesis of genetics and biochemistry, from the wish to understand the composition, spatial structure, and functions of the hereditary material (see, e.g., Olby 1974). In all three cases a number of previously disconnected disciplines converged to tackle such emergents as feedback loops and the coordination of a large number of units, which are characteristic of multi-component systems, such as cells and factories.

An even more recent case of hybridization is the emergence of evolutionary developmental biology, or evo-devo. This new science, with its own journal – *Evolution and Development* – is currently all the rage (see, e.g., Maynard Smith et al. 1985; Gould 1992; Raff 2000; Arthur 2002; Wilkins 2002). Evo-devo emerged from the wish to account for both the evolution and the conservation of developmental pathways (mechanisms) and, in particular, for the origin of both speciation and stasis. Since evolutionary novelties emerge in the course of individual development, phylogeny must somehow emerge from ontogeny – even if not exactly the way Ernst Haeckel had imagined in the 1860s. And because the nascent science seeks to uncover the mechanisms underlying the patterns of phylogenetic branching, it constitutes a missing link between genetics and evolutionary biology. Another such missing link is the study of phenotypic plasticity, or the ability of a genome to sculpt different phenotypes under distinct environmental conditions. This study too requires the cooperation of several disciplines, mainly genetics, organismic biology, systematics, evolutionary biology, and ecology.

As a consequence of the evo-devo synthesis, the traditional picture of evolution as only a sieve that eliminates the unfit genic variations is seen to have two flaws (Arthur 1997). It is one-sided, because it stresses destruction (death and extinction) at the expense of construction (emergence); and it is simplistic, because selection acts directly on whole organisms, not on genes. The traditional schema, 'Mutation → New genes → New species,' is therefore being completed nowadays to yield 'Mutation → New genes → New proteins → New developmental patterns → New phenotypes → New organism–environment relations → New species.' Evo-devo studies the intermediate process leading from new genes (which emerge mainly through mutation and recombination), and the corresponding new proteins, to new phenotypes. Moreover, the study of phenotypic plasticity emphasizes the role of the environment in gene expression and suppression.

Thus, the new picture puts the individual organism back at the centre of biology between molecule and population, as well as in the midst of its habitat. It also replaces the instant-adult fiction of classical evolutionary biology and population genetics with the real developing organism, which it regards as the site of qualitative novelty. In sum, the emergence of evo-devo has supplied not just a host of new important findings, but also a novel approach to both development and evolution. And it has reinforced the links between the two parent disciplines, as well as with these and genetics, ecology, and paleontology.

In general, the blurring of interdisciplinary borders occurs as the result of either of two moves: hybridization and phagocytation. The outcome of the first process is the emergence of a new interdiscipline, such as political geography, evolutionary developmental biology, or cognitive neuroscience. The second process, phagocytation, ensues in the inclusion of one discipline into another. Such inclusion can be legitimate, as in the case of the reduction of optics to electromagnetism; or illegitimate, as in the cases of molecular and structural biology.

The reduction of optics to electromagnetic theory towards the end of the nineteenth century was legitimate because it was found that light waves are electromagnetic waves. By contrast, molecular and structural biology do not belong in biology but in biochemistry and biophysics respectively. This is because biology proper, like life, only starts at the cell level. Indeed, molecular and structural biology study molecules, not living beings. In particular, structural biology investigates the folding of proteins, a physical process. The name 'biology' seems to be only a gimmick designed to attract students and research funds at a time when biology is the star science, and physics an impoverished brahmin. This case should alert us to the importance of the philosophical analysis of cross-disciplinarity.

3 Caution against Premature Unification

A word of caution is in order: Premature marriage among disciplines should be avoided, for it risks being barren. An examination of a few famous premature amalgamations may exhibit some of the necessary conditions for a synthesis of disciplines to succeed. Let us take a quick look at three such cases: phrenology, evolutionary psychology, and quantum cosmology.

It is well known that phrenologists attempted to diagnose psychological traits through examining bumps on the scalp, on the assumptions that such bumps are reliable indicators of brain development, and that mental functions are localized. Although eventually neuroscience vindicated a moderate version of localizationism (see chapter 12), its postulation by phrenology was purely speculative; and of course touching the scalp is useless for identifying brain subsystems.

The first seed of evolutionary psychology was planted, or rather dropped, by Charles Darwin in his work *The Descent of Man and Selection in Relation to Sex* (1871). However, this attempt failed to attract interest: psychology was still in its infancy, as well as far removed from biology, and ethology had not been born yet. This failure teaches an obvious lesson: For a marriage of two disciplines to be fruitful, the two must have attained some maturity.

The next attempt was made one century later (see Barkow, Cosmides, and Tooby 1992). Although (or because) it became instantly popular, this attempt too failed, though for different reasons. The main reasons for this particular failure are the following. First, the scientific method was not observed: Ingenious if simplistic stories about the origin of nearly all social items were offered in place of hard evidence. Second, the fusion consisted in a reduction of psychology to biology, and of the latter to genetics. As noted in chapter 9, this was a double error, because the brain is shaped not only by genes and the natural environment, but also by development (ontogeny) and the social environment.

Our last case of a failed synthesis will be quantum cosmology (see, e.g., Atkatz 1994). This theory is claimed to be a synthesis of quantum theory and relativistic cosmology. But this claim is debatable, because the theory deliberately violates the main conservation theorems included in the founding theories. (In particular, it violates the quantum-theoretical theorems of conservation of energy, and of linear and angular momentum, as well as the relativistic theorem of conservation of the stress-energy tensor.) Consequently, quantum cosmology contradicts the axioms that entail those theorems. It only uses the convenient parts of the founding theories. Therefore, this theory is not the synthesis it claims to be.

Worse, quantum cosmology includes decidedly wacky ideas, such as that the universe came into existence by tunnelling through a potential barrier. There are two problems with this idea. The first is that nothingness cannot traverse anything. The second is that the said potential barrier happens to be only a mathematical construct with a shape resembling a real potential barrier. Indeed, far from representing a physical field, as an ordinary potential barrier is supposed to, the said barrier is a function of the cosmic scale function (the radius of the universe in a closed model). Furthermore, it has a null gradient, since it only depends on time. And it does not even have the dimension of an energy. Hence, it makes no sense to calculate the tunnelling probability – which should equal 1 anyway, since the universe exists. An even more outrageous assumption that has been made is that, before coming into existence, the universe is (or is not) free to choose the curvature it will take (Coule and Martin 2000). We are not told what or who predetermines or makes the decision. But at least we are reassured that there are certain obstacles to creating a universe in the laboratory.

What is the difference between quantum cosmology and science fiction?

And how about declaring a moratorium on quantum cosmology until we find out at least whether the universe is spatially finite (closed) or infinite (open)? In sum, it seems that the necessary conditions for a successful fusion of quantum theory with classical cosmology have not yet been met.

The main lesson of the above examples is that, in amalgamating two research fields or theories, one should keep their distinctive and best-confirmed principles, since no embryo will develop in an empty shell. Some sacrifices may of course be required. But only the obviously false components can be safely dropped. For example, the so-called Modern Synthesis (of evolutionary biology with genetics) did not keep the hypothesis of inheritance of acquired characters, which Darwin had inherited from Lamarck. And no future synthesis of evolutionary biology with psychology should keep any of the patently false assumptions of computational cognitive psychology, such as the detachment of cognition from emotion, and the strict compartmentalization (modularity) of mental functions. (Recall chapter 11.)

The above examples suggest adopting the following methodological principle: The synthesis of fields or theories must retain their bulk. The word 'bulk' is deliberately vague. We should not specify a priori the percentage of principles of the founder theories that may have to be sacrificed in effecting their synthesis. Obviously, however, the best corroborated among them should take precedence over the others.

4 Why Both Processes Are Required

There are two main reasons for splitting disciplines. One is epistemological: as the knowledge of a subject increases in coverage and depth, it requires increasing specialization and therefore it becomes less accessible to students in other fields. For instance, although a chemical reaction is both a physical and a chemical process, it may be tackled either by chemists or by physicists. (However, only chemical physicists, or physical chemists, will produce the most complete account.) A second reason, or rather cause, is social: every specialization involves the formation of a community with interests that may conflict with those of other fiefdoms. In some cases, this social division masks an underlying unity. This seems to be the case with chemical physics and physical chemistry.

Some fields of inquiry have split, only to reunite later on, when the existence of cross-disciplinary problems was realized. Well-known examples are evolutionary and developmental biology, psychology and biology, and economics and sociology. These were cases of reversible divergence. The cases of irreversible divergence are far less numerous. In fact, I only know of one such case: that of mathematics and physics, which diverged in the nineteenth century with the emergence of abstract algebra and set theory. Even as recently as one century ago, rational mechanics was taught by mathematicians. (Many of them rejected relativistic mechanics, believing that mechanics, being rational, is an a priori science and thus impregnable to experiment.)

The reason for convergence is less obvious but no less compelling. The real world is one, and many processes, though different in some features, share others. Thus, the study of spontaneous self-organization processes, such as the emergence of macrophysical order out of microphysical disorder – as in solidification, the onset of ferromagnetism, and the clumping of cells that occurs in morphogenesis – have revealed some common general principles. This has prompted the formation of synergistics, a multidiscipline (see Haken 1989).

Just as the explanation of emergence often calls for the convergence of two or more disciplines, so convergence may in turn explain or even predict emergence. For example, Alan Turing, of Turing machine fame, predicted in 1952 the existence of chemical waves. He did so in studying chemical reactions diffusing through a medium, such as a liquid. That is, he coupled the equations of chemical kinematics with Fourier's diffusion equation. Quite independently, B.P. Belousov and A.M. Zhabotinsky produced such waves in the laboratory (see, e.g., Glansdorff and Prigogine 1971; Ball 2001). The study of reaction-diffusion systems is now a thriving chapter of chemistry.

At other times, fragmentation results from shallowness. It suffices to recall three historical episodes. Until the seventeenth century, mechanics was divided into two disconnected branches: terrestrial and celestial (or astronomy). Only the work of Galileo suggested that the matter involved was one, a point Newton proved when he built the first successful scientific theory that could legitimately claim to embrace the whole universe – though not the totality of processes. It took another two centuries to join mechanics to thermology, a fusion that produced the thermodynamic laws met by all macrophysical systems, regardless of their composition. Third example: Before the acceptance of the materialist thesis that mental events are brain events, the study of the mind and that of the brain ran parallel courses. Cognitive neuroscience is only now firmly established.

As for the epistemological reason for convergence, it was given by David Hilbert, who stated (1935 [1918]: 151) that it is not enough that the formulas of a physical theory be mutually consistent: they must also not contradict the propositions of a neighbouring field. For example, mechanics must be compatible with electromagnetic theory. I have called this the condition of *external consistency*, and have extended it to all the scientific theories (Bunge 1967a). Thagard (1978) calls it 'explanatory coherence.' For example, a psychological theory should cohere with the relevant neuroscience; and a linguistic theory should jibe with the bulk of psychology and sociology. Likewise, an economic theory should match the relevant sociology; thus, a theory of economic transactions should take into account that they occur among members of a social network rather than among mutually independent agents.

External consistency may be regarded as a condition of scientificity. The isolation of a discipline, as is the case with parapsychology and psychoanalysis, is a mark of pseudo-science (Bunge 1983b). In turn, the reason for this requirement is that the parcelling of the system of human knowledge is partly conventional. Consequently, this system should not be pictured in analogy with a map of sovereign nations, but rather as a rosette of partially overlapping petals.

The preceding considerations have an important bearing on the question of the relation between the depth and the breadth of a study. The received view on this matter is that these features of knowledge are inversely related to one another. (More precisely, the relation holds between the intension or sense of a predicate and its extension or coverage: see, e.g., Bunge 1974b.) However, the clear lead of cross-disciplinary over unidisciplinary strategies in tackling multifaceted and multilevel problems challenges the received view. We have seen, indeed, that at least in these cases breadth is necessary to attain depth.

Let us finally see how the information revolution is affecting the existing division of scientific labour. At first sight, the widening and tightening of the global communication network should strengthen interdisciplinary bonds. In fact, the Internet has facilitated enormously the formation of international 'collaboratories,' as well as the search for information that used to be conceptually and geographically distant from the specialist's topics. After all, two documents picked up at random on the web are on average only nineteen clicks away (Albert, Jeong, and Barabási 1999). The world of knowledge is thus a small world – at least potentially.

However, Van Alstyne and Brynjolfsson (1996) have argued that the very same mechanism can also help to balkanize science, in strengthening the ties among investigators within highly specialized fields, thus corralling them rather than prompting them to jump over the fence. In other words, conceptual convergence would be masked or even frustrated by excessive social cohesion. However, whether the Web promotes insularity or universality depends largely on individual interests, which are partly shaped by one's philosophical perspective. Whence the potential of philosophy to either promote or hinder the integration of science. Which suggests one more test for evaluating philosophies: Do they favour or obstruct the unification of knowledge and thereby its advancement?

5 Logic and Semantics of Integration

The problem of the fusion of scientific theories and entire research fields has been the object of far too few philosophical analyses (e.g., Darden and Maull 1977; Bunge 1983b). Let us therefore examine some of the major ideas involved. If the precursor disciplines are represented by ellipses, their merger may be represented as the intersection between those ellipses. And the entire system of the sciences can then be represented as a rosette of hundreds of partially overlapping petals. Obviously, however, not every juxtaposition of disciplines will result in their legitimate union. For example, geosociology, quantum anthropology, and molecular economics are hopeless monsters.

Of course, in addition to clear-cut cases there are also borderline ones. One of these may be ecophysiology, or the study of physiological processes under environmental stresses, such as droughts, food scarcity, and intense air pollution. Indeed, perhaps the only difference between ecophysiology and traditional physiology is that, whereas the latter feigned that the environment is roughly constant, ecophysiology takes environmental changes into account. So, it may not really be a synthesis of two disciplines, but rather a correction of standard physiology. To admit ecophysiology as a genuine new interscience, we must await the emergence of a hybrid theory containing both physiological and ecological laws, as well as bridges among key concepts of both disciplines.

I submit that any two disciplines, however alien they might be to one another, can be connected via a suitable number of links. For instance, sociology reaches out to the earth sciences through the intermediary of resource management – a social technology. And politology makes contact with pure mathematics via cultural policy-making, since mathematical production calls for funds as well as brains.

Let us sketch the necessary conditions for the fertile union of two previously separate disciplines, to form either a multidiscipline M_{12} or an interdiscipline I_{12}. Call D_1 and D_2 the precursors of the hybrid discipline (whether multi or inter). Next, call $\mathcal{R}(D_1)$ and $\mathcal{R}(D_2)$ the respective reference classes (or collections of entities they investigate). Further, let $C(D_1)$ and $C(D_2)$ designate the sets of technical (or specific) concepts of the respective disciplines. These concepts occur in the following two conventions.

Definition 17.1 M_{12} is the *multidiscipline* constituted by the disciplines D_1 and D_2 if

(a) $\mathcal{R}(M_{12}) = \mathcal{R}(D_1) \cup \mathcal{R}(D_2)$ is non-empty;

(b) there is a non-empty set $G(M_{12})$ of glue (or bridge or clamp) formulas in $D_1 \cup D_2$, wherein concepts of both $C(D_1)$ and $C(D_2)$ occur.

Definition 17.2 I_{12} is the *interdiscipline* lying between the disciplines D_1 and D_2 if

(a) $\mathcal{R}(I_{12}) = \mathcal{R}(D_1) \cap \mathcal{R}(D_2)$ is non-empty;
(b) there is a non-empty set $G(I_{12})$ of glue (or bridge or clamp) formulas in $D_1 \cap D_2$, wherein concepts of both $C(D_1)$ and $C(D_2)$ occur.

It will be noticed that, whereas $\cup$ (logical addition) is the trademark of multidisciplinarity, that of interdisciplinarity is $\cap$ (logical intersection). $\mathcal{R}$, the semantic concept of reference, is exactified elsewhere (Bunge 1974a), as is the epistemological concept of a research field (Bunge 1983b). I take it that the latter is logically prior to the sociological concept of a discipline, because people join those laboratories, institutes, or departments where the subjects that interest them are cultivated. This is why whoever undertakes to study, say, the political-sociology community presupposes some idea of political sociology as a branch of knowledge. Tell me what you study and I may find out the professional network where you belong.

6 Glue

The idea of a specific or technical concept, such as that of social structure in sociology, productivity in economics, power in politology, and revolution in historiography, is so familiar that it hardly needs comment. By contrast, the concept of a glue (or clamp or bridge) formula can do with some clarification, all the more since the expression 'bridge formula' has been used with a different meaning in the philosophical literature on reduction. Let the following examples suffice to sketch the intuitive idea.

Biophysics   The hydrophobic components of a protein are ensconced in its folds.
Evolutionary developmental biology   Evolutionary novelty emerges during development.
Psychophysics   The felt intensity of a physical stimulus is a power function of the latter.
Cognitive neuroscience   Learning consists in the emergence of new systems of neurons (Hebb).
Social cognitive neuroscience   The reaction of the amygdala to negative social stimuli depends on the subject's personality (extraverted or intraverted).
Neurolinguistics   A severe stroke in a language centre causes speech impediments.

Social medicine Morbidity and poverty are positively correlated.
Social psychology Physiological stress is the more intense, the lower the status in a social system.
Economic demography The pace of economic development is inversely proportional to the birth rate.
Socio-economics Labour productivity is inversely correlated to income inequality.
Political sociology An effect of the welfare state is to moderate the political left.
Political economy Economic globalization gives rise to undemocratic systems of supranational control and consultation.
Legal sociology As societies become more complex, criminal law becomes less important than civil law (Durkheim).

These examples suggest that a glue (or bridge) formula may be either a law or an indicator hypothesis or 'operational definition' – that is, an observable-unobservable link. Of course, it is up to the researcher to invent suitable glue formulas for the disciplines that he endeavours to synthesize. But at least a standard result in mathematical logic guarantees their existence under certain conditions. This is the interpolation theorem in elementary logic (see, e.g., Adamowicz and Zbierski 1997). Adapted to our problem, it reads as follows:

Theorem 17.1 Let A and B be formulas in the disciplines D_1 and D_2 respectively. If $A \Rightarrow B$ holds in the union $D_1 \cup D_2$, then there is a proposition C in the intersection $D_1 \cap D_2$ such that $A \Rightarrow C$ holds in D_1, and $C \Rightarrow B$ holds in D_2.

This theorem is pertinent to our subject for two reasons. To begin with, it relates multidisciplinarity $(D_1 \cup D_2)$ to interdisciplinarity $(D_1 \cap D_2)$. Second, it guarantees that the former entails the latter – which may sound rather counterintuitive. That is, we have proved

Theorem 17.2 If D_1 and D_2 are disciplines satisfying Definition 17.2 and Theorem 17.1, then their intersection is non-empty: $I_{12} = D_1 \cap D_2 \neq \emptyset$.

Needless to say, this is not a recipe for crafting interdisciplines: it just encourages the quest for them, in showing that certain hybrids are conceptually possible. Whether the suitable glue (or bridge) formulas will be found depends not only upon the researcher's ingenuity: it depends primarily on whether the formulas in question represent bridges in the real world.

Theorem 17.2 suggests the following measure of the closeness (or degree of interdisciplinarity) between two disciplines D_1 and D_2:

$\kappa = |D_1 \cap D_2|^2/|D_1||D_2|,$

where $|D|$ stands for the numerosity of the set D. The value of κ ranges between 0 for totally alien fields, such as particle physics and art history, and 1 for identical fields. Needless to say, the measure κ is easier defined than estimated in particular cases.

In sum, a broad research field is likely to consist not only of disciplines and subdisciplines, but also of multidisciplines and interdisciplines. For example, linguistics contains not only general linguistics (or the general study of grammar), but also such interdisciplines as neurolinguistics, psycholinguistics, sociolinguistics, and historical linguistics. The more bridges, the better entrenched and the more useful.

7 Integrated Sciences and Technologies

Let us now introduce a last convention:
Definition 17.3 A multidiscipline or interdiscipline is
(i) either a *multiscience* or an *interscience* if it is scientific;
(ii) either a *multitechnology or* an *intertechnology* if it is technological.

For a multidiscipline or an interdiscipline to be scientific (or technological), it is necessary, though insufficient, that the pertinent glue formulas be reasonably well confirmed, or at least testable in principle. Another necessary condition of scientificity is that at least one of the precursors of the hybrid discipline be scientific (or technological). That this condition is insufficient is proved by the existence of astrology (a hybrid of astronomy and folk psychology) and psychohistory (a cross of history with psychoanalysis).

In sum, pedigree is not enough: any claim to the scientific status of a hybrid discipline must be established independently. That is, the hybrid in question must be shown to abide by the scientific method, and to be consistent with the bulk of the relevant (scientific or technological) knowledge, as well as with the naturalistic and realistic outlook of science. As we argued in chapter 9, human sociobiology and its currently fashionable successors, evolutionary psychology and evolutionary medicine, fail to meet these conditions because they are mostly fantasies. These stories are off the mark if only because they ignore all the economic, political, and cultural variables. Examples: The claims that all males, in all societies, attempt to maximize the dispersion of their genes; that our brains have been 'designed' to cope with the conditions prevailing about 100,000 years ago in the East African savanna; that religiosity is in the genome; and that sadness and even depression are healthy adaptations.

Finally, Theorem 17.2 may be rewritten as the following programmatic methodological:

Thesis 17.1 Given any two disciplines with some common referents, a third discipline can be interpolated between them.

This thesis is unfalsifiable, because any failure to fashion an interdiscipline may be blamed on shortcomings of the people who have made the attempt. However, like other unfalsifiable theses, such as those of the lawfulness and knowability of the world, and the perfectibility of man, the thesis in question is heuristically fertile. (Incidentally, not only philosophy but also science in the making teems with programmatic hypotheses, such as 'Variable y depends upon variable x,' where the form of the dependence is to be determined by research. And some of these programmatic hypotheses may be unfalsifiable in real time.)

A first corollary of the above thesis is that, at least in principle, there are no independent sciences or technologies. To put it in positive terms, all the sciences and technologies are interdependent to some extent. This feature is so important that it deserves being raised to the status of a norm involved in the very definition of scientificity. I have called it the condition of *external consistency*, and have assigned it the same weight as internal consistency, precision, and testability (Bunge 1967a). External consistency is much more than mere compatibility, which is met trivially by any two fields that have no common referents, as is the case of geography and virology. External consistency consists in partial overlap. For example, an economic theory must respect the principle of energy conservation; on the other hand, it may violate the second principle of thermodynamics, of increasing disorder, since this only holds for closed systems. In short, every genuine science overlaps partially with other sciences. Strictly speaking, there are no unidisciplinary fields.

It follows that, if a field of knowledge is disjunct from all the sciences, then it is non-scientific. For instance, graphology, parapsychology and psychoanalysis are alien to, and even at odds with, experimental psychology and neuroscience. This suffices to dismiss them as pseudo-sciences. Much the same holds for neo-classical microeconomics and the game-theoretic models in politology, because they take pride in being self-reliant. Tell me which company you fail to keep, and I will tell you where you are bound to fail.

How far should one go in seeking to fuse disciplines? Edward O. Wilson (1998), the eminent zoologist and founder of contemporary sociobiology, has argued for the 'consilience' (unification) of all the branches of knowledge. I agree that this goal is desirable, attainable, and increasingly close. But I disagree with Wilson's thesis that such unity must be attained through reduction. In particular, I disagree that every social convention has biological roots and can thus be explained by biology.

A major objection to the project of unifying social science through reduction to biology is this: There is overwhelming evidence for the hypothesis that social norms are social inventions, and that these are adopted or imposed, reformed or rejected, by people regardless of their biological similarities or differences (recall chapter 9). Just think of the enormous and swift changes in social mores that have occurred in the developed countries during the last century as a result of political and cultural movements: the political empowerment of women; the tolerance of homosexuality; the acceptance of love-making for the sake of pleasure rather than just reproduction; the abolition of the death penalty; the spread of the welfare state; the massification of the universities; the worldwide expansion of the fast-food industry; and the replacement of the printed press by television as the main medium of information and entertainment. None of these social changes can be traced back to either genic alterations or natural selection.

The consilience defended in the present book does not erase the differences between disciplines, because it does not flatten the world to either the physical or the cultural levels.

Concluding Remarks

The qualitative richness of the universe is so patent that, at first sight, the unity of the sciences looks like an impossible dream – as Dupré (1993) and others have claimed. But of course such diversity has always spurred philosophers and scientists to search for underlying unities, both ontological and epistemological. For example, far from being dualist, cognitive neuroscience conceives of mental processes as brain processes; and physiological social psychology supplements the biological account of the mental with a consideration of the social inputs and outputs of cognition and emotion. Besides, in both research fields the scientific method is being practised, even if there is no consensus concerning its nature. In sum, the multiplication of the sciences of the mind has been accompanied by their increasing convergence.

Arguably, the ultimate root of the unification movements is the materiality, systemicity, lawfulness, and knowability of the universe. Thus, diversity in the details is consistent with overall unity, just as an account of particulars is complementary to the search for pattern: contrary to what the hermeneuticists contend, all of the sciences are both idiographic and nomothetic, and use the scientific method (see Bunge 1996).

However, the basic unity of both the world and science does not entail the success of radical reductionism. For example, the reduction of chemistry to atomic physics via quantum mechanics is so far only partial, if only because it requires additional concepts, such as that of covalent bond, as well as additional assumptions, such as the basic rate equation of classical chemical kinetics. Likewise, the reduction of psychology to neuroscience is ontological (in identifying the mental and the neural) but not epistemological, since cognitive neuroscience is a typical interdiscipline that employs such psychological concepts unknown to biology as those of attention and perception, pleasure and fear, empathy and intention, personality and neuroticism (see Bunge 1990).

Another example is the recent completion of the sequencing of the human genome, hyped as the deciphering of the Book of Life, and hailed as a victory of radical reductionism. Although this is undoubtedly a remarkable achievement of Big Science, it was no such philosophical victory, because the gap between genotype and phenotype is still to be filled. Indeed, the participation of proteins in intracellular processes, and the intercellular interactions that make organs and the whole organism work, are still far from being well known, as even Sydney Brenner (2000) has warned. (This is the same eminent molecular biologist who, in 1982, had prophesied that, if he were given the complete sequence of an organism's genome and a powerful computer, he could compute the whole organism.)

In short, here as elsewhere, knowledge of the parts is necessary but insufficient to explain how they combine into a whole with (emergent) properties of its own. This is why reduction, though certainly desirable whenever feasible, must be supplemented with integration: because the world is both one in some respects and diverse in others, and because it is not a pile of simples but a system of systems. Here, as elsewhere, epistemology and methodology must fit ontology, and the latter must cohere with science.

To conclude. One century ago, in his epoch-making address to the Second International Congress of Mathematicians, the great David Hilbert (1901) pointed out that, far from being isolated, every mathematical problem belongs to some problem chain – a chain that may trespass previously erected barriers. And he closed his address by stating that mathematics is 'an indivisible whole, an organism whose viability is conditioned by the interdependence of its parts.' The thrust of the present chapter is that the same holds for all the sciences: Their health depends on their being a part of a single conceptual system – though not a seamless one – even if their respective research communities are socially separate. No one can network with a million investigators, but anyone can detect or craft a conceptual bridge to a neighboring discipline, particularly if he adopts the systemic approach and the scientific method, and uses the language of all the sciences, namely mathematics. Such convergence processes are often necessitated by the discovery of emergents, and in turn they may spawn the search for further convergents.

Glossary

Agathonism The moral philosophy whose peculiar maxims are 'Enjoy life and help live,' and 'Every right implies a duty and conversely.' A sort of merger of deontologism and utilitarianism.
Axiology Value theory.
Background knowledge What is known so far, the starting point of a research project.
Cause and effect The cause of an event is the event that is both necessary and sufficient for its occurrence.
Computationism The programmatic hypothesis according to which natural or social processes of certain kinds are computational or algorithmic.
Conditional Proposition of the form 'If A, then B,' or 'A $\Rightarrow$ B' for short.
Constructivism Ontological: The world is a human (individual or social) construction. Epistemological: All ideas are constructs, and none derives directly from perception.
Convergence (of research lines) Merger of two or more previously separate lines of inquiry. Examples: cognitive neuroscience and socio-economics.
Counter-Enlightenment Reaction against the Enlightenment. Characteristics: irrationalism and social conservatism. Icons: Hegel, Nietzsche, and Heidegger.
Counter-example Exception to an alleged law or rule.
Datum A piece of information, such as 'The world population currently exceeds six billion people.' A scientific datum is evidence for or against some scientific hypothesis or theory.
Definition Convention whereby one concept or term is set identical with other concepts or terms, as in '2 $=_{df}$ 1 + 1' and 'Emergent property $=_{df}$ Property of a whole that its constituents lack.' Not to be confused with either description or assumption.

Deontologism The moral philosophy that enjoins us to do our duties regardless of consequences for self or others. Rivals: utilitarianism and agathonism.

Divergence (of research lines) Splitting of a research line into two or more specialties.

Dualism The family of doctrines according to which there are two equally basic but mutually irreducible kinds, such as matter and mind, or nature and society. A particular case of pluralism.

Emergence Qualitative novelty. A property of systems.

Empiricism The epistemological doctrine according to which experience is the source and test of every idea.

Enlightenment tradition Respect for reason, inquiry, truth, and human rights, especially liberty. First articulated in the eighteenth century.

Epistemic Having to do with knowledge, as in 'Data are the epistemic counterparts of facts.'

Epistemology The philosophical study of cognition and its product, knowledge. It can be descriptive or prescriptive (normative). Main schools: scepticism, intuitionism, empiricism, conventionalism, fictionism, rationalism, and ratio-empiricism.

Ethics The philosophical study of moral norms, or rules of right conduct.

Event Change of state of a concrete thing.

Evidence A set of data relevant to some hypothesis.

Exactification Conversion of a fuzzy construct into an exact one. Example: likely $\to$ probable.

Existence Conceptual: belonging to a conceptual context, such as a theory. Material: being part of the real world, or having energy. Example: Something exists: $(\exists x)Ex$.

Existential quantifier Exactification of the concept of someness, as in 'Some objects are systems,' or $\exists x Sx$ for short. In mathematics, 'some' and 'exists' are interchangeable.

Explanation Description of a mechanism.

Fact State or change of state of a material thing.

Fallibilism The family of epistemologies that asserts that most if not all of our knowledge is subject to revision.

Functionalism The view that only function matters, and 'substrate' (matter) does not. Prevalent in cognitive psychology before the emergence of cognitive neuroscience.

Glue Formula that bridges two of more disciplines to form an interdiscipline. Example: 'Longevity and income are positively correlated' is one of the glues between human biology and socio-economics.

Hermeneutics The school of thought according to which (a) symbols are all-important or even the only existents, and (b) social facts are understood by grasping intuitively the intentions or goals of the actors involved in them.

Holism The family of doctrines according to which things come in unanalysable wholes. The ontological partner of intuitionism.

Hypothesis A proposition, whether particular or general, that asserts more than any of the data relevant to it.

Idealism, philosophical The ontology according to which ideas pre-exist and dominate everything else. Not to be mistaken for altruism.

Individualism The view that the basic constituents of the universe are separate individuals.

Interdiscipline The discipline formed by a combination (not just a juxtaposition) of two or more disciplines.

Intuitionism The family of epistemological doctrines according to which intuition is superior to any other kind of cognition, in being total and immediate. The epistemological partner of holism.

Law Regularity or pattern. Objective law: inherent in things, hence a property of theirs. Conceptual law: formula that captures an objective pattern. Sometimes called 'nomological statement' to avoid confusion with the pattern it is expected to represent.

Lawfulness, principle of The principle, presupposed tacitly by all the sciences, that everything, even chance, satisfies laws.

Level (of organization) Collection of things characterized by a system of properties. Examples: physical, chemical, biological, social, and technological levels.

Logic The common part of logic and mathematics, centred in logical form and deducibility regardless of the content of the propositions in question. Logic can be ordinary or non-standard. Ordinary or classical logic is the theory of deduction inherent in all of the factual sciences.

Macro-reduction Reduction to higher-level entities. Example: explanation of individual behaviour in terms of environmental stimuli.

Material Capable of changing by itself, having energy.

Materialism The family of naturalist ontologies according to which all existents are material. Two main versions: vulgar or physicalist ('Everything is physical'), and emergentist ('There are several kinds of matter: physical, chemical, living, social, and technical'). According to materialism, the predicates 'is material' and 'is real' are coextensive (apply correctly to the same things), even though they are not cointensive (i.e., do not mean the same).

Meaning, semantic   A property of constructs. Reference or denotation together with sense or connotation. Not to be confused with the 'meaning' of an action, i.e., its goal.
Mechanism   The totality of processes that make a system tick, such as metabolism in the case of organisms, and litigation in that of courthouses.
Meliorism   The family of epistemologies that assert the corrigibility of epistemic error.
Method   A standardized procedure for doing something.
Method, scientific   Background knowledge → Problem → Hypothesis or technique → Test → Evaluation of hypothesis or technique → Revision of background knowledge.
Methodology   The study of methods: a branch of epistemology. Not to be confused with 'method.'
Micro-reduction   Reduction to lower-level entities. Example: the great-man theory of history.
Mind   Theological view: The immaterial soul, that may or may not use the brain, hence may be understood without the help of neuroscience. Scientific view: The family of specific functions of highly developed brains, whence it can only be explained by cognitive and affective neuroscience.
Mind-body problem   The question of the nature of mental functions and their relation to the brain. Main answers: psychoneural dualism and materialist monism (either physicalist or emergentist).
Module   Unit on a certain level.
Monism, substance   The family of philosophies that assert that all things are of a single basic kind – material, ideal, or other.
Multidiscipline   Set of disciplines engaged in studying multifaceted facts.
Naturalism   The family of ontologies that reject the belief in supernatural agencies.
Norm   Ideal pattern of behaviour consecrated in a social group on the belief, true or false, that it favours the interests of some or all of its members.
Objectivism   The view that it is possible and desirable to account for things regardless of the way they may affect us.
Ontology   The philosophical study of being and becoming.
Phenomenalism   The philosophical view that there are only phenomena (appearances to someone), or that only these can be known. Phenomenalism opposes realism.
Phenomenon   Perception of a fact, in contradistinction to the fact in itself. For instance, colour as distinct from wavelength. Phenomena happen only in brains.
Pluralism   The family of philosophies that assert the plurality of basic kinds

(substance pluralism), basic properties (property pluralism), or both. Emergentist materialism is property-pluralist but substance-monist: one substance, many properties.

Pragmatism The epistemological doctrine according to which practice is the origin, test, and value of all ideas.

Probability Quantitation of chance or randomness. A property of states or changes of state, not of propositions. Not to be confused with qualitative possibility, likelihood, or plausibility.

Process A non-instantaneous change of state of a thing, a sequence of states of a thing. Formalizable as a sequence of values of a state function, i.e., a trajectory in the state space of the things of some kind.

Property Feature or trait of an item, such as the area of a closed figure, the energy of a body, the age of an organism, and the structure of a society.

Proposition A statement, such as 'I am reading,' capable of being true or false to some degree. Not to be confused with 'proposal.'

Pseudo-science A body of belief presented as scientific although it is inconsistent with the bulk of antecedent scientific knowledge.

Ratio-empiricism Merger of rationalism and empiricism.

Rational-choice theory The theory according to which all actors prefer, or should prefer, the action leading to the outcome that maximizes their expected utilities. (The expected utility of an outcome equals the product of its probability by its utility.)

Rationalism The epistemological doctrine according to which cognition involves reasoning. Radical rationalism: Reason is sufficient for knowing. Moderate rationalism: Reason is necessary for understanding.

Real Existing in and by itself, that is, whether or not it is perceived or conceived of by someone.

Realism The philosophy according to which the external world exists independently of the inquirer, and can be known at least in part. Realism opposes radical scepticism, conventionalism, phenomenalism, and idealism.

Reduction Conceptual operation whereby an item is claimed or shown to be identical with another or included in it, or to be either an aggregate, a combination, or an average of other items.

Relativism, epistemological. The view according to which all ideas are relative to the subject or to a social group, hence none are cross-cultural, let alone universally valid.

Rule Standardized procedure or human behaviour pattern, as in 'The rule of incest avoidance is universal.'

Scepticism The family of epistemologies that either deny the possibility of certain knowledge (radical scepticism) or else affirm that we should doubt before believing – and often even after having adopted a belief (moderate scepticism).

Scientism The epistemological view according to which whatever can be known is best investigated scientifically. Scientism joins ratio-empiricism with the scientific method. Not to be confused with the disregard for supra-physical features.

Semantics The study of meaning and truth. Not to be confused with 'a mere matter of words.'

State (of a thing) The totality of properties of a thing at a given instant.

State space The set of all the possible states that things of a given kind can be in. Conceptualizable as the n-dimensional Cartesian space spanned by an n-dimensional state function such as the human development index.

Structure of a system The totality of relations among the system's components, and among these and the system's environment.

Subjective A process that occurs inside a subject's head. The main object of study of psychology.

Subjectivism The family of philosophies according to which everything is in a subject's mind (subjective idealism), or nothing can be known objectively.

Submergence The disappearance of one or more properties of a thing or system due to its metamorphosis into other things, as in the loss of mass when an electron fuses with an anti-electron (positron) to produce a photon.

System A complex object whose constituents are held together by strong bonds – logical, physical, biological, or social – and possessing global (emergent) properties that their parts lack.

Systemics The branch of ontology concerned with the general features of concrete systems. Some of its chapters are the theory of Turing machines, cybernetics (the general theory of control), and synergistics (the general theory of self-assembly). Most of the contributors to systemics have been scientists, engineers, or both.

Systemism Ontology: Everything is either a system or a component of a system. Epistemology: Every piece of knowledge is or ought to become a member of a conceptual system, such as a theory.

Tautology A truth of logic, such as 'The conjunction of a proposition and its denial is false.'

Technology Body of knowledge invented and utilized to design, produce, or maintain artefacts, whether physical (like computers), biological (like cultivated rice), or social (like hospitals).

Testability, empirical The ability of a hypothesis or a theory to be confirmed or falsified to some extent by empirical data. A characteristic of science and scientific philosophy in contradistinction to theology and esotericism.

Theory Hypothetico-deductive system: A system of propositions every one of which implies or is implied by others.

Truth, factual The adequacy of ideas to the facts they represent, as with 'It rains' when it is actually raining.

Truth, formal The coherence of an idea with a previously accepted body of ideas. Examples: tautologies and theorems.

Truth, partial The less than perfect adequacy of an idea to its referent, as in '$\pi = 3$.'

Truth value One of various degrees of adequacy (true, false, half-true, etc.) that a proposition may have or acquire.

Universal quantifier Prefix 'for all,' as in '$\forall x$ (x is real if and only if x is changeable).'

Utilitarianism The moral philosophy that enjoins us to maximize the expected utility or gain for self or others. Rivals: deontologism and agathonism.

Utility Subjective value.

Verstehen Intuitive understanding. The 'method' of hermeneutics.

For more, see Bunge 2003a.

References

Adamowicz, Zofia, and Pawel Zbierski. 1997. *Logic of Mathematics: A Modern Course of Classical Logic*. New York: John Wiley & Sons.

Adams, Marie, Zvonimir Dogic, Sarah L. Keller, and Seth Fraden. 1998. 'Entropically Driven Microphase Transitions in Mixtures of Colloidal Rods and Spheres.' *Nature* 393: 249–351.

Adorno, Theodor, and Max Horkheimer. 1972. *Dialectic of Enlightenment*. New York: Herder & Herder.

Agassi, Joseph. 1987. 'Methodological Individualism and Institutional Individualism.' In J. Agassi and I.C. Jarvie, eds, *Rationality: The Critical View*, 119–50. Dordrecht: Martinus Nijhoff.

Agazzi, Evandro, ed. 1991. *The Problem of Reductionism in the Sciences*. Dordrecht, Boston: Kluwer.

Agnati, L.F., K. Fuxe, C. Nicholson, and S. Syková, eds. 2000. *Volume Transmission Revisited, Progress in Brain Research*, vol. 125. Amsterdam: Elsevier.

Albert, Hans. 1994. *Kritik der reinen Hermeneutik*. Tübingen: J.C.B. Mohr (Paul Siebeck).

Albert, Réka, Hawoong Jeong, and Albert-Lászlo Barabási. 1999. 'Diameter of the World-Wide Web.' *Nature* 401: 130.

Anderson, Adam K., and Elizabeth A. Phelps. 2001. 'Lesions of the Human Amygdala Impair Enhanced Perception of Emotionally Salient Events.' *Nature* 411: 305–9.

Arkin, Adam, and John Ross. 1994. 'Computational Functions in Biochemical Reaction Networks.' *Biophysical Journal* 67: 560–78.

Armstrong, E., and D. Falk, eds. 1982. *Primate Brain Evolution: Methods and Concepts*. New York: Plenum.

Arthur, Wallace. 1997. *The Origin of Animal Body Plans: A Study in Evolutionary Developmental Biology*. Cambridge: Cambridge University Press.

– 2002. 'The Emerging Conceptual Framework of Evolutionary Developmental Biology.' *Nature* 415: 757–64.

Asfaw, Berhane, W. Henry Gilbert, Yonas Beyene, William K. Hart, Paul B. Renne, Giday WoldeGabriel, Elisabeth S. Vrba, and Tim D. White. 2002. 'Remains of *Homo erectus* from Bouri, Middle Awash, Ethiopia.' *Nature* 416: 317–20.

Ashby, W. Ross. 1963 [1956]. *An Introduction to Cybernetics*. New York: John Wiley & Sons.

Athearn, Daniel. 1994. *Scientific Nihilism: On the Loss and Recovery of Physical Explanation*. Albany: State University of New York Press.

Atkatz, David. 1994. 'Quantum Cosmology for Pedestrians.' *American Journal of Physics* 62: 619–27.

Baldi, S. 1998. 'Normative versus Social Constructivist Processes in the Allocation of Citations: A Network-analytic Model.' *American Sociological Review* 63: 829–46.

Ball, Philip. 2001. *The Self-Made Tapestry: Pattern Formation in Nature*. Oxford: Oxford University Press.

Barbas, Helen. 1995. 'The Anatomical Basis of Cognitive–Emotional Interactions in the Primate Prefrontal Cortex.' *Neuroscience Behavioral Reviews* 19(3): 499–510.

Barkow, Jerome H., Leda Cosmides, and John Tooby, eds. 1992. *The Adapted Mind: Evolutionary Psychology and the Generation of Culture*. Cambridge, MA: MIT Press.

Bartlett, M.S. 1975. *Probability, Statistics and Time: A Collection of Essays*. London: Chapman & Hall.

Barton, R.A., and P.H. Harvey. 2000. 'Mosaic Evolution of Brain Structure in Mammals.' *Nature* 405: 1055–58.

Bates, Elizabeth, and Judith C. Goodman. 1999. 'On the Emergence of Grammar from the Lexicon.' In MacWinney, ed., 29–79.

Baudrillard, Jean. 1973. *Le miroir de la production*. Paris: Casterman.

Bechtel, William, ed. 1986. *Integrating Scientific Disciplines*. Dordrecht, Boston: Martinus Nijhoff.

Becker, Gary S. 1976. *The Economic Approach to Human Behavior*. Chicago: University of Chicago Press.

Berlinski, David. 1976. *On Systems Analysis*. Cambridge, MA: MIT Press.

Bernard, Claude.1865. *Introduction à l'étude de la médecine expérimentale*. Paris: Flammarion, 1952.

Berry, Margaret. 1975. *An Introduction to Systemic Linguistics*. 2 vols. New York: St Martin's Press.

Bertalanffy, Ludwig von. 1950. 'An Outline of General System Theory.' *British Journal for the Philosophy of Science* 1: 139–64.

– 1968. *General System Theory*. New York: George Braziller.

Bhabha, H.K. 1990. *Nation and Narration*. London: Routledge.

Bindra, Dalbir. 1976. *A Theory of Intelligent Behavior*. New York: Wiley.

Black, Donald. 1976. *The Behavior of Law*. New York: Academic Press.

Blair, R.J.R., J.S. Morris, C.D. Frith, D.I. Perreu, and R.J. Dolan. 1999. 'Dissociable Neural Responses to Facial Expressions of Sadness and Anger.' *Brain* 122: 883–93.

Bliss, Michael. 2000. *The Discovery of Insulin*. 3rd ed. Toronto: University of Toronto Press.

Blitz, David. 1992. *Emergent Evolution: Qualitative Novelty and the Levels of Reality*. Dordrecht, Boston: D. Reidel.

Bloom, Harold, Paul de Man, Jacques Derrida, Geoffrey Hartman, and J. Hillis Miller. 1990. *Deconstruction and Criticism*. New York: Continuum.

Blumer, Harold. 1969. *Symbolic Interactionism*. Englewood Cliffs, NJ: Prentice-Hall.

Bohr, Niels. 1958. *Atomic Physics and Human Knowledge*. New York: John Wiley & Sons.

Bolzano, Bernard. 1950 [1851]. *Paradoxes of the Infinite*. Trans. D.A. Steele. London: Routledge & Kegan Paul.

Booth, J.W., P. James, and H. Meadwell, eds. 1993. *Politics and Rationality*. Cambridge: Cambridge University Press.

Boring, Edwin G. 1950. A *History of Experimental Psychology*. 2d ed. New York: Appleton-Century-Crofts.

Botvinick, Matthew, Leigh E. Nystrom, Kate Fissell, Cameron S. Carter, and Jonathan D. Cohen. 1999. 'Conflict Monitoring versus Selection-for-action in Anterior Cingulate Cortex.' *Nature* 402: 179–81.

Boudon, Raymond. 1974. *Education, Opportunity, and Social Inequality*. New York: Wiley.

– 1980. *The Crisis in Sociology*. New York: Columbia University Press.

– 1981. *The Logic of Social Action*. London, Boston, and Henley, UK: Routledge & Kegan Paul.

– 1998. *Études sur les sociologues classiques*. Paris: Quadrige / Presses Universitaires de France.

– 2001. *The Origin of Values: Sociology and Philosophy of Beliefs*. New Brunswick, NJ: Transaction Publishers.

– 2002. 'Sociology That Really Matters.' *European Sociological Review* 18: 371–8.

– 2003. *Y-a-t-il encore une sociologie?* Paris: Odile Jacob.

Bourricaud, François. 1977. *L'individualisme institutionnel: Essai sur la sociologie de Talcott Parsons*. Paris: Presses Universitaires de France.

Braithwaite, Richard B. 1953. *Scientific Explanation*. Cambridge: Cambridge University Press.

Braudel, Fernand. 1969. *Écrits sur l'histoire*. Paris: Flammarion.

– 1972 [1949]. *The Mediterranean and the Mediterranean World in the Age of Philip II*. 2nd ed. Trans. S. Reynolds. New York: Harper & Row.

Brenner, Sydney. 2000. 'The End of the Beginning.' *Science* 287: 2173–4.

Broad, C[harlie] D[unbar]. 1925. *The Mind and Its Place in Nature*. London: Routledge & Kegan Paul.

Buck, R.C. 1956. 'On the Logic of General Behavior Systems Theory.' In Herbert Feigl and Michael Scriven, eds, *Minnesota Studies in the Philosophy of Science*, vol. 1: 223–38. Minneapolis: University of Minnesota Press.

Bunge, Mario. 1951. 'What Is Chance?' *Science & Society* 15: 209–31.

– 1956. 'Do Computers Think?' *British Journal for the Philosophy of Science* 7: 139–8; 7: 212–19.

– 1959a. *Causality: The Place of the Causal Principle in Modern Science*. Cambridge, MA: Harvard University Press. Rev. New York: Dover, 1979.

– 1959b. *Metascientific Queries*. Evanston, IL: Charles C. Thomas.

– 1962a. *Intuition and Science*. Englewood Cliffs, NJ: Prentice-Hall. Repr. Westport, CT: Greenwood Press, 1975.

– 1962b. 'Cosmology and Magic.' *The Monist* 44: 116–41.

– 1963. *The Myth of Simplicity*. Englewood Cliffs, NJ: Prentice-Hall.

– 1964. 'Phenomenological Theories.' In M. Bunge, ed., *The Critical Approach: Essays in Honor of Karl Popper*, 234–54. New York: Free Press.

– 1967a. *Scientific Research*. 2 vols. Berlin, Heidelberg, New York: Springer-Verlag. Rev. as *Philosophy of Science*. 2 vols. New Brunswick, NJ: Transaction, 1998.

– 1967b. *Foundations of Physics*. Berlin, Heidelberg, New York: Springer-Verlag.

– 1971. 'Is Scientific Metaphysics Possible?' *Journal of Philosophy* 68: 507–20.

– 1973a. *Philosophy of Physics*. Dordrecht, Boston: D. Reidel.

– 1973b. *Method, Model, and Matter*. Dordrecht, Boston: D. Reidel.

– 1974a. *Treatise on Basic Philosophy*, vol. 1: *Sense and Reference*. Dordrecht, Boston: D. Reidel.

– 1974b. *Treatise on Basic Philosophy*, vol. 2: *Interpretation and Truth*. Dordrecht, Boston: D. Reidel.

– 1976. 'Possibility and Probability.' In W.L. Harper and C.A. Hooker, eds, *Foundations of Probability Theory, Statistical Inference, and Statistical Theories of Science*, vol. 3: 17–34. Dordrecht, Boston: D. Reidel.

– 1977a. *Treatise on Basic Philosophy*, vol. 3: *The Furniture of the World*. Dordrecht, Boston: D. Reidel.

– 1977b. 'Levels and Reduction.' *American Journal of Physiology: Regulatory, Integrative and Comparative Physiology* 2: 75–82.

– 1979a. *Treatise on Basic Philosophy*, vol. 4: *A World of Systems*. Dordrecht, Boston: D. Reidel.

– 1979b. 'The Mind-Body Problem in an Evolutionary Perspective.' In *Brain and Mind: Ciba Foundation Symposium* 69: 53–63. Amsterdam: Excerpta Medica.

– 1980. *The Mind-Body Problem*. Oxford: Pergamon Press.

- 1981a. *Scientific Materialism.* Dordrecht, Boston: D. Reidel.
- 1981b. 'Four Concepts of Probability.' *Applied Mathematical Modelling* 5: 306–12.
- 1982. 'Is Chemistry a Branch of Physics?' *Zeitschrift für Allgemeine Wissenschaftstheorie* 13: 209–33.
- 1983a. *Treatise on Basic Philosophy,* vol. 5: *Exploring the World.* Dordrecht, Boston: D. Reidel.
- 1983b. *Treatise on Basic Philosophy,* vol. 6: *Understanding the World.* Dordrecht, Boston: D. Reidel.
- 1984. 'Philosophical Problems in Linguistics.' *Erkenntnis* 21: 107–73.
- 1985a. *Treatise on Basic Philosophy,* vol. 7: *Philosophy of Science and Technology,* part 1: *Formal and Physical Sciences.* Dordrecht, Boston: D. Reidel.
- 1985b. *Treatise on Basic Philosophy,* vol. 7: *Philosophy of Science and Technology,* part 2: *Life Science, Social Science, and Technology.* Dordrecht, Boston: D. Reidel.
- 1986. 'A Philosopher Looks at the Current Debate on Language Acquisition.' In I. Gopnik and M. Gopnik, eds, *From Models to Modules,* 229–39. Norwood, NJ: Ablex Pub. Co.
- 1989. *Treatise on Basic Philosophy,* vol. 8: *Ethics: The Good and the Right.* Dordrecht, Boston: D. Reidel.
- 1990. 'What Kind of Discipline Is Psychology: Autonomous or Dependent, Humanistic or Scientific, Biological or Sociological?' *New Ideas in Psychology* 8: 121–37.
- 1991. The Power and Limits of Reduction.' In Agazzi, ed., 31–49.
- 1996. *Finding Philosophy in Social Science.* New Haven, CT: Yale University Press.
- 1997a [1980]. *Ciencia, técnica y desarrollo.* 2nd ed. Buenos Aires: Sudamericana.
- 1997b. 'Moderate Mathematical Fictionism.' In E. Agazzi and G. Darwas, eds, *Philosophy of Mathematics Today,* 51–71. Dordrecht, Boston: Kluwer Academic.
- 1997c. 'Mechanism and Explanation.' *Philosophy of the Social Sciences* 27: 410–65.
- 1998. *Social Science under Debate.* Toronto: University of Toronto Press.
- 1999. *The Sociology-Philosophy Connection.* New Brunswick, NJ: Transaction.
- 2000a. 'Systemism: The Alternative to Individualism and Holism.' *Journal of Socio-Economics* 29: 147–57.
- 2000b. 'Ten Modes of Individualism – None of Which Works – and Their Alternatives.' *Philosophy of the Social Sciences* 30: 384–406.
- 2000c. 'Energy: Between Physics and Metaphysics.' *Science and Education* 9: 457–61.
- 2000d. 'Physicians Ignore Philosophy at Their Risk – and Ours.' *Facta philosophica* 2: 149–60.
- 2001a. *Philosophy in Crisis: The Need for Reconstruction.* Amherst, NY: Prometheus Books.

- 2001b. 'Systems and Emergence, Rationality and Imprecision, Free-wheeling and Evidence, Science and Ideology: Social Science and Its Philosophy according to van den Berg.' *Philosophy of the Social Sciences* 3: 404–23.
- 2003a. *Philosophical Dictionary.* 2nd ed. Amherst, NY: Prometheus Books.
- 2003b. 'How Does It Work?' *Philosophy of the Social Sciences* forthcoming.

Bunge, Mario, and Rubén Ardila. 1987. *Philosophy of Psychology.* New York: Springer-Verlag.

Bunge, Silvia A., T. Klingberg, R.B. Jacobsen, and J.D.E. Gabrieli. 2000. 'A Resource Model of the Neural Basis of Executive Working Memory.' *Proceedings of the National Academy of Sciences* 97(7): 3573–8.

Buss, David M., Marie G. Haselton, Todd K. Shackelford, April L. Bleske, and Jerome C. Wakefield. 1998. 'Adaptations, Exaptations, and Spandrels.' *American Psychologist* 53: 533–48.

Cabanac, Michel. 1999. 'Emotion and Phylogeny.' *Japanese Journal of Physiology* 49: 1–10.

Cacioppo, John T., and Richard E. Petty, eds. 1983. *Social Psychophysiology: A Sourcebook.* New York: Guilford Press.

Card, D. 1995. *Myth and Measurement: The New Economics of the Minimum Wage.* Princeton, NJ: Princeton University Press.

Carnap, R. 1938. 'Logical Foundations of the Unity of Science.' In *International Encyclopedia of Unified Science*, vol. 1, no. 1: 42–62. Chicago: University of Chicago Press.

Cavalli-Sforza, Luigi Luca, Paolo Menozzi, and Alberto Piazza. 1994. *The History and Geography of Human Genes.* Princeton, NJ: Princeton University Press.

Changeux, Jean-Pierre. 1997. *Neuronal Man: The Biology of Mind.* Princeton, NJ: Princeton University Press.

Chomsky, Noam. 1984. *Modular Approaches to the Study of the Mind.* San Diego: San Diego State University Press.

- 1995. 'Language and Nature.' *Mind* 104: 1–61.

Churchland, Patricia S., and Terrence J. Sejnowski. 1993. *The Computational Brain.* Cambridge, MA: MIT Press.

Churchman, C. West, Russell Ackoff, and E. Leonard Arnoff. 1957. *Introduction to Operations Research.* New York: John Wiley & Sons.

Clemens, T.S. 1990. *Science vs. Religion.* Buffalo, NY: Prometheus Books.

Clutton-Brock, Tim. 2002. 'Breeding Together: Kin Selection and Mutualism in Cooperative Vertebrates.' *Science* 296: 69–72.

Cockburn, Andrew. 1998. 'Evolution of Helping Behavior in Cooperatively Breeding Birds.' *Annual Review of Ecology and Systematics* 29: 141–77.

Coleman, James S. 1964. *Introduction to Mathematical Sociology.* Glencoe, IL: The Free Press.

- 1990. *Foundations of Social Theory.* Cambridge, MA: Belknap Press of Harvard University Press.
Collins, Randall. 1998. *The Sociology of Philosophies: A Global Theory of Intellectual Change.* Cambridge, MA: Belknap Press of Harvard University Press.
Collins, W. Andrew, Eleanor E. Maccoby, Laurence Steinberg, E. Mavis Hetherington, and Marc H. Bornstein. 2000. 'Contemporary Research on Parenting.' *American Psychologist* 55: 218–32.
Cook, Karen S., and Russell Hardin. 2001. 'Norms of Cooperativeness and Networks of Trust.' In Hechter and Opp, eds, 327–47.
Corballis, Michael C. 2002. *From Hand to Mouth: The Origins of Language.* Princeton, NJ: Princeton University Press.
Cosmides, Leda, and John Tooby. 1987. 'From Evolution to Behavior: Evolutionary Psychology as the Missing Link.' In John Dupré, ed., *The Latest on the Best: Essays on Evolution and Optimality,* 277–306. Cambridge, MA: MIT Press.
- 1992. 'Cognitive Adaptations of Social Exchange. In Barkow et al., eds, 163–228.
Coule, D.H., and Jérôme Martin. 2000. 'Quantum Cosmology and Open Universes.' *Physical Review* D 61: 063501.
Cramér, Harald. 1946. *Mathematical Methods of Statistics.* Princeton, NJ: Princeton University Press.
Crews, Frederick. 1998. *Unauthorized Freud: Doubters Confront a Legend.* New York: Penguin Books.
Curtis, James E., Douglas E. Baer, and Edward G. Grab. 2001. 'Nations of Joiners: Explaining Voluntary Association Membership in Democratic Societies.' *American Sociological Review* 66: 783–805.
D'Abro, A. 1939. *The Decline of Mechanism (in Modern Physics).* New York: D. Van Nostrand.
Damasio, Antonio R. 1994. *Descartes' Error: Emotion, Reason, and the Human Brain.* New York: Putnam.
- 2000. *The Feeling of What Happens.* London: Heinemann.
Damasio, Antonio R., Anne Harrington, Jerome Kagan, Bruce S. McEwen, Henry Moss, and Rashid Shaikh, eds. 2001. *Unity of Knowledge: The Convergence of Natural and Human Science.* New York: New York Academy of Sciences.
Darden, Lindley, and Nancy L. Maull. 1977. 'Interfield Theories.' *Philosophy of Science* 44: 43–64.
Dawkins, Richard. 1976. *The Selfish Gene.* New York: Oxford University Press.
Deblock, Christian, and Dorval Brunelle. 2000. 'Globalization and New Normative Frameworks: The Multilateral Agreement on Investment.' In Guy Lachapelle and John Trent, eds, *Globalization, Governance and Identity: The Emergence of New Partnerships,* 83–126. Montreal: Presses de l'Université de Montréal.

de Finetti, Bruno. 1937. La prévision: Ses lois logiques, ses sources subjectives.' *Annales de l'Institut Henri Poincaré* 7: 1–68.

Dehaene, Stanislas, Michael Kerszberg, and Jean-Pierre Changeux. 1998. 'A Neuronal Model of a Global Workspace in Effortless Cognitive Tasks.' *Proceedings of the National Academy of Sciences of the USA* 95: 14529–34.

Dehaene, Stanislas, and Lionel Naccache. 2001. 'Towards a Cognitive Neuroscience of Consciousness: Basic Evidence and a Workspace Framework.' *Cognition* 79: 1–37.

Dennett, Daniel C. 1991. *Consciousness Explained*. Boston: Little, Brown.

– 1995. *Darwin's Dangerous Idea: Evolution and the Meanings of Life*. New York: Simon & Schuster.

d'Holbach, Paul-Henry Thiry. 1773. *Système social*. 3 vols. Repr. Hildesheim, New York: Georg Olms, 1969.

Diamond, Jared. 1997. *Guns, Germs, and Steel*. New York: W.W. Norton.

Dijksterhuis, Eduard Jon. 1961. *The Mechanization of the World Picture*. Oxford: Clarendon Press.

Dillinger, Mike. 1990. 'On the Concept of a Language.' In Weingartner and Dorn, eds, 5–28.

Dilthey, Wilhelm. 1959 [1883]. *Einleitung in die Geisteswissenschaften*. In *Gesammelte Schriften*, 4: 318–31. Stuttgart: Teubner.

Dixon, W.J., and T. Boswell. 1996. 'Dependency, Disarticulation, and Denominator Effects: Another Look at Foreign Capital Penetration.' *American Journal of Sociology* 102: 543–62.

Donald, Merlin. 1991. *Origins of the Modern Mind*. Cambridge, MA: Harvard University Press.

Dover, Gabriel. 2000. *Dear Mr Darwin: Letters on the Evolution of Life and Human Nature*. London: Weidenfeld & Nicolson.

Drosdowski, Günther. 1990. *Deutsch-Sprache in einem geteilten Land*. Manheim: Dudenverlag.

Dubrovsky, Bernardo. 2002. 'Evolutionary Psychiatry: Adaptationist and Nonadaptationist Conceptualizations.' *Progress in Neuro-Psychopharmacology & Biological Psychiatry* 26: 1–19.

Du Pasquier, L.-Gustave. 1926. *Le calcul des probabilités: Son évolution mathématique et philosophique*. Paris: Hermann.

Dupré, John. 1993. *The Disorder of Things*. Cambridge, MA: Harvard University Press.

Eddy, Charles. 1982. 'Probabilistic Reasoning in Clinical Medicine: Problems and Opportunities.' In Kahneman, Slovic, and Tversky, eds, 249–67.

Eddy, David M., and Charles H. Clanton. 1982. 'The Art of Diagnosis.' *New England Journal of Medicine* 306: 1263–8.

Engels, Frederick. 1954 [1878]. *Anti-Dühring*. Moscow: Foreign Languages Publishing House.

Everett, Hugh, III. 1957. '"Relative state" Formulation of Quantum Mechanics.' *Reviews of Modern Physics* 29: 454–62.
Fehr, Ernst, and Simon Gächter. 2000. 'Cooperation and Punishment in Public Goods Experiments.' *American Economic Review* 90: 980–94.
Fiske, Donald W., and Richard A. Shweder, eds. 1986. *Metatheory in Social Science: Pluralisms and Subjectivities*. Chicago: University of Chicago Press.
Fleck, Ludwik. 1979 [1935]. *Genesis and Development of a Scientific Fact*. Foreword T.S. Kuhn. Chicago, London: University of Chicago Press.
Fodor, Jerry A. 1975. 'The Mind-Body Problem.' *Scientific American* 244(1): 114–23.
– 1983. *The Modularity of the Mind*. Cambridge, MA: MIT Press.
Fogel, Robert W. 1994. 'Economic Growth, Population Theory, and Physiology: The Bearing of Long-term Processes on the Making of Economic Policy.' *American Economic Review* 84: 369–95.
Franklin, James. 2001. *The Science of Conjecture*. Baltimore, MD: Johns Hopkins University Press.
Franklin, Ursula. 1990. *The Real World of Technology*. Montreal, Toronto: CBC Enterprises.
Fréchet, Maurice. 1955. *Les mathématiques et le concret*. Paris: Presses Universitaires de France.
Fredrickson, George M. 2002. *Racism: A Short History*. Princeton, NJ: Princeton University Press.
Freud, Sigmund. 1938 [1900]. *The Interpretation of Dreams*. In A.A. Brill, ed., *The Basic Writings of Sigmund Freud*. New York: Modern Library.
Gadamer, Hans-Georg. 1976. *Philosophical Hermeneutics*. Berkeley: University of California Press.
Galbraith, James K., and Maureen Berner, eds. 2001. *Inequality and Industrial Change: A Global View*. Cambridge: Cambridge University Press.
Galison, Peter, and David J. Stump, eds. 1996. *The Disunity of Science: Boundaries, Contexts and Power*. Stanford, CA: Stanford University Press.
Gallagher, H.L., F. Happé, N. Brunswick, P.C. Fletcher, U. Frith, and C.D. Frith. 2000. 'Reading the Mind in Cartoons and Stories: An fMRI Study of "Theory of Mind" in Verbal and Nonverbal Tasks.' *Neuropsychologia* 38: 11–21.
Gazzaniga, Michael S., ed. 2000. *The New Cognitive Neurosciences*. Cambridge, MA: MIT Press.
Geertz, Clifford. 1973. *The Interpretation of Cultures*. New York: Basic Books.
Giddens, Anthony. 1984. *The Constitution of Society: Outline of the Theory of Structuration*. Cambridge, UK: Polity Press.
Gilbert, Scott F., John M. Opitz, and Rudof A. Raff. 1996. 'Resynthesizing Evolutionary and Developmental Biology.' *Developmental Biology* 173: 357–72.
Gilfillan, S.C. 1970. *The Sociology of Invention*. 2nd ed. Cambridge, MA: MIT Press.

Gini, Corrado. 1952. *Patologia economica*. 5th ed. Turin: UTET.
Glansdorff, P., and I. Prigogine. 1971. *Thermodynamic Theory of Structure, Stability and Fluctuations*. London: Wiley-Interscience.
Good, Irving John. 1950. *Probability and the Weighing of Evidence*. New York: Charles Griffin.
Gould, Stephen J. 1977. *Ontogeny and Phylogeny*. Cambridge, MA: Belknap Press of Harvard University Press.
– 1992. 'Ontogeny and Phylogeny – Revisited and Reunited.' *BioEssays* 14: 275–79.
– 2002. *The Structure of Evolutionary Theory*. Cambridge, MA: Harvard University Press.
Gould, Stephen J., and Richard C. Lewontin. 1979. 'The Spandrels of San Marco and the Panglossian Paradigm: A Critique of the Adaptationist Programme.' *Proceedings of the Royal Society of London*, series B, 205: 581–98.
Graham, Loren R. 1993. *The Ghost of the Executed Engineer*. Cambridge, MA: Harvard University Press.
Greenfield, Susan. 2000. *The Private Life of the Brain*. London: Penguin Press.
Griffin, Donald R. 2001. *Animal Minds: Beyond Cognition and Consciousness*. Chicago: University of Chicago Press.
Gross, Paul R., and Norman Levitt. 1994. *Higher Superstition: The Academic Left and Its Quarrels with Science*. Baltimore, MD: Johns Hopkins University Press.
Habermas, Jürgen. 1967. *Zur Logik der Sozialwissenschaften*. Frankfurt: Suhrkamp.
Haken, H. 1989. 'Synergistics: An Overview.' *Reports on Progress in Physics* 52: 515–53.
Halliday, [Michael] A.K. 1985. *An Introduction to Functional Grammar*. London: Edward Arnold.
Hamilton, William D. 1964. 'The Genetical Evolution of Social Behavior.' *Journal of Theoretical Biology* 7: 1–16, 17–52.
Handy, Rollo, and Paul Kurtz. 1964. *A Current Appraisal of the Behavioral Sciences*. Great Barrington, MA: Behavioral Research Council.
Harding, Susan. 1986. *The Science Question in Feminism*. Ithaca, NY: Cornell University Press.
Hardy, G.H. 1967. *A Mathematician's Apology*. Foreword C.P. Snow. Cambridge: Cambridge University Press.
Harrington, Anne. 1996. *Reenchanted Science: Holism in German Culture from Wilhelm II to Hitler*. Princeton, NJ: Princeton University Press.
Hartmann, Nicolai. 1949. *Neue Wege der Ontologie*. 3d ed. Stuttgart: Kohlhammer Verlag.
Hauser, Marc D., Noam Chomsky, and W. Tecumseh Fitch. 2002. 'The Faculty of Language: What Is It, Who Has It, and How Did It Evolve?' *Science* 298: 1569–79.
Hebb, Donald O. 1980. *Essay on Mind*. Hillsdale, NJ: Erlbaum.

Hebb, D.O., W.E. Lambert, and G.R. Tucker. 1971. 'Language, Thought and Experience.' *Modern Language Journal* 55: 212–22.

Hechter, Michael, and Elizabeth Borland. 2001. 'National Self-determination: The Emergence of an International Norm.' In Hechter and Opp, eds, 186–233.

Hechter, Michael, and Dieter Opp, eds. 2001. *Social Norms*. New York: Russell Sage Foundation.

Hedström, Peter, and Richard Swedberg, eds. 1998. *Social Mechanisms: An Analytical Approach to Social Theory*. Cambridge: Cambridge University Press.

Helmuth, Laura. 2001. 'From the Mouths (and Hands) of Babes.' *Science* 293: 1758–9.

Hempel, Carl G. 1965. *Aspects of Scientific Explanation*. New York: The Free Press.

Hilbert, David. 1901. 'Mathematische Probleme.' Repr. in *Gesammelte Abhandlungen*, 3: 290–329. Berlin: Julius Springer, 1935.

– 1918. 'Axiomatisches Denken.' Repr. in *Gesammelte Abhandlungen*, 3: 146–56. Berlin: Julius Springer, 1935.

Hill, A[rchibald] V[ivian]. 1956. 'Why Biophysics?' *Science* 124: 1233–7.

Hintikka, Jaakko. 1999. 'The Emperor's New Intuitions.' *Journal of Philosophy* 96: 127–47.

Hirschman, Albert O. 1981. *Essays in Trespassing: Economics to Politics and Beyond*. Cambridge: Cambridge University Press.

– 1990. 'The Case against 'One thing at a time.' *World Development* 18: 1119–20.

Hoffman, Ralph E., and Thomas H. McGlashan. 1999. 'Using a Speech Perception Neural Network Simulation to Explore Normal Neurodevelopment and Hallucinated "Voices" in Schizophrenia.' *Progress in Brain Research* 121: 311–25.

Hogarth, Robin M., and Melvin W. Reder. 1987. *Rational Choice: The Contrast between Economics and Psychology*. Chicago: University of Chicago Press.

Holland, John H. 1998. *Emergence: From Chaos to Order*. Oxford: Oxford University Press.

Homans, George Caspar. 1974. *Social Behavior: Its Elementary Forms*. Rev. ed. New York: Harcourt Brace Jovanovich.

Hubel, David H., and Torsten N. Wiesel. 1968. 'Receptive Fields, Binocular Interaction, and Functional Architecture in the Cat's Visual Cortex.' *Journal of Physiology* 160· 515–24.

Huelsenbeck, John P., Fredrik Ronquist, Rasmus Nielsen, and Jonathan P. Bollback. 2001. 'Bayesian Inference of Phylogeny and Its Impact on Evolutionary Biology.' *Science* 294: 2310–14.

Hughes, G.E., and M.J. Cresswell. 1968. *An Introduction to Modal Logic*. London: Methuen.

Humphreys, Paul. 1985. 'Why Propensities Cannot Be Probabilities.' *Philosophical Review* 94: 557–70.

Husserl, Edmund. 1970 [1936]. *The Crisis of European Sciences and Transcendental Phenomenology*. Evanston, IL: Northwestern University Press.

Huxley, T.H., and Julian Huxley. 1947. *Evolution and Ethics: 1893–1943*. London: Pilot Press.

James, Patrick. 2002. *International Relations and Scientific Progress: Structural Realism Reconsidered*. Columbus: Ohio State University Press.

Jeffrey, Harold. 1937. *Scientific Inference*. 2nd ed. Cambridge: Cambridge University Press.

Johansson, S. Ryan. 2000. 'Macro and Micro Perspectives on Mortality History.' *Historical Methods* 33: 59–72.

Johnson, Jeffrey G., Patricia Cohen, Elizabeth M. Smalles, Stephanie Jasen, and Judith S. Brook. 2002. 'Television Viewing and Aggressive Behavior during Adolescence and Adulthood.' *Science* 295: 2468–71.

Kahneman, Daniel, Paul Slovic, and Amos Tversky. 1982. *Judgment under Uncertainty: Heuristics and Biases*. Cambridge: Cambridge University Press.

Kahneman, Daniel, and Amos Tversky. 1982. 'Subjective Probability: A Judgment of representativeness. In Kahneman, Slovic, and Tversky, eds, 32–47.

Kanazawa, Satoshi, and Mary C. Still. 2001. 'The Emergence of Marriage Norms: An Evolutionary Psychological Perspective.' In Hechter and Opp, eds, 274–304.

Kary, Michael, and Martin Mahner. 2002. 'How Would You Know If You Synthesized a Thinking Thing?' *Minds and Machines* 12: 61–86.

Kauffman, Stuart A. 1993. *The Origins of Order: Self-Organization and Selection in Evolution*. New York: Oxford University Press.

Kempthorne, Oscar. 1978. 'Logical, Epistemological and Statistical Aspects of Nature-Nurture Data Interpretation.' *Biometrica* 34: 1–23.

Keynes, John Maynard. 1973 [1936]. *The General Theory of Employment, Interest and Money*. In *Collected Writings*, vol. 7. London: Macmillan and Cambridge University Press.

Kim, Jaegwon. 1978. 'Supervenience and Nonlogical Incommensurability.' *American Philosophical Quarterly* 15: 149–56.

Kincaid, Harold. 1997. *Individualism and the Unity of Science: Essays on Reduction, Explanation, and the Special Sciences*. Lanham, MD: Rowman & Littlefield.

Kitcher, Philip. 1985. *Vaulting Ambition: Sociobiology and the Quest for Human Nature*. Cambridge, MA: MIT Press.

Klir, George J., ed. 1972. *Trends in General Systems Theory*. New York: Wiley-Interscience.

Koerner, E.F. Konrad. 1973. *Ferdinand de Saussure: Origin and Development of His Linguistic Thought in Western Studies of Language. A Contribution to the History and Theory of Linguistics*. Braunschweig: Vieweg.

Kolmogorov, A[lexander] N. 1956 [1933]. *Foundations of the Theory of Probability*. New York: Chelsea Pub. Co.

Kolnai, Aurel. 1938. *The War against the West*. London: Victor Gollancz.
Koshland, Daniel E., Jr. 2002. 'The Seven Pillars of Life.' *Science* 295: 2215–16.
Kosslyn, Stephen M., and Olivier Koenig. 1995. *Wet Mind: The New Cognitive Neuroscience*. New York: The Free Press.
Kraft, Julius. 1957. *Von Husserl zu Heidegger: Kritik der phänomenologischen Philosophie*. 2nd ed. Frankfurt am Main: Öffentliches Leben.
Kreiman, Gabriel, Christof Koch, and Itzhak Fried. 2000. Imagery Neurons in the Human Brain.' *Nature* 408: 357–61.
Kronz, Frederick, and Justin Tiehen. 2002. 'Emergence and Quantum Mechanics.' *Philosophy of Science* 69: 324–47.
Kuhl, Patricia K. 2000. 'Language, Mind, and Brain: Experience Alters Perception.' In Michael S. Gazzaniga, ed., *The New Cognitive Neurosciences*, 2nd ed., 99–115. Cambridge, MA: MIT Press.
Kurtz, Paul. 1986. *The Transcendental Temptation: A Critique of Religion and the Paranormal*. Buffalo, NY: Prometheus Books.
Kuznets, Simon. 1955. 'Economic Growth and Income Inequality.' *American Economic Review* 65: 1–28.
Kyburg, Henry E., Jr. 1961. *Probability and the Logic of Rational Belief*. Middletown, CT: Wesleyan University Press.
Labov, William. 1972. *Sociolinguistic Patterns*. Philadelphia: University of Pennsylvania Press.
Lacan, Jacques. 1966. *Écrits*. Paris: Éditions du Seuil.
Lanczos, Cornelius. 1949. *The Variational Principles of Mechanics*. Toronto: University of Toronto Press.
Lange, Friedrich Albert. 1905 [1866]. *Geschichte des Materialismus*. 2 vols. Leipzig: Philipp Reclam.
Lashley, Karl S. 1949. 'Persistent Problems in the Evolution of Mind.' *Quarterly Review of Biology* 24: 28–42.
Laszlo, Ervin. 1972. *Introduction to Systems Philosophy*. New York: Gordon and Breach.
Latour, Bruno, and Steve Woolgar. 1986. *Laboratory Life: The Construction of Scientific Facts*. Rev. ed. Princeton, NJ: Princeton University Press.
Lepore, Ernest, and Zenon Pylyshyn, eds. 1999. *What Is Cognitive Science?* Malden, CA: Blackwell.
Lerner, David, ed. 1963. *Parts and Wholes*. New York: Free Press.
Lévy, M. 1979. 'Relations entre chimie et physique.' *Epistemologia* 2: 337–70.
Lewes, George Henry. 1874. *Problems of Life and Mind*. London: Truebner.
Lewis, David. 1986. *On the Plurality of Worlds*. Oxford: Basil Blackwell.
Lewontin, Richard. 2000. *It ain't necessarily so*. New York: New York Review Books.
Llinás, Rodolfo. 2001. *I of the Vortex: From Neuron to Self*. Cambridge. MA: MIT Press.

Lloyd, Elisabeth A. 1999. 'Evolutionary Psychology: The Burdens of Proof.' *Biology and Philosophy* 14: 211–33.
Lloyd Morgan, Conwy. 1923. *Emergent Evolution*. London: Williams & Norgate.
Lomnitz, Larissa A. 1977. *Networks and Marginality: Life in a Mexican Shantytown*. San Francisco: Academic Press.
Looijen, Rick C. 2000. *Holism and Reductionism in Biology and Ecology*. Dordrecht, Boston: Kluwer Academic.
Lovejoy, Arthur O. 1953 [1936]. *The Great Chain of Being*. Cambridge, MA: Harvard University Press.
Lowe, E.J. 2002. *A Survey of Metaphysics*. Oxford: Oxford University Press.
Luhmann, Niklas. 1987. *Soziale Systeme: Grundrisse einer allgemeinen Theorie*. Frankfurt: Suhrkamp.
Machamer, Peter K., Lindley Darden, and Carl F. Craver. 2000. 'Thinking about Mechanisms.' *Philosophy of Science* 67: 1–25.
Machamer, Peter K., Rick Grush, and Peter McLaughlin, eds. 2001. *Theory and Method in the Neurosciences*. Pittsburgh: University of Pittsburgh Press.
MacWhinney, Brian, ed. 1999. *The Emergence of Language*. Mahwah, NJ: Lawrence Erlbaum Associates.
Mahner, Martin, ed. 2001. *Scientific Realism: Selected Essays by Mario Bunge*. Amherst, NY: Prometheus Books.
Mahner, Martin, and Mario Bunge. 1996a. 'Is Religious Education Compatible with Science Education?' *Science & Education* 5: 101–23.
– 1996b. 'The Incompatibility of Science and Religion Sustained: A Reply to Our Critics.' *Science & Education* 5: 189–99.
– 1997. *Foundations of Biophilosophy*. Berlin, New York: Springer-Verlag.
– 2001. 'Function and Functionalism: A Synthetic Perspective.' *Philosophy of Science* 68: 75–94.
Mann, Michael. 1986. *The Sources of Social Power*, vol. I: *A History of Power from the Beginning to A.D. 1760*. Cambridge: Cambridge University Press.
– 1993. *The Sources of Social Power*, vol. II: *The Rise of Classes and Nation-States*. Cambridge: Cambridge University Press.
March, James G., and Herbert A. Simon. 1958. *Organizations*. New York: John Wiley.
Marr, David. 1982. *Vision*. San Francisco: W.H. Freeman.
Martin, Michael. 2000. *The Uses of Understanding in Social Science: Verstehen*. New Brunswick, NJ: Transaction Publishers.
Martinich, A.P., ed. 1996. *The Philosophy of Language*. 3rd ed. New York: Oxford University Press.
Marx, [Karl,] [Frederick] Engels, and [Vladimir Illich] Lenin. 1977. *On Dialectical Materialism*. Moscow: Progress.
Massey, Douglas S. 2002. 'A Brief History of Human Society: The Origin and Role of Emotion in Social Life.' *American Sociological Review* 67: 1–29.

Maynard Smith, J., R. Burian, S. Kauffman, P. Alberch, J. Campbell, B. Goodwin, R. Lande, D. Raup, and L. Wolpert. 1985. 'Developmental Constraints and Evolution.' *Quarterly Review of Biology* 60: 265–87.
Mayr, Ernst. 1982. *The Growth of Biological Thought*. Cambridge, MA: Harvard University Press.
Medawar, Peter. 1984. *The Limits of Science*. New York: Harper & Row.
Melvill Jones, G., A. Berthoz, and B. Segal. 1984. 'Adaptive Modifications of the Vestibulo-ocular Reflex by Mental Effort in Darkness.' *Experimental Brain Research* 56: 149–53.
Merton, Robert K. 1968 [1957]. *Social Theory and Social Structure*. New York: The Free Press.
– 1973. *The Sociology of Science: Theoretical and Empirical Investigations*. Chicago: University of Chicago Press.
– 1976. *Sociological Ambivalence and Other Essays*. New York: The Free Press.
– 2001 [1938]. *Science, Technology & Society in Seventeenth-century England*. New York: Howard Fertig.
Mesulam, M.-Marsel. 1990. 'Large-scale Neurocognitive Networks and Distributed Proccessing for Attention, Language, and Memory.' *Annals of Neurology* 28: 597–613.
Miikkulainen, Risto, and Marshall R. Mayberry III. 1999. 'Disambiguation and Grammar as Emergent Soft Constraints. In McWhinney, ed., 153–76.
Mill, John Stuart. 1952 [1843, 1875]. *A System of Logic*. London: Longmans, Green.
Milsum, John H., ed. 1968. *Positive Fedback: A General Systems Approach to Positive/Negative Feedback and Mutual Causality*. Oxford: Pergamon Press.
Moessinger, Pierre. 1988. *La psychologie morale*. Paris: Presses Universitaires de France.
Morris, J.S., A. Öhman, and R.J. Dolan. 1998. 'Conscious and Unconscious Emotional Learning in the Human Amygdala.' *Nature* 393: 467–70.
Morrison, Margaret. 2000. *Unifying Scientific Theories*. Cambridge: Cambridge University Press.
Moser, Paul K., ed. 1990. *Rationality in Action: Contemporary Approaches*. Cambridge: Cambridge University Press.
Mountcastle, Vernon. 1998. *Perceptual Neuroscience: The Cerebral Cortex*. Cambridge, MA: Harvard University Press.
Murphy, Edmond A. 1997. *The Logic of Medicine*. 2nd ed. Baltimore, London: Johns Hopkins University Press.
Nagel, Ernest. 1961. *The Structure of Science: Problems in the Logic of Scientific Explanation*. New York: Harcourt, Brace & World.
Navarro, Vicente, ed. 2002. *The Political Economy of Social Inequalities: Consequences for Health and Quality of Life*. Amityville, NY: Baywood Pub. Co.
Negroponte, Nicholas. 1996. *Being Digital*. New York: Vintage Books.

Nesse, Randolph M., and George C. Williams. 1994. *Why We Get Sick: The New Science of Darwinian Medicine.* New York: Vintage.

Neurath, Otto. 1931. 'Soziologie im Physikalismus.' In R. Haller and H. Rutte, eds, *Gesammelte philosophische und methodologische Schriften*, vol. 2: 533–62. Vienna: Hölder-Pichler-Tempsky, 1981.

– 1944. 'Foundations of the Social Sciences.' In *International Encyclopedia of Unified Science*, vol. 2, no. 1. Chicago: University of Chicago Press.

Newmeyer, Frederick J., ed. 1988. *Linguistics: The Cambridge Survey*, vol. 4: *Language: The Socio-cultural Context*. Cambridge: Cambridge University Press.

Novikoff, Alex B. 1945. 'The Concept of Integrative Levels in Biology.' *Science* 101: 209–15.

Ocampo, José Antonio, and Lance Taylor. 1998. 'Trade Liberalization in Developing Economies: Modest Benefits but Problems with Productivity Growth, Macroprices, and Income Distribution.' *Economic Journal* 108: 1523–46.

Ochsner, Kevin N., Silvia A. Bunge, James J. Gross, and John D.E. Gabrieli. 2002. 'Rethinking Feeling: An fMRI Study of the Cognitive Regulation of Emotion.' *Journal of Cognitive Neuroscience* 14: 1215–29.

Ochsner, Kevin N., and Matthew D. Lieberman. 2001. 'The Emergence of Social Cognitive Neuroscience. *American Psychologist* 56: 717–34.

Olby, Robert. 1974. *The Path to the Double Helix.* Seattle: University of Washington Press.

Olson, Mancur. 1971. *The Logic of Collective Action: Public Gods and the Theory of Groups.* 2nd ed. Cambridge, MA: Harvard University Press.

Outhwaite, William. 1987. *New Philosophies of Social Science: Realism, Hermeneutics, and Critical Theory.* London: Macmillan.

Pareto, Vilfredo. 1935 [1916]. *A Treatise on General Sociology.* 4 vols. New York: Harcourt, Brace & Co.

Peirce, Charles S. 1965 [1982–93]. *Scientific Metaphysics.* In C. Hartshorne and P. Weiss, eds, Collected Papers, vol. 6. Cambridge, MA: Harvard University Press.

Petroski, Henry. 1992. *To Engineer Is Human: The Role of Failure in Successful Design.* New York: Vintage.

Pfisterer, Andrea R., and Bernhard Schmid. 2002. 'Diversity-dependent Production Can Decrease the Stability of Ecosystem Functioning.' *Nature* 416: 84–6.

Piaget, Jean. 1965. *Études sociologiques.* Genève: Librairie Droz.

Pickel, Andreas. 2001. 'Between Social Science and Social Technology.' *Philosophy of the Social Sciences* 31: 459–87.

Pinker, Steven. 1997. *How the Mind Works.* New York: W.W. Norton.

Polya, George. 1954. *Mathematics and Plausible Reasoning.* 2 vols. Princeton, NJ: Princeton University Press.

Popper, Karl R. 1945. *The Open Society and Its Enemies.* London: George Routledge & Sons.

- 1957. 'The Propensity Interpretation of the Calculus of Probability, and the Quantum Theory.' In S. Körner, ed., *Observation and Interpretation*, 65–70, 88–9.
- 1959 [1935]. *The Logic of Scientific Discovery*. London: Heineman.
- 1970. 'A Realist View of Logic, Physics, and History.' In Wolfgang Yourgrau and Allen D. Breck, eds, *Physics, Logic, and History*, 1–30. New York, London: Plenum.
- 1974. 'Scientific Reduction and the Essential Incompleteness of All Science.' In Francisco Ayala and Theodosius Dobzhansky, eds, *Studies in the Philosophy of Biology*, 259–84. Berkeley, Los Angeles: University of California Press.
- 1985 [1967]. 'The Rationality Principle.' In David Miller, ed., *Popper Selections*, 357–65. Princeton, NJ: Princeton Univesity Press.

Portes, Alejandro. 2000. 'The Hidden Abode: Sociology as Analysis of the Unexpected.' *American Sociological Review* 65: 1–18.

Premack, David, and Guy Woodruff. 1978. 'Does the Chimpanzee Have a Theory of Mind?' *Behavioral and Brain Sciences* 1: 515–26.

Pylyshyn, Zenon. 1999. 'What's in Your Mind?' In Lepore and Pylyshyn, eds, 1–25.

Raff, Rudolf. A. 2000. 'Evo-devo: The Evolution of a New Discipline.' *Nature Rev. Genetics* 1: 74–9.

Ramsey, Frank Plumpton. 1931. *The Foundation of Mathematics*. London: Kegan Paul.

Rayner, Keith, Barbara R. Foorman, Charles A. Perfetti, David Pesetsky, and Mark S. Seidenberg. 2002. 'How Should Reading Be Taught?' *Scientific American* 286(3): 84–91.

Renfrew, Colin, and Ezra B.W. Zubrow, eds. 1994. *The Ancient Mind: Elements of Cognitive Archaeology*. Cambridge: Cambridge University Press.

Ricoeur, Paul. 1981. *Hermeneutics and the Human Sciences: Essays on Language, Action and Interpretation*. Ed. and trans. John B. Thompson. Cambridge: Cambridge University Press.

Rilling, James K., David A. Gutman, Thorsten R. Zeh, Giuseppe Pagnoni, Gregory S. Berns, and Clinton D. Kiets. 2002. 'A Neural Basis for Social Cooperation.' *Neuron* 35: 395–405.

Rodríguez, Alejandro, and Dani Rodrik. 2001. 'Trade Policy and Economic Growth: A Skeptic's Guide to the Cross-national Evidence.' In B. Bemanke and K. Rogoff, eds, *NBER Macroeconomic Annual 2000*. Cambridge, MA: MIT Press.

Rorty, Richard. 1979. *Philosophy and the Mirror of Nature*. Princeton, NJ: Princeton University Press.

Ruppin, Eytan, James A. Reggia, and Dennis Glanzman, eds. 1999. 'Understanding Brain and Cognitive Disorders: The Computational Perspective.' *Progress in Brain Research* 121: ix–xv.

Saarinen, Esa, ed. 1980. *Conceptual Issues in Ecology*. Dordrecht, Boston: D.D. Reidel.

Sachs, Ignacy. 2000. *Understanding Development*. Oxford: Oxford University Press.

Sampson, Robert J., Stephen W. Raudenbush, and Felton Earls. 1997. 'Neighborhoods

and Violent Crime: A Multilevel Study of Collective Efficacy.' *Science* 277: 918–24.
Savage, Leonard J. 1954. *The Foundation of Statistics.* New York: John Wiley & Sons.
Schacter, Daniel. 2001. *The Seven Sins of Memory: How the Mind Forgets and Remembers.* Boston: Houghton Mifflin.
Schaffner, K. 1969. 'The Watson-Crick Model and Reductionism.' *British Journal for the Philosophy of Science* 20: 325–48.
Schelling, Thomas C. 1978. *Micromotives and Macrobehavior.* New York: W.W. Norton.
Schönwandt, Walter L. 2002. *Planung in der Krise? Theoretische Orientierungen für Architektur, Stadt-und Raumplanung.* Stuttgart: W. Kohlhammer.
Schutz, Alfred. 1967 [1932]. *The Phenomenology of the Social World.* Evanston, IL: Northwestern University Press.
Searle, John R. 1980. 'Minds, Brains and Programs.' *Behavioral and Brain Sciences* 3: 417–24.
– 1997. *The Mystery of Consciousness.* New York: New York Review Books.
Sellars, Roy Wood. 1922. *Evolutionary Naturalism.* Chicago: Open Court.
Sellars, Roy Wood, V.J. McGill, and Marvin Farber, eds. 1949. *Philosophy for the Future: The Quest for Modern Materialism.* New York: Macmillan.
Sherif, Muzafer, and Carolyn W. Sherif, eds. 1969. *Interdisciplinary Relationships in the Social Sciences.* Chicago: Aldine.
Smelser, Neil J., and Richard Swedberg, eds. 1994. *The Handbook of Economic Sociology.* Princeton, NJ: Princeton University Press; New York: Russell Sage Foundation.
Smoluchowski, Marian. 1918. 'Über den Begriff des Zugalls und den Ursprung der Wahrscheinlichkeitsgesetze in der Physik.' *Naturwissenschaften* 6: 253–63.
Sober, Elliott. 2000. *Philosophy of Biology.* 2nd ed. Boulder, CO: Westview Press.
Sober, Elliott, and David Sloan Wilson. 1998. *Unto Others.* Cambridge, MA: Harvard University Press.
Soros, George. 1998. *The Crisis of Global Capitalism. (The Open Society Endangered.)* New York: Public Affairs.
Stenger, Victor J. 1995. *The Unconscious Quantum: Metaphysics in Modern Physics and Cosmology.* Amherst, NY: Prometheus Books.
Sternberg, Robert J. 1985. *A Triarchic Theory of Human Intelligence.* New York: Cambridge University Press.
Stiglitz, Joseph E. 2001. '*Quis custodiet custodes*?: Corporate Governance Failures in the Transition.' In Joseph E. Stiglitz and Pierre-Alain Muet, eds, *Governance, Equity, and Global Markets,* 22–54. Oxford: Oxford University Press.
Stinchcombe, Arthur L. 1968. *Constructing Social Theories.* Chicago: University of Chicago Press.
Stove, David. 1995. *Darwinian Fairytales.* Aldershot, UK: Avebury Ashgate Publications.

Streeten, Paul. 2001. *Globalisation: Threat or Opportunity?* Copenhagen: Copenhagen Business School Press.
Suppes, Patrick. 1970. *A Probabilistic Theory of Causality. Acta Philosophica Fennica* XXIV. Amsterdam: North Holland.
Sutherland, Stuart. 1995. *Irrationality: Why We Don't Think Straight*. New Brunswick, NJ: Rutgers University Press.
Swedberg, Richard. 1990. *Economics and Sociology: Redefining Boundaries. Conversations with Economists and Sociologists*. Princeton, NJ: Princeton University Press.
Taylor, Charles. 1971. 'Interpretation and the Sciences of Man.' *Review of Metaphysics* 25: 3–51.
Tegmark, Max. 2002. 'Measuring Spacetime: From the Big Bang to Black Holes.' *Science* 296: 1427–33.
Thagard, Paul. 1978. 'The Best Explanation: Criteria for Theory Choice.' *Journal of Philosophy* 75: 76–92.
Thompson Klein, Julie. 1990. *Interdisciplinarity: History, Theory, and Practice.* Detroit: Wayne State University Press.
Tilly, Charles. 1998. *Durable Inequality*. Berkeley: University of California Press.
– 2001. 'Mechanisms in Political Processes.' *Annual Reviews of Political Science* 4: 21–41.
Tocqueville, Alexis de. 1998 [1856]. *The Old Regime and the French Revolution*, vol. 1. Trans. A.S. Kahan. Chicago: University of Chicago Press.
Tootell, Roger B.H., Eugene Switkes, Martin B. Silverman, and Susan L. Hamilton. 1998. 'Functional Anatomy of Macaque Striate Cortex II. Retinotopic Organization.' *Journal of Neuroscience* 8: 1531–68.
Torrey, E. Fuller. 1992. *Freudian Fraud: The Malignant Effect of Freud's Theory on American Thought and Culture*. New York: HarperCollins.
Treisman, Ann M., and G. Gelade. 1980. 'A Feature-integration Theory of Attention.' *Cognitive Psychology* 12: 97–136.
Trigger, Bruce. 2003a. *Artifacts & Ideas: Essays in Archaeologys*. New Brunswick, NJ: Transaction Publishers.
– 2003b. *Understanding Early Civilizations*. Cambridge: Cambridge University Press.
Tsebelis, George. 1990. 'Penalty Has No Impact in Crime: A Game-theoretic Analysis.' *Rationality and Society* 2: 255–86.
Turnbull, Colin M. 1972. *The Mountain People*. New York: Simon & Schuster.
Tversky, Amos, and Daniel Kahneman. 1974. 'Judgment under Uncertainty: Heuristics and Biases.' *Science* 185: 1124–31.
– 1982. 'Availability: A Heuristic for Judging Frequency and Probability.' In Kahneman, Slovic, and Tversky, eds, 163–78.
Urbain, G., P. Job, G. Allard, and G. Champetier. 1939. *Traité de chimie générale*. Paris: Hermann.

Van Alstyne, Marshall, and Erik Brynjolfsson. 1996. 'Could the Internet Balkanize Science?' *Science* 274: 1479–80.
Van den Berg, Axel. 2001. 'The Social Sciences according to Bunge.' *Philosophy of the Social Sciences* 31: 83–103.
Vanderburg, Willem H. 2000. *The Labyrinth of Technology*. Toronto: University of Toronto Press.
Van der Dennen, Johan M.G., David Smillie, and Daniel R. Wilson, eds. 1999. *The Darwinian Heritage and Sociobiology*. Westport, CT: Praeger.
Vaughan, Susan C., Randall D. Marshall, Roger A. McKinnon, Roger Vaughan, Lisa Mellman, and Steven P. Roose. 2000. 'Can We Do Psychoanalytic Outcome Research? A Feasibility Study.' *International Journal of Psychoanalysis* 81: 513–27.
Venn, John. 1888. *The Logic of Chance*. 3rd ed. London: Macmillan & Co.
Ville, Jean. 1938. *Étude critique de la notion de collectif*. Paris: Gauthier-Villars.
Von Mises, Richard. 1931. *Wahrscheinlichkeitsrechnung*. Vienna: Springer-Verlag.
Von Schelting, Alexander. 1934. *Max Webers Wissenschaftslehre*. Tübingen: Mohr.
Von Wright, Georg Henrik. 1971. *Explanation and Understanding*. Ithaca, NY: Cornell University Press.
Vygotsky, L[ev] S. 1978. *Mind in Society: The Development of Higher Psychological Processes*. Cambridge, MA: Harvard University Press.
Waal, Frans B.M. de. 1996. *Good Natured: The Origins of Right and Good in Humans and Other Animals*. Cambridge, MA: Harvard University Press.
Warren, W. Preston, ed. 1970. *Principles of Emergent Realism: Philosophical Essays by Roy Wood Sellars*. St Louis, MO: Warren H. Green.
Warrington, Elizabeth K., and Rosalen A. McCarthy. 1987. 'Categories of Knowledge.' *Brain* 110: 1273–96.
Weber, Max. 1976 [1922]. *Wirtschaft und Gesellschaft*. 5th ed., 3 vols. Tübingen: J.C.B. Mohr (Paul Siebeck).
Weingartner, Paul, and Georg J.W. Dorn, eds. *Studies on Mario Bunge's Treatise*. Amsterdam, Atlanta: Rodopi.
Weiss, Paul A., ed. 1971. *Hierarchically Organized Systems in Theory and Practice*. New York: Hafner Pub. Co.
Weissman, David. 2000. *A Social Ontology*. New Haven, CT: Yale University Press.
Wertheimer, Max. 1912. 'Experimentelle Studien über das Sehen von Bewegungen.' *Zeitschrift für Psychologie* 61: 161–265.
West, Stuart A., Ido Pen, and Ashleigh S. Griffin. 2002. 'Cooperation and Competition between Relatives.' *Science* 296: 72–5.
Whewell, William. 1847. *The Philosophy of the Inductive Sciences*. Rev. ed., 2 vols. London: John W. Parker; repr. London: Frank Cass, 1967.
Whyte, Lancelot Law, Albert G. Wilson, and Donna Wilson, eds. 1969. *Hierarchical Levels*. New York: American Elsevier, 1969

Wiener, Norbert. 1948. *Cybernetics: Control and Communication in the Animal and the Machine*. New York: John Wiley & Sons.
– 1993. *Invention: The Care and Feeding of Ideas*. Cambridge, MA: MIT Press.
Wilkins, Adam S. 2002. *The Evolution of Developmental Pathways*. Sunderland, MA: Sinauer Associates.
Williams, George C. 1966. *Adaptation and Natural Selection*. Princeton, NJ: Princeton University Press.
Wilson, Edward O. 1975. *Sociobiology: The New Synthesis*. Cambridge, MA: Belknap Press of Harvard University Press.
– 1998. *Consilience: The Unity of Knowledge*. New York: Knopf.
Winch, Peter. 1958. *The Idea of a Social Science and Its Reduction to Philosophy*. London: Routledge.
Winfree, Arthur T. 1980. *The Geometry of Biological Time*. New York: Springer-Verlag.
Wolf, Arthur P. 1995. *Sexual Attraction and Childhood Association*. Stanford, CA: Stanford University Press.
Wolpert, Lewis. 1992. *The Unnatural Nature of Science*. London: Faber & Faber.
Wulff, Henrik R. 1981. *Rational Diagnosis and Treatment: An Introduction to Clinical Decision-Making*. 2nd ed. Oxford: Blackwell Scientific Publications.
Zakhariev, Boris N., and Boris M. Chabanov. 1997. 'New Situation in Quantum Mechanics (Wonderful Potentials from the Inverse Problem).' *Inverse Problems* 13: R47–R79.
Zatorre, Robert J., and Isabelle Peretz, eds. 2001. *The Biological Foundations of Music. Annals of the New York Academy of Sciences*, vol. 930.
Zeki, Semir. 1980. 'The Representation of Colours in the Cerebral Cortex.' *Nature* 284: 412–18.
– 1993. *A Vision of the Brain*. Oxford: Blackwell.

Index of Names

Ackoff, Russell 271
Adamowicz, Zofia 279
Adams, Marie 28
Adorno, Theodor 197
Agassi, Joseph 105
Agnati, L.F. 182
Albert, Hans 198
Albert, Réka 276
Alembert, Jean le Rond d' 138
Anderson, Adam K. 188
Annales school 268
Aquinas, Thomas 83, 196
Archimedes 241
Ardila, Rubén 183, 188
Aristotle 239
Arkin, Adam 150
Arnoff, E. Leonard 271
Arthur, Wallace 271
Asfaw, Berhane 158
Ash, S.E. 193
Ashby, W. Ross 36, 42
Athearn, Daniel 24
Atkatz, David 273

Bacon, Francis 236
Baldi, S. 120
Ball, Philip 28, 275

Barabasi, Albert-Laszlo 276
Barbas, Helen 159
Barkow, Jerome H. 180, 273
Bartlett, M.S. 229
Barton, R.A. 187
Bates, Elizabeth 66
Baudrillard, Jean 197, 210
Bechtel, William 270
Becker, Gary S. 124, 164, 197
Belousov, B.P. 275
Bergson, Henri 66
Berlinski, David 42
Bernard, Claude 139, 251
Berner, Maureen 23, 72, 104, 174
Berry, Margaret 65
Bhabha, H.K. 153
Bindra, Dalbir 51, 190
Black, Donald 87
Bliss, Michael 251
Blitz, David 16, 147
Bloom, Harold 197
Blumer, Harold 197
Bohr, Niels 138
Boltzmann, Ludwig 235
Bolzano, Bernhard 16
Bondi, Hermann 31
Booth, J.W. 197

Bopp, Franz 61
Boring, Edwin G. 17
Boswell, T. 174
Botvinick, Matthew 189
Boudon, Raymond 70, 112, 118, 120, 169, 172, 209, 210
Bourbaki, Nicholas 89
Bourricaud, François 105
Braithwaite, Richard 24
Braudel, Fernand 168, 170, 172
Brenner, Sydney 283
Bridgman, Percy W. 152
Broad, C.D. 13
Brunelle, Dorval 174
Brynjolfsson, Erik 276
Buck, R.C. 42
Bunge, Silvia A. 190–1
Burke, Edmund 113
Buss, David M. 157

Cabanac, Michel 157
Cacioppo, John T. 163, 193
Calvino, Italo 218
Card, D. 121
Carnap, Rudolf 137, 152
Cavalli-Sforza, Luigi Luca 65, 174
Changeux, Jean-Pierre 190
Chomsky, Noam 55, 56, 63–5, 252
Churchland, Patricia S. 151, 183
Churchman, C. West 271
Clanton, Charles H. 232, 260
Claparède, Édouard 185
Clutton-Brock, Tim 155
Cockburn, Andrew 155
Coleman, James S. 16, 70, 74, 115, 120, 163, 164, 197, 210
Collins, Randall 117, 122–3
Collins, W. Andrew 144
Comte, Auguste 83, 113, 247
Cook, Karen S. 163

Corballis, Michael C. 63
Cosmides, Leda 157, 158, 180, 273
Coule, D.H. 273
Cournot, Antoine-Augustin 235
Cramér, Harald 222
Craver, Carl F. 24
Cresswell, M.J. 215
Crews, Frederick 163
Curtis, James E. 156

D'Abro, A. 132
Dahl, Robert 210
Damasio, Antonio R. 159, 191
Darden, Lindley 24, 277
Darwin, Charles 156–8, 272
Davidson, Donald 14
Dawkins, Richard 133, 141
Deblock, Christian 174
Debreu, Gerard 106
Deep Blue 77
de Finetti, Bruno 226
Dehaene, Stanislas 190
Dennett, Daniel C. 151, 157, 180
Derrida, Jacques 61
Descartes, René 131-2
Dewey, John 83
Diamond, Jared 202
Dillinger, Mike 62
Dilthey, Wilhelm 114, 162, 166, 192, 197, 198, 200
Dixon, W.J. 174
Djerassi, Carl 173
Donald, Merlin 55, 157
Dover, Gabriel 143
Drosdowski, Günther 68
Dubrovsky, Benard 143, 160
Du Pasquier, L.-Gustave 227, 235
Dupré, John 270, 282
Durkheim, Émile 83, 103, 113, 169, 210, 279

Eddy, Charles 232
Eddy, David M. 260–1
Einstein, Albert 235
Encyclopedists 122
Engels, Friedrich 16, 146
Euler, Leonhard 182
Everett, Hugh, III 216

Falk, D. 187
Faraday, Michael 119
Fehr, Ernst 155
Fiske, Donald W. 197
Fleck, Ludwik 253
Fleming, Alexander 263–4
Florey, Howard 264
Fodor, Jerry A. 180
Fogel, Robert W. 176
Foucault, Michel 117, 165
Franklin, James 228
Fréchet, Maurice 235
Fredrickson, George M. 156
Freud, Sigmund 163, 197

Gächter, Simon 155
Gadamer, Hans-Georg 197
Galbraith, James K. 23, 72, 104, 174–5
Galen 185
Galilei, Galileo 275
Galison, Peter 270
Gauss, Carl Friedrich 241
Gazzaniga, Michael 191
Geertz, Clifford 166, 197, 210
Gelade, G. 67
Giddens, Anthony 74, 124
Gilfillan, S.C. 75, 119
Gini, Corrado 113
Glansdorff, P. 275
Goethe, Johann Wolfgang 82
Goffman, Erving 210
Goodman, Judith C. 66

Goody, Jack 193
Gorbachev, Mikhail 125
Gould, Stephen J. 49, 157, 159, 226, 271
Graham, Loren R. 77
Gramsci, Antonio 16
Greenspan, Alan 165
Gross, Paul R. 67

Habermas, Jürgen 197
Haeckel, Ernst 271
Haken, H. 275
Haldane, J.B.S. 216
Halliday, Michael A.K. 65
Hamilton, William D. 154
Handy, Rollo 169
Hardin, Russell 163
Harding, Susan 165
Hardy, G.H. 42
Harrington, Anne 118
Hartmann, Nicolai 16
Harvey, P.H. 187
Harvey, William 251
Hauser, Marc D. 63
Hayek, F.A. 156
Hebb, Donald O. 51, 64, 278
Hechter, Michael 197
Hegel, Friedrich 22, 113, 198
Heidegger, Martin 90, 198
Helmuth, Laura 63
Hempel, Carl G. 13, 24
Henderson, Lawrence J. 47
Henry, Joseph 119
Hilbert, David 222-3, 275, 283
Hill, A.V. 163
Hintikka, Jaakko 66
Hirschman, Albert O. 124, 177
Hobbes, Thomas 124, 177
Hoffman, Ralph E. 184
Hogarth, Robin M. 164, 197
Holbach, Paul-Henry Thiry d' 41

318   Index of Names

Holland, John H. 15
Homans, George Caspar 106, 162, 197
Horkheimer, Max 197
Hoyle, Fred 31
Huelsenbeck, John P. 228
Hughes, G.E. 215
Hume, David 120, 156
Humphreys, Paul 224
Husserl, Edmund 55, 66, 198
Huxley, Julian 162
Huxley, Thomas H. 16

Ibn Khaldûn 83, 113, 171, 210

James, William 84
Jeffrey, Harold 226
Jeong, Hawoong 276
Johansson, S. Ryan 116
Johnson, Jeffrey G. 163

Kahneman, Daniel 165, 207, 214, 227, 231
Kanazawa, Satoshi 161
Kant, Immanuel 80, 166, 249
Kary, Michael 151–2
Kempthorne, Oscar 14
Kerszberg, Michael 190
Keynes, John Maynard 106, 210
Kim, Jaegwon 14
Kincaid, Harold 167, 270
Kitcher, Philip 143
Klir, George J. 42
Koenig, Olivier 184
Koerner, E.F. Konrad 61
Kolmogorov, Alexander N. 223
Kolnai, Aurel 118
Koshland, Daniel 159
Kosslyn, Stephen M. 184
Kraft, Julius 66
Kreiman, Gabriel 186
Kripke, Saul 215

Kronz, Frederick 44
Kuhl, Patricia 64
Kurtz, Paul 169
Kuznets, Simon 175

Labov, William 65
Lacan, Jacques 197
Lanczos, Cornelius 215
Lange, Friedrich Albert 146
Lashley, Karl 157
Laszlo, Ervin 42
Latour, Bruno 120, 153, 165, 196
Leibniz, Gottfried Wilhelm 84, 238
Lenin, Vladimir I. 146
Leontief, Wassily 106
Lepore, Ernest 181
Lerner, David 9
Levitt, Norman 67
Lewes, George Henry 16
Lewis, David 215–16
Lewontin, Richard 143, 159
Lieberman, Matthew D. 163, 193
Llinás, Rodolfo 191
Lloyd, Elisabeth A. 160
Locke, John 113, 249
Lombroso, Cesare 154
Lomnitz, Larissa A. 144
Looijen, Rick C. 140
Lorente de No, Rafael 187
Lovejoy, Arthur O. 9
Lowe, E.J. 26, 217
Lucretius 30–1
Luhmann, Niklas 87
Luria, Alexander Romanovich 193
Luther, Martin 247

Machamer, Peter 24, 183
MacWhinney, Brian 16, 64
Mahner, Martin xiii, 15, 46–7, 49, 151–2, 212

Malinowski, Bronislaw 113
Mann, Michael 172
March, James G. 207
Marr, David 180, 183
Martin, Jerôme 273
Martin, Michael 197, 199
Martinich, A.P. 54
Marx, Karl 72, 83, 101-2, 113–14, 196, 201, 207
Massey, Douglas S. 159
Maull, Nancy L. 277
Maxwell, James Clerk 235
Mayberry, Marhall 182
Maynard Smith, John 39, 271
Mayr, Ernst 14
McCarthy, Rosalen H. 186
McGlashan, Thomas H. 184
Medawar, Peter 270
Meinong, Alexius 218
Melvill Jones, G. 190
Melzack, Ronald 19
Merton, Robert K. 70, 93, 99, 107, 156, 160, 173, 200, 210, 249
Miikkulainen, Risto 182
Mill, John Stuart 16, 24, 162, 198
Milsum, John 42
Moessinger, Pierre 155
Moore, George E. 14
Morgan, Conway Lloyd 16, 157
Morris, J.S. 189
Morrison, Margaret 270
Moser, Paul K. 164, 197
Mountcastle, Vernon 50, 180
Murphy, Edmond A. 232, 255
Müller, Adam 113

Naccache, Lionel 190
Nagel, Ernest 13
Navarro, Vicente 176
Negroponte, Nicholas 151

Nesse, Randolph M. 160
Neurath, Otto 152
Newton, Isaac 26, 43, 82, 275
Nietzsche, Friedrich 198
Novikoff, Alex B. 16, 147

Ocampo, José Antonio 104, 174
Ochsner, Kevin N. 163, 193-4
Odum, Eugene 140
Olby, Robert 271
Olson, Mancur 164, 197
Opp, Dieter 197
Outhwaite, William 197

Palchinsky, Peter 77
Pareto, Vilfredo 172, 198–9
Parmenides 89
Parsons, Talcott 70, 113
Pascal, Blaise 91
Peirce, Charles S. 83, 220
Peretz, Isabelle 186
Petroski, Henry 76
Petty, Richard E. 163, 193
Pfisterer, Andrea R. 141
Phelps, Elizabeth A. 188
Piaget, Jean 17, 65
Pickel, Andreas 125, 176, 208
Pinker, Steven 63, 65, 180
Plato 83, 180
Poincaré, Henri 231
Polya, George 232
Popper, Karl R. 14, 24, 105, 118, 121, 146, 175, 199, 201, 205, 235, 241, 243
Portes, Alejandro 73
Premack, David 202
Prigogine, Ilya 275
Putnam, Hilary 166
Pylyshyn, Zenon 181

Quesnay, François 106

Raff, Rudolf A. 271
Ramsey, Frank Plumpton 226
Rawls, John 249
Rayner, Keith 67
Reder, Melvin W. 164, 197
Reichenbach, Hans 243
Renfrew, Colin 157
Renyi, A. 223
Rickert, Heinrich 200
Ricoeur, Paul 197
Rilling, James K. 155
Rodríguez, Alejandro 104
Rodrik, Dani 104
Rorty, Richard 196
Ross, John 150
Rousseau, Jean-Jacques 100, 249
Routley, Richard 218
Ruppin, Eytan 183
Russell, Bertrand 91

Saarinen, Esa 141
Sachs, Ignacy 175
Saussure, Ferdinand de 61–3
Savage, Leonard 226
Schachter, Daniel 159
Schaffner, Kenneth 139
Schelling, Thomas C. 106, 164, 197
Schmid, Bernhard 16
Schönwandt, Walter L. 122
Schutz, Alfred 197
Searle, John R. 182, 192
Sejnowski, Terrence J. 151, 183
Sellars, Roy Wood 16, 147
Sherif, Muzafer 169, 193
Shweder, Richard A. 197
Simberloff, Daniel 140
Simon, Herbert A. 20
Smith, Adam 113, 198
Smoluchowski, Marian 235
Sober, Elliott 154, 156

Soros, George 121, 175
Stenger, Victor J. 31
Sternberg, Robert J. 144
Stevenson, Robert Louis 100
Stiglitz, Joseph E. 174
Still, Mary C. 161
Stinchcombe, Arthur L. 71
Stove, David 142-3
Streeten, Paul 23
Suppes, Patrick 226
Sutherland, Stuart 66
Swedberg, Richard 172

Taylor, Charles 197
Taylor, Lance 104, 174
Tegmark, Max 31
Tench, Watkin 203
Tesla, Nikola 119
Thucydides 210
Tiehen, Justin 44
Tilly, Charles 74, 125, 208
Tocqueville, Alexis de 100, 118, 122, 171, 210
Tooby, John 157–8, 180, 273
Torrey, E. Fuller 163
Treisman, Ann M. 67
Trigger, Bruce G. 5, 112, 157, 166
Tsebelis, George 209
Turing, Alan 275
Turnbull Colin M. 155
Tversky, Amos 214, 227, 231

Van Alstyne, Marshall 276
Van den Berg, Axel 177
Vanderburg, Willem H. 73
Van der Dennen, Johan M.G. 154
Vaughan, Susan C. 163
Venn, John 221
Vienna Circle 152, 243
Ville, Jean 222

Virchow, Rudolf 251
Von Bertalanffy, Ludwig 42
Von Mises, Richard 221–2
Von Schelting, Alexander 115, 200
Von Wright, Georg Henrik 196
Vygotsky, Lev S. 164

Waal, Frans de 155
Warren, W. Preston 147
Warrington, Elizabeth K. 186
Weber, Max 84, 114, 172, 198, 200, 201, 210
Weiss, Paul A. 42
Weissman, David 16, 112
Wertheimer, Max 16
West, Stuart A. 155
Wheeler, John A. 151
Whewell, William 151
Whitehead, Alfred North 84
Wiener, Norbert 16, 36, 119, 270

Wilkins, Adam S. 49, 271
Williams, George C. 158, 160
Wilson, David Sloan 154, 156
Wilson, Edward O. 133, 141, 154, 281
Winch, Peter 196
Winfree, Arthur 18
Wittgenstein, Ludwig 91, 196, 238
Wolf, Arthur P. 163
Wolpert, Lewis 120
Woodruff, Guy 202
Woolgar, Steve 120, 153, 196
Wulff, Henrik R. 232

Yunus, Muhammed 76

Zatorre, Robert J. 186
Zbierski, Pawel 279
Zeki, Semir 192, 194
Zhabotinsky, A.M. 275
Zubrow, Ezra 157

Index of Subjects

Action 69, 123; theory 98–9
Adaptation 158
Adaptationism 156, 159
Aggregate 10–11
AI. *See* Artificial Intelligence
AL. *See* Artificial Life project
Algorithm 158, 160, 181, 193
Altruism 154–247
Analysis, conceptual 89, 90–1, 130, 195. *See also* Reduction
Anarchism 100, 104
Antibiotics 264–5
Anti-reductionism 131
Aphasia 55
Approximation theory 241, 246
Architecture of a system. *See* Structure
Argument 12, 87, 254–9
Artefact 63, 77–8, 151–2, 185; statistical 115
Artificial Intelligence 77, 151, 191–3
Artificial Life project 46, 151
Association 10–1
Associationist psychology 11
Atomism, logical 91
Attitude 162–3
Axiology 97–8. *See also* Value theory

Bayesianism 214, 226–32, 259–61. *See also* Probability, subjective
Bayes's theorem 228, 259–60
Behaviour 201–6
Behavioural science 163
Behaviourism 36, 158, 163, 181
B-E-P-C square 169–71
Big Bang 31
Bijection 239
Biochemistry 139
Biofunction 49
Biologism 154–62
Biology 45, 139–41; molecular 271–2; structural 272
Biospecies 48–9
Biosystem 46–8
Black box 23, 36
Block universe 41
Body systems 251
Bond 19–20, 36
Bottom-up strategy 9, 94, 135, 200
Boudon-Coleman diagram 76, 116, 118, 208
Boundary 36, 92
Brain imaging 189
Breadth, conceptual 276

Bridge hypothesis 144–5. *See also* Glue formula

Category, mathematical 8
Causation 170, 190, 192, 226, 232
CESM model of a system 35–7
Chain of being 45. *See also* Hierarchy
Chance 218–32, 260. *See also* Probability; Propensity; Randomness
Chemistry 21, 139, 275, 282–3
Chinese room 182
Choice 107, 161
Class 48
Classification 57
Code, moral 248
Cognition 92, 181
Cognitive neuroscience 50–1, 191–3; social 163, 268
Cognitive science 191–3
Coherence, explanatory 275. *See also* Consistency, external
Cohesiveness 103–4
Cold War 125
Collection 35–6
Combination 10
Commmunication system 67
Community, scientific 120, 274
Competition 119–20, 155
Composition of a system 11, 35
Computationism 150–2, 159–60, 181–5
Computer 185
Computer science 185
Concatenation 10–11, 53. *See also* Association
Concept 56
Condorcet Jury Theorem 227
Confusion 213–15, 243
Connectionism 183–5
Connectivity 215
Consciousness 189–91

Conservation principles 30–1, 273
Consilience 281–2
Consistency, external 31, 275–6, 281. *See also* Coherence, explanatory
Constructivism: epistemological 117; psychological 65–6; social 102, 114, 200, 253
Context 57
Contradiction 243
Convention: semiotic 55, 60; social – *see* Norm
Convergence; functional 179, 195; stealthy 196–212
Convergence of disciplines xi, 3–5, 129, 268–83
Cooperation 20, 119, 155
Copenhagen interpretation 131
Correspondence theory of truth 238–41
Cosmology 30–1, 273–4
Cost-benefit algorithm 160
Counter-Enlightenment 123, 197
Counter-norm 99, 120
Covering-law model 14, 95. *See also* Subsumption
Creationism 224–5
Credibility 226–32
Crime 188, 208–9
Crisis, normative 32
Critical theory 196
Cross-disciplinarity 169, 171. *See also* Interdisciplinarity; Multidisciplinarity
Culturalism 166. *See also* Hermeneutics
Cybernetics 36, 270

Darwinism xii. *See also* Evolutionary biology
Datum 214
Decision rule 261–2
Decision theory 231, 161–3
Deducibility 217

Definability 217
Definition 144–5; operational – *see* Indicator hypothesis
Denotation 56, 58
Descartes's error 159
Designation 56, 58
Determinism 79
Development: biological 18; national 72, 174–5
Diagnosis, medical 253–9
Dialectics 22
Discipline 277–8
Disconnection syndrome 186
Disease 160, 252–3
Disposition 224. *See also* Propensity
Distance, alethic 244
Divergence of disciplines 268–70, 274–5
DNA 139, 141, 154–5, 225
Drake formula 225
Dry/wet psychology 184
Dualism, psychoneural 181, 191
Dynamicism 79

Ecology 140–1
Economism 164–5. *See also* Economic imperialism
Egalitarianism 248–9
Electrodynamics 13
Emergence xi, 3–5, 12–14, 17, 23–31. *See also* Novelty, qualitative
Emergent 17
Emergentism 38. *See also* Systemism
Emotion 52, 157, 159, 191, 194
Empiricism 221
Endostructure 36, 62
Energy 12, 18, 20, 43
Engineering 77; social 175, 206 – *see also* Policy, social
Enlightenment 41, 197
Entanglement 44

Environmentalism 39, 131
Environment of a system 35, 130
Enzyme 45–6
Epidemiology 259–60
Epistemology 3, 137, 148; probabilistic 228–32
Equivalence, functional 29
Error 241–5, 266
Ethics 99, 162, 247–8
Event 27
Everything, theory of 149–50
Evidence 160
Evo-devo. *See* Evolutionary developmental biology
Evolution 15, 18, 158, 271–2; mosaic 187; social 159, 174; unit of 49
Evolutionary biology xii, 132, 154, 270
Evolutionary developmental biology 4, 271–2
Evolutionary psychology. *See* Psychology, evolutionary
Evolutionism, philosophical 79
Ex nihilo 17, 30–1, 273
Exactness 48
Exaptation 159
Existence 215–6
Experiment 258–9
Explanation 22–3, 275; mechanismic 95; reductive 145
Extensionalism 88–9
Extension of a predicate 88–9
Externalism 117, 172

Fact 74, 123, 169, 176, 197, 214–19, 238–40
Faculty, psychological 180. *See also* Module
Fallacy, individualist 86
Falsificationism 243
Feasibility 219–20

Feedback 106
Feminism, academic 165
Fictionism 215
Fractal 12
Fragmentation of social science 168–9
Freedom 23–4
Frequency, relative 220–3, 259–60
Function, specific 49, 88–9
Functionalism 30, 151, 183
Functional system 188

Geisteswissenschaften 71, 166, 192
Gene 19, 24, 45, 64
Geneticism 133, 154–5
Genetics 19, 24, 110, 139–40
Gestalt psychology 188
Globalization 23, 104, 211
Glue formula 278–9. *See also* Bridge hypothesis
Gradualism 18
Grammar 54–5, 64–5

Hamilton's variational principle 43–4
Happiness 99
Hermeneutics 153, 166, 196–206
History 18, 38; human 100; of ideas 117
Holism 38, 40, 42, 74, 100–5; axiological 103–4; epistemological 91, 102; logical 100; methodological 94, 103; ontological 87; semantic 89–90, 101–2
Holoindividualism 105–6
Human Genome 132
Hybrid, philosophical 105–6
Hybridization of disciplines 272, 280
Hypothesis 154–9, 206, 232–4

Iatrophilosophy 251
Idealism, philosophical 23, 192
Identity theory 192
Ideology 248–9

Imperialism: economic 164–5 – *see also* Economism, linguistic 152–3
Impossibility 218
Indicator hypothesis 152, 253–9
Individholism 105–6
Individual 82, 84–5
Individualism 38, 40, 74, 82–101; axiological 97–8; epistemological 91–5; historical 100; institutional 105–6; logical 87–9; methodological 93–4, 199–201; ontological 85–7; political 100; semantic 89–90
Individualist's Dilemma 110
Information, genetic 47
Informationism 181–3. *See also* Computationism
Innatism 63, 66, 158
Insulin, discovery of 251
Integration of disciplines 168–78, 268, 277–82. *See also* Convergence; Cross-disciplinarity; Interdisciplinarity; Multidisciplinarity
Interaction 94, 105–6
Inter-level relation 135–6
Interdisciplinarity 176–7; degree of 279–80. *See also* Cross-disciplinarity; Multidisciplinarity
Interdiscipline 277–82
Internalism 117
International Monetary Fund 104–5
Interno-externalism 64
Interpretation: of an action – see *Verstehen*; psychological 197, 201 – *see also* Theory of Mind; semantic 58–9
Interscience 280
Intertechnology 280
Intuition 102, 226, 228
Intuitionism 66, 102
Invention 75–6, 119, 200; social 76–7, 142

IQ 143–4
Issue, practical 109, 114, 121

Juxtaposition. *See* Association

Kind, natural 48, 85. *See also* Species, biological
Kin selection 141–2

Language 34, 55–65, 238
Lattice 90
Law, natural or social 85, 217–19, 257
Lawfulness, principle of 217
Law statement 91
Learning 95, 135, 247
Level: hierarchy 15, 133; lexical 54; of organization 9–10, 15, 78, 133–4, 251
Libertarianism 100
Life 46–8
Likelihood 220–1, 224, 259–61
Likert scale 220
Linguistics 54, 63
Localizationism 185–7
Logic: classical 241; modal 213–16, 219; quantum 103, 240

Machine 16, 47, 77, 179, 193
Machinism 46. *See also* Artificial Life project; Computationism
Macro-micro 9, 135
Macro-reduction 131
Macro-reductionism. *See* Holism
Market 106, 119, 200
Marriage norm 124, 161
Marxism 164, 238. *See also* Materialism, dialectics
Materialism 33, 146, 192; dialectical 146–7; emergentist 79, 147; vulgar – *see* Physicalism

Mathematics 42, 61, 108–9, 153, 182, 215, 237–8
Maximization 203, 207, 261
Maybe 213–36
Meaning, semantic 55, 58–9, 243
Meaningfulness 243
Measurement 216, 225
Mechanics: analytical 44; classical 241; quantum 92, 103, 138; statistical 135; vectorial 43–4, 90, 138
Mechanism 20–4, 35, 46, 74, 95, 132, 206–9, 256–9
Mechanismic. *See* Explanation
Mechanistic 46
Medicine 160, 231–2, 250–67
Membership relation 88
Mereology 11, 85
Metaphysics 26, 220. *See also* Ontology
Method, scientific 129–30, 202, 250–1
Micro-macro 133
Micro-micro 135
Micro-reduction 131–3
Mind 49–52, 179–9; theory of 202–3
Module 9–10, 180
Modus ponens 254
Money 59
Monty Hall problem 230
Multidisciplinarity 171–6. *See also* Cross-disciplinarity; Interdisciplinarity
Multidiscipline 277–82
Multiscience 280

Narrative 115
Nativism. *See* Innatism
Naturalism 110, 112. *See also* Materialism
Nature/culture dualism 80
Nazism 118
Necessity: logical 216–17; physical 218–19

Negation 243
Network 215, 183–5; neural 183–5; social – *see* System, social
NGO 156
Nominalism 85
Norm 156, 161, 172, 282
Novelty 11, 16

Observability 20
Ontology 3, 19, 26, 42, 79–80, 108, 137, 148
Organization. *See* Structure

Paradox 229–30
Part–whole relation 10–11, 48, 82, 101
Penicillin, discovery of 164–5
Perception, social 160
Phagocytation of disciplines 272
Phenotype 271, 283
Philosophy 173, 276
Phonics 66–7
Phrenology 272
Physicalism 146, 149–50. *See also* Materialism, vulgar
Planning 73, 122
Plasticity 50–1, 72
Platonism 245
Plausibility 232–5, 260–1
Pleistocene 158
Pluralism 79
Plurality of worlds 216
Policy, social 121
Politicism 165–6
Politics 103, 118
Possibility 213–36; logical 214–17; real 217–20
Possible world 215–16
Postmodernism 173
Praxiology. *See* Action theory
Predicate 56, 88

Premature unification 272–4
Probability 221–7, 243–4; objective 221–6; subjective 226–32 – *see also* Bayesianism
Problem 91–2, 203–6; inverse 30, 203–6, 253–7, 258; moral 99
Process 16–18, 252
Program 158. *See also* Algorithm
Propensity 224–6. *See also* Disposition; Possibility, real
Property 14, 17, 56
Proposition 53, 56, 234, 238, 260
Proteome 133
Pseudo-science 281
Psychiatry 160
Psychoanalysis 163
Psychologism 162–4
Psychology 49–50, 163, 180, 230–1; evolutionary 143, 156–62, 272–3; Gestalt 188; social 193
Psychon 51
Public health 176
Purpose 49

Quantum physics 150. *See also* Mechanics, quantum

Racism 156
R&D 266
Randomness 222, 224–5
Rational-choice theory 107, 159, 164–5, 187–206
Rationalism 83
Rationality 197
Realism 80, 112, 212, 238
Reality 219
Reduction 96, 107, 129–48
Reductionism 131, 146–7
Reduction statement 152–3
Reference 55–7, 60, 89, 240, 277–8

Refutabilty 31
Relation 82, 100
Relativism, cultural 102
Reliability 244
Representation 240
Reproductionism 157
Revolution, French 122
Romanticism, philosophical 197–8. *See also* Counter-Enlightenment
Rule 265

Saltationism 18
Sampling 229
Science 172–4; biomedical 250–65; social 112–26, 168–78
Sectoral approach 41
Self-assembly 13
Selfishness 28
Semantics 55
Semigroup 10–11
Sense 60
Sentence 53, 238. *See also* Proposition
Set theory 87
Sign 58–9
Signification 58
Simulation, computer 151–2, 184
Situation, logic of 105
Social cognitive neuroscience 163, 268
Society 70–81, 106, 113
Sociobiology 141–3, 154–6
Socio-economics 172
Socio-linguistics 67
Sociologism 164
Soviet Empire 32, 125, 175–6
Spandrel 159
Species, biological 48–9
Speech 62–3, 65–7, 187
State: function 37, 252–3; mental 51; space 37–8, 252–3; of a thing 37, 152–3
Statistical physics 135, 219, 223, 135

Statistics, descriptive 115, 259
Structuralism 39
Structure of a system 19, 35, 74
Subjective 201–3
Submergence xi, 15, 17, 31–3
Subsumption 23. *See also* Covering-law model
Supervenience 14. *See also* Emergence
Syllogism, proportional 229
Symbol 59, 153, 164, 181–2
Symptom 254–61
Syncategorematic 53
Synergistics 275
Synergy 179
Synonymy 58
Synthesis 195. *See also* Convergence
System 28–9; of actions 70; artificial 34; conceptual 33–4; functional 188; material 43; mathematical 42; natural 33–4; nervous 50; semiotic 33–4, 53–67; social 34, 112–26; technical 33–4; types of 33–4
Systemic approach 40–52, 71–3, 114–15
Systematics 48
Systemism 38, 42. *See also* Systemic approach

Technology 70–1, 263–6; philosophical 97–110; social 72–3, 178
Testability, empirical 57
Theorem 233
Theory 42, 88, 257; reduction 145–6
Therapy 258–9, 264
Thermodynamics 95, 275
Thomas's theorem 159–60
Thought 240
Top-down strategy 9, 94, 136, 200
Translucent box 36
Trespassing across disciplines 124
Truth 237–49; factual 238–49; formal

237–8; partial 241–5; relative 245; test 145–6; value 56, 91, 241
Typology 22

Uncertainty 231–2, 258
Unconscious 189
Unfalsifiability 281
Unidisciplinarity 170, 175
Unity of the sciences 270. *See also* Convergence
Universal 85
Universe 40, 274
Utilitarianism 99
Utility 261–2

Value 98, 156

Variational principle 43–4
Verisimilitude 232. *See also* Plausibility
Vérité de fait 238. *See also* Truth, factual
Vérité de raison 238. *See also* Truth, formal
Verstehen 196, 200, 202–3, 209
Violence 176–7
Virtual 215
Vitalism 46

Washington consensus 210
Whole 12, 84. *See also* System
Wholeness 43–4
Workspace 190
World, possible 216